LIBRARY
World Conservation Monitoring Centre
DS 5399

Plant conservation in the Mediterranean area

Geobotany 7

Series Editor

M.J.A. WERGER

1985 **DR W. JUNK PUBLISHERS**
a member of the KLUWER ACADEMIC PUBLISHERS GROUP
DORDRECHT / BOSTON / LANCASTER

Plant conservation in the Mediterranean area

edited by

C. GÓMEZ-CAMPO

1985 **DR W. JUNK PUBLISHERS**
a member of the KLUWER ACADEMIC PUBLISHERS GROUP
DORDRECHT / BOSTON / LANCASTER

Distributors

for the United States and Canada: Kluwer Academic Publishers, 190 Old Derby Street, Hingham, MA 02043, USA
for the UK and Ireland: Kluwer Academic Publishers, MTP Press Limited, Falcon House, Queen Square, Lancaster LA1 1RN, UK
for all other countries: Kluwer Academic Publishers Group, Distribution Center, P.O. Box 322, 3300 AH Dordrecht, The Netherlands

Library of Congress Cataloging in Publication Data

Main entry under title:

Plant conservation in the Mediterranean area.

(Geobotany ; 7)
1. Plant conservation--Mediterranean Region.
2. Rare plants--Mediterranean Region. 3. Endangered species--Mediterranean Region. I. Gómez-Campo, C.
II. Series.
QK86.M47P58 1985 639.9'9'091822 84-27817
ISBN 90-6193-523-7

ISBN 90-6193-523-7 (this volume)
ISBN 90-6193-895-3 (series)

Copyright

© 1985 by Dr W. Junk Publishers, Dordrecht.

All rights reserved. No part of this publication may be reproduced, stored in a retrieval system, or transmitted in any form or by any means, mechanical, photocopying, recording, or otherwise, without the prior written permission of the publishers,
Dr W. Junk Publishers, P.O. Box 163, 3300 AD Dordrecht, The Netherlands.

PRINTED IN THE NETHERLANDS

Contents

Foreword

The Mediterranean basin harbours one of the richest floras in the world outside the tropics. A combination of the characteristic Mediterranean climate and the varied geology has produced a diverse vegetation which man has, in turn, dramatically modified through his agricultural activities - by clearing trees and scrub for crops, grazing of sheep and goats, terracing of the slopes, irrigation and the deliberate and accidental use of fire. The characteristic landscapes that we associate today with the Mediterranean - the terraced hillsides, orchards, vineyards, olive groves, scattered forests of oaks, pines, cedars and firs, and the extensive scrublands or matorral - are largely man-made or modified.

These same landscapes have played a considerable rôle in the development of man's civilization and the various peoples of the Mediterranean have been shaped, in Di Castri's words, into a single cultural unit with a common denominator of shared values, based largely on a close integration with the environment.

Alas, this pleasant region of the world fulfils so many of man's inspirations is fragile and under great threat through pollution, excessive urbanization, deforestation and tourism, all leading to the loss of habitats and risks to the survival of many species.

The Mediterranean basin was one of the first to suffer the effects of massive deforestation and the need for conservation was voiced as long ago as the 6th century BC by Plato, and a graphic account is given in J V Thurgood's 'Man and the Mediterranean Forest. A history of resource depletion'.

This present volume which has been organized and edited by Professor César Gómez-Campo, a champion of Mediterranean conservation, is particularly timely as it focuses attention on the serious problems facing us in this area, giving case histories of plants for many of the countries concerned, and suggesting possible solutions.

Many of the earliest botanical studies were made in countries bordering the Mediterranean and it may be recalled that the first botanic gardens were established in cities such as Pisa, Padova, Firenze and Montpellier. In more recent times, botany has not found much favour in many Mediterranean countries and has gone through a period of decline compared with more northern countries. This is particularly unfortunate given the richness of the flora and the importance of the region as the source of many important crop plants and their relatives.

It is to be hoped that this valuable book will help draw attention to these issues and stimulate the necessary conservation action to maintain the rich genetic plant resources of this region.

August 1984 V.H. Heywood

Acknowledgement

The editor wishes to acknowledge the important role played by Prof. Vernon H. Heywood and his staff of Plant Science Laboratories (University of Reading) in helping to obtain the proper English style and making useful suggestions to improve the text.

List of contributors

Avishai, Michael. Jerusalem Botanical Gardens, Jerusalem, Israel

Baytop, Turhan. Department of Pharmacology, University of Istanbul, Istanbul, Turkey

Boulos, Loutfy. National Research Centre, Dokki, Cairo, Egypt

Demiriz, Hüsnü. Department of Botany, University of Istanbul, Istanbul, Turkey

Filipello, Sebastiano. Istituto di Botanica, Università di Pavia, Pavia, Italy

Gardini-Peccenini, Simonetta. Instituto di Botanica, Università di Pavia, Pavia, Italy

Gómez-Campo, César. Departamento de Biología, Escuela T.S. Ing. Agrónomos, Universidad Politécnica, E-28040 Madrid, Spain

Leon, Christine. IUCN Conservation Monitoring Centre, The Royal Botanic Gardens, Kew, Richmond, Surrey, UK

Lucas, Gren. The Herbarium, The Royal Botanic Gardens, Kew, Richmond, Surrey, UK

Malato-Beliz, Joâo. Dep. de Biología Analítica, Estacão Nacional de Melhoramento de Plantas, Elvas, Portugal

Mathez, Jöel. Institut de Botanique, Université des Sciences et Techniques de Languedoc, Montpellier, France

Papanicolaou, Kostas. Institute of Systematic Botany, University of Copenhagen, 140 Gothersgade, DK-1123 Copenhagen K, Denmark

Pons, Armand. Faculté des Sciences et Techniques de Saint-Jérôme, Rue Henri Poincaré, F-13397 Marseille Cedex 13, France

Quézel, Pierre. Faculté des Sciences et Techniques de Saint-Jerôme, Rue Henri Poincaré, F-13397 Marseille Cedex 13, France.

Raynaud, Christian. Institut de Botanique, Université des Sciences et Techniques du Languedoc, Montpellier, France

Ruiz de la Torre, Juan. Departamento de Botánica, Escuela T.S. Ing. de Montes, Universidad Politécnica, E-28040 Madrid, Spain

Snogerup, Sven. The Botanical Museum, Ö. Vallgatan 18, S-223 61 Lund, Sweden

Strid, Arne. Institute of Systematic Botany, University of Copenhagen, 140 Gothersgade, DK-1123 Copenhagen K, Denmark

Synge, Hugh. IUCN Conservation Monitoring Centre, The Royal Botanic Gardens, Kew, Richmond, Surrey, UK

Part I

General

CHAPTER 1

The conservation of Mediterranean plants: Principles and problems

C. GÓMEZ-CAMPO

1. Conservation

In theory, conservation could be applied to any level of organization or information, e.g. genic, organismal, ecological, etc. Since some individual genes have been already isolated, the time may come when large collections of them might be stored for the future. But, for the time being, genes are conserved in large groups, either as organisms, populations or ecosystems. A seed bank of cultivars of *Lupinus albus* or the efforts being made by many zoos to keep Siberian tigers, are both efforts directed at the preservation of specific taxa. Since individuals are subsets of the genetic pool of the species they belong to, such a pool needs to be sampled through a sufficient number of individuals. But genes in an organism are associated with each other through multiple interactions as are the indivuals of a species or the different species within an ecosystem. Thus, when conservation is aimed at the highest information level – the ecosystem – not only the components themselves but also their reciprocal relations are conserved. Thus it is frequently stated that the most logical and economic method to conserve any biological entity is within the ecosystem of which it forms a part.

To be precise, conservation should be distinguished from preservation. Conservation consists of keeping organisms under the same conditions as in nature, following the same paths of evolution. Preservation is to keep them essentially in the same state as when they were taken from their natural habitat (Simmons et al., 1976; Frankel & Soulé, 1981). An integral reserve may be an example of the first, a seed bank of the second. But the distinction is perhaps too academic, and both terms are very often used synonymously. In fact, strict conservation 'as things were before human influence started' is an illusory concept. Even within the limited space of an integral reserve, evolution will never proceed in exactly the same manner as if man were not around.

Man first became aware of his critical role in the extinction of other species through his own observations on the animal kingdom. As a hunter during most of his existence as a species, it is logical to think that his earliest contacts with this problem were through the depletion of the populations of his prospective prey. It is said that during the past 300 years, the extinction of at least 100 species of terrestrial vertebrates has coincided with the age of geographical exploration and discovery. For other zoological groups data are scarcer and more incomplete, but it is believed that extinction rates have also been high. Today, in spite of many protective measures in force, it is thought that increasing human presence and pressure are still a direct or indirect threat for a large number of animals. More than one thousand species and subspecies are listed as threatened in the IUCN 'Red Data Book' (IUCN, 966).

It may be unnecessary to emphasize that the real problem is not the very fact of extinctions, but the

Gómez-Campo, C. (ed.), Plant conservation in the Mediterranean area.
© 1985, Dr W. Junk Publishers, Dordrecht. ISBN 90 6193 523 7.

high rate of them that human action had induced. Extinction has been a common and natural phenomenon during the course of evolution. If we try to compute how many Mesozoic species or even genera are living today, only a handfull will be found. If the same average rate of extinction prevailing in the past one hundred million years were applied to historic times, only one or two extinctions would have occurred in the last millenium. However, the actual situation is quite different.

2. Conservation of plants

For many decades, plants have constituted the neglected side of conservation. Man has been always aware of the sharp regression of forests and other natural plant communities as a consequence of agriculture, grazing, fire, wood extraction, etc., but he has always considered this clearance and cultivation as a triumph rather than as a source of concern. Only when the destruction of natural ecosystems has become increasingly associated with economic shortages (as of wood, hunting, recreational areas, etc.) has a recognition of the problem occurred to the ordinary citizen. But for many people, the need to conserve indivual species of plants is still today an obscure and unimportant problem.

Anyone who has tried to explain the problem to an audience of non-botanists is aware of the difficulties it presents. There are plenty of reasons for conservation (Melville, 1970; Heslop-Harrison 1974), but most of them are difficult to convey to the public. To say that plants were present in the world much before man himself and thus it should be recognized that they have a certain right to survive, may sound too philosophical. To say that plants are fundamental to the function of the biosphere, that the stability of the biosphere itself is narrowly linked to its genetic diversity, and that many species becoming extinct means a step backwards in the stability of our environment, may sound exaggerated and only partly comprehensible. To say that plant diversity is a legacy we have received and we are ethically obliged to transmit to our descendents, may perhaps be met with some doubts as to whether our descendents will really miss this weed or that dwarf shurb. To say that every plant, as a participant in the general process of evolution, merits a scientific value, may be received with some indifference; for many people, the scientific study of a species is completed the moment it is discovered and described.

One would then need to resort to utilitarian reasons which are usually the best understood. In effect, modern man is more and more turning his eyes towards wild plants as sources for oils, protein, fibre, medicinal or aromatic substances, and even recently, for energy. Domestication of plants did not finish many centuries ago, but is still an active process today and there are strong indications that it may be increasingly important in the future. Wild plants may also be a precious source of valuable genetic information to incorporate into crop species through well-directed breeding programmes (Bennett & Frankel, 1970; Frankel & Hawkes, 1975).

Since most botanists, ecologists and other biologists are convinced of the need to protect plant species, it seems clear that reaching public attention should be a continuing objective, not only because it contributes to education generally, but also because increasing public awareness of a problem paves the way to governmental recognition and thus increases the possibility of effective action. The need to protect animal species is now a very popular idea and successful campaigns have been carried out in many countries through the mass media, but the case of plants is clearly lagging far behind. However, as plants are immobile and much linked to geography, extreme care is necessary to avoid the case where publicity itself makes things worse. Unscrupulous collectors will be always present even if a small proportion, so that the exact location of the rarest or most vulnerable taxa should never be disclosed in too much detail. The mountain or even the nearest town may be sufficient, but detailed co-ordinates or the number of kilometres from a recognizable point by road would be excessive.

Official interest in plant conservation on a world scale started within the IUCN (International Union for the Conservation of Nature and Natural Resources). Its 7th Technical Meeting in 1959 was devoted to the conservation of plants, but it took more than a decade before the TPC (Threatened Plant Committee) was founded within that organization. By 1975, both the International Botanic Congress in Lenigrad and the First OPTIMA (Organization for the Taxonomic Investigation of the Mediterranean Area) Congress in Herakleion, showed a special interest and put much emphasis on plant conservation. The same year, a conference devoted to this subject was held in Royal Botanic Gardens, Kew. In the agreed resolutions (Simmons et. al. 1976) the main priorities for conservation of plant species on a world scale were clearly established. Other follow-up conferences were held in Kew and Cambridge (Synge & Townsend, 1979; Synge, 1981). On the other side, TPC recently became TPU (Threatened Plants Unit) within the I.U.C.N. dependent C.M.C. (Conservation Monitoring Centre) and it continues on as an international body to stimulate and coordinate activities in plant conservation.

Priorities established in the Kew conference of 1975 have been later on maintained and emphasized in several occasions, for example in the UNESCO-UNEP-WWF document 'World Conservation Strategy' (1980).

The first priority refers to the tropical rain forest whose surface area is reportedly dwindling at a rate of 2-3% each year (UNESCO/UNEP-FAO, 1978; Myers, 1981). The great floristic diversity of this biome and the ecological implications its dissapearance may have for the whole biosphere, make this problem very urgent. At the same time, there are several obstacles to rapid solutions. Apart from its strong economic interests, there are few botanic institutions or botanic gardens in the tropics that could co-operate by creating conservation-oriented plant collections or by providing technical support in the establishment of suitable genetic reserves. To make things more difficult, seeds of many rain forest trees are 'recalcitrant' and therefore not suitable for low temperature storage. Resolutions of the Kew conference urged: (1) 'that a strong network of nature reserves and conservation oriented gardens should be established throughout the tropics both through the strengthening and development of existing foundations and the creation of new ones where the need exists, (2) that institutions of temperate countries should offer all possible help in the programme through technical aid training and the secondment of personnel, and (3) that this aim should be pursued through the IUCN to ensure good coordination and proper understanding of the importance of the work for the tropical countries themselves and for the whole of mankind'.

The second priority 'ex aequo' corresponds to 'the conservation of threatened floras, particularly of islands and those parts of the world with Mediterranean or similar climates since both are often inhabited by very large numbers of narrowly endemic species of plants endangered by human activities'. Though the text might be taken more widely, it is evident that the Mediterranean region and the islands therein contained remain at the very core of this resolution.

3. Conservation of plants in the Mediterranean area

The floristic wealth of the Mediterranean region is very high and obvious but so far it is not possible to give an accurate estimate. The estimated total number of present species (25,000) begins to acquire significance when it is referred to surface area or compared to other regions of the world (see Chapter 3 by Quézel). The number of endemics in each country also provide good comparative figures. In this recpect it might be instructive to note that four out of five European endemics of that kind do correspond to the Mediterranean region. While the endemics of North and Central European countries are scored by fives or tens, those of Mediterranean countries are very often scored by hundreds. North African and East Asian Mediterranean countries are no less rich (Lucas, 1980). The Med Checklist being prepared under the auspices of OPTIMA (Organization for the Phyto-Taxonomic Investigation of the Mediterranean Area) is expected to provide more accurate data in the near future (Greuter, Burdet & Long, 1984).

The fact that many of the Mediterranean endemics are 'narrow' endemics is highly relevant to conservation policies and, as such, deserves comment. Firstly we are supposed to determine which plant species should be conserved, and it seems obvious that a good correlation exists between threatened plants and narrow endemics. But narrowness may be only geographical, only ecological or, more frequently, a combination of both. The approach provided by the endemics to a single country (as in TPU lists) is fairly satisfactory in the first instance, since a high percentage of threatened species fall within that category. However, several alterations will be necessary if we wish to improve the list and to make it one of prospective threatened plants with direct application in a conservation programme.

The first modification would consist of detecting those species which simultaneously belong to two or three countries but still occupy a relatively small area. These should be searched for in borderline areas, especially when these are mountainous as the Alps or the Pyrenees. Secondly, the islands should in general be treated separately from the mainland, as has already been done in the TPU lists (Anonymous, 1983) extracted from 'Flora Europaea' (Tutin et al., 1964–80). But this may in turn represent a source of inaccuracies, i.e. for those taxa which are common to the adjoining mainland but very poorly represented there. These cases should also be detected and included. For the larger Mediterranean countries it might be useful to assign the endemics to natural sub-regions and then to select those species which are only present in one or two of these sub-regions, say, to select priority cases amongst the most locally distributed endemics. By following this procedure, many widely distributed endemics, which are far from being threatened, can be set momentarily aside and much effort saved (see Chapter 4 by Gómez-Campo & Malato-Beliz on the Iberian Peninsula). The sub-regions should be natural biogeographic divisions and not political or administrative areas.

Ecological criteria should also be added to the purely geographic ones. Though they very often run parallel, this may not be always so. For plants adapted to very strict and vulnerable ecological habitats, a particular hazard can be present even if they belong to the floras of two or more different countries. As an example we could take those species living on humid, dripping and almost vertical rocks (the phytosociological division *Adiantea*) which are both local and rare in the Mediterranean region. On the other hand, rareness not only derives from successful adaptation to extreme or narrow habitats but may also be associated with a low but sufficient competitiveness in widely distributed non-extreme habitats. Plants which are widely distributed but always few in number belong to this group (Drury, 1974). This could also supply some candidates for the lists of threatened species. Finally, the existance of actual, observed decline in the populations of rare species provides another criterion for the inclusion of such species in the lists (Walters, 1976; Aymonin, 1980).

Many aspects of island biogeography (MacArthur & Wilson, 1967; Bramwell, 1979) are of much relevance in the Mediterranean region, not only due to the existence of many true islands in the Mediterranean Sea, but also because of the isolated situations of many of the mainland species. Rocky environments, humid zones, etc. are very often many kilometres away from other similar habitats, and tiny distant populations occur in agreement with that pattern. As pointed by Smith (1976), in relatively small islands with uniform habitat, the entire population may be a single interbreeding unit with a narrow homogeneous specialization, since genetic interchange with other neighbouring populations is lacking. Under these conditions – a small gene pool, reduced interspecific competition, inbreeding and small size of the population – such species become highly vulnerable to extinction. The importance of the breeding system and population aspects in conservation is discussed by Jain (1975a and b) or Bradshaw & Doody (1978).

The problem of the origin of the Mediterranean flora is not only zasic but has deep implications with regard to our conservation attitudes. This is why two chapters of this book (by Quézel and by Pons and Quézel) are almost entirely devoted to providing the scientific background to our understanding of its origin. In short, apportionment by other ancient floras plus a large series of more recent paleogeographic and paleoclimatic events have resulted in the present high diversity. Valuable relict taxa (paleoendemics) now co-exist with many other more recently evolved ones (neoendemics). This wealth poses one of the greatest difficulties for conservation. Lists of plants to be protected soon become very long, and much more so if infra-specific taxa are also included. In our opinion they should be, because they often may hold the clue to interesting or unresolved evolutionary problems. Moreover, imperfectly known taxa are also rather abundant and pose a number of doubts as to whether they should be included or not. Again we believe they should, at least until their taxonomic status becomes more clearly established.

Though due attention is merited by the taxa themselves, conservation becomes more logical when directed to syntaxa or to ecosystems because a number of species can be thus conserved together within their natural habitat and with a minimum of energy. However, simplistic views of conservation as an absence of human intervention in ecosystems may be completely misleading for the Mediterranean region (see Chapter 12 by Ruiz de la Torre). A basic fact is provided by the practical non-existence of endemics within Mediterranean climax or quasi-climax ecosystems. The Mediterranean endemic seems to have envolved under conditions of frequent disturbances and/or strong natural exploitation and they are mostly adapted either to subseral stages (steppes, low matorral, etc.) or to more stable situations where a strong limiting factor exists (as in rocky environments where the scarcity of available soil acts as a natural exploiting factor). Thus any conservation policy needs to avoid being entirely passive. A continuous monitoring to check the persistence of the conditions of existence of the ecosystem to be preserved will be necessary. Opportune and well-calibrated interventions to secure that persistence will, in many cases, be an essential requisite. On the contrary, a strict non-intervention policy will induce the ecological succession to go ahead and to approach a climax vegetation with the probable extinction of the endemic species we wish to protect.

These ideas might sound somewhere heterodox to many people who associate conservation to non-intervention. However, they should be taken into account when trying to conserve many Mediterranean species. It is a kind of paradox; Mediterranean endemics look as if they were more or less man-adapted since the beginning. They are not certainly man-adapted because most of them antedate humans by many hundreds of thousands or millions of years. But we could well say that they show a certain pre-adaptation to man as a consequence of their evolution under conditions of frequent disturbances, rapid climatic changes, naturally exploiting factors, etc.

Perhaps one of the most basic problems in the conservation of Mediterranean endemics lies in the fact that, while we need to intervene, we simultaneously lack ideas on how and when we must intervene. In other words, we know very little about the structure, function and conditions of stability of natural ecosystems in general and of Mediterranean ones in particular to be able to act with full success. Not in vain' the UNESCO 'Man and Biosphere' programme has put the greatest emphasis in stimulating basic studies on the natural biomes of the earth as a means to predict human impacts, and to ensure proper management either for exploitation or for conservation (UNESCO, 1971).

In an article by Greuter (1979), the 'what', 'where', 'why' and 'how' of the conservation of Mediterranean species are preliminarily discussed from the point of view of a plant taxonomist. This book pretends to treat the same subject in a wider and more general way. Were it produced a few years later, then more emphasis would have been put on the solutions and less on the problems themselves. But the right solutions will be derived from a sensible appreciation of the problems and this is still very imperfect today. Consequently, several chapters are devoted to the problems and prospects of plant-conservation in different Mediterranean subregions under the personal view of each author and with the help of a number of case histories. These are intended to provide material for further thought and a pool of information to be shared by botanists of different countries. As the efficiency of conservation is to be benefited by international co-operation (Myers, 1976), this comparative approach may be initially expected to stimulate at least an exchange of ideas. In its final chapters, the book additionally pretends to be a source of concepts and ideas on what to do in a subject which is both urgent and difficult.

References

Anonymous (1983). List of rare, threatened and endemic plants in Europe. TPU, IUCN-CMC. Strasbourg: Council of Europe.

Aymonin, G. (1980). Quelques considerations sur la notion de plante en danger. Bull. Soc. Bot. France 127: 105-110.

Bennett, E. & Frankel, O.H. (Eds.) (1970). Genetic resources in plants. Oxford and Edinburgh Blackwell.

Bradshaw, M.E. & Doody, J.P. (1978) Plant populations studies and their relevance to nature conservation. Biol. Conserv. 14: 223-242.

Bramwell, D. (1979). Plants and Islands. London: Academic Press.

Drury, W.H. (1974). Rare species. Biol. Conserv. 6: 162-169.

Frankel, O.H. & Hawkes, J.G. (Eds.) (1975). Crop Genetic Resources for Today and Tomorrow. Cambridge: Cambridge University Press.

Frankel, O.H. & Soulé, M.E. (1981). Conservation and Evolution. Cambridge: University Press.

Greuter, W. (1979). Mediterranean conservation as viewed by a plant taxonomist. Webbia 34: 87-99.

Greuter, W., Burdet, H.M. & Long, G. (1984). Med Checklist. Geneva: OPTIMA.

Heslop-Harrison, J. (1974). The plant kingdom: An exhaustible resource? Trans. Bot. Soc. Edinburgh 42: 1-15.

I.U.C.N. (1966). Red Data Book. Morges, Switzerland: IUCN.

Jain, S.K. (1975a). Population structure and the effects of breeding systems. In: Frankel, O.H. & Hawkes, J.G. (Eds.), Crop Genetic Resources for Today and Tomorrow; pp. 15-36. Cambridge: Cambridge University Press.

Jain, S.K. (1975b). Genetic reserves. In: Frankel, O.H. & Hawkes, J.G. (Eds.), Crop Genetic Resources for Today and for Tomorrow, pp. 379-396. Cambridge: Cambridge University Press.

Lucas, G. L. (1980). First preliminary draft of the list of rare, threatened and endemic plants for the countries of North Africa and the Middle East. IUCN-TPC. Kew: Royal Botanic Gardens.

MacArthur, R.H. & Wilson, E.O. (1967). The Theory of Island Biogeography. Princeton, N.J.: Princeton University Press.

Melville, R. (1970): Plant conservation and the Red Book. Biol. Conserv. 2: 185-188.

Myers, N. (1976). An expanded approach to the problem of dissapearing species. Science 193: 198-202.

Myers, N. (1981). Conservation needs and opportunities in Tropical Moist Forests. In: Synge, M. (Ed.), The Biological Aspects of Rare Plant Conservation, pp. 141-154. Chichester: Wiley.

Simmons, J.B. et al. (Eds.) (1976). Conservation of threatened plants, NATO Conference Series. New York and London: Plenum Press.

Smith, R.L. (1976). Ecological genesis of endangered species: The philosophy of preservation. Ann. Rev. Ecol. Syst. 7: 33-55.

Synge, H. (Ed.) (1981). The biological aspects of rare Plant Conservation. Chichester: Wiley.

Synge, H. & Townsend, H. (1976). Survival or Extinction. Kew: Bentham-Moxon and Royal Botanic Gardens.

Tutin, T.J. et al. (1964-1980). Flora Europaea. Cambridge: Cambridge University Press.

UNESCO (1971). Final Report. First Session of the International Coordinating Council of the Programme on Man and the Biosphere. Paris: UNESCO.

UNESCO-UNEP-FAO (1978). Tropical Forest Ecosystems. Paris: UNESCO.

UNESCO-UNEP-WWF (1980). World Conservation Strategy. Paris: UNESCO.

Walters, S.M. (1976). The conservation of threatened vascular plants in Europe. Biol. Conserv. 10: 31-42.

Departamento de Biología
Escuela T.S. Ing. Agrónomos
Universidad Politécnica
E-28040 Madrid, Spain

CHAPTER 2

Definition of the Mediterranean region and the origin of its flora

P. QUÉZEL

1. Introduction

The Mediterranean region might easily give the impression of forming a geographical unity. It is formed by similar territories that surround an almost enclosed large sea and extends from east to west along a zone within middle latitudes. Actually this is a false impression because the region includes a very wide range of ecological conditions and a similar diversity in the history and biogeographical significance of the component flora. These variations have been further accentuated by human action which dates from ancient times.

A geographical as well as climatic or floristic definition of the Mediterranean region is not easy to propose. Different authors have offered differing interpretations, and it would be pretentious to even try to solve such controversial problems in this paper. However, the past few years have seen much progress in the knowledge and interpretation of the Mediterranean climate and vegetation. This enables us to propose a more coherent synthesis of the Mediterranean world that will lead to a better comprehension of its floristic, ecological and historical significance.

2. The limits of the Mediterranean region

For more than a century the problem of the limits of the Mediterranean region has been studied either by floristic methods, vegetation structure analysis, climatic interpretations or, finally, bioclimatic methods.

2.1 Floristic criteria

The use of certain species to define precisely the extent and limits of the Mediterranean region can be traced back to the nineteenth century. Drude (1884) was the first to try to characterize this region by the presence of *Quercus ilex L.* In France, Durand & Flahault (1886) used the presence of olive tree cultivation as their criterion, a position that has also been recently adopted by some geographers (Birot & Gabert, 1964).

This method has been criticized by Emberger (1943), on the basis of his own observations in North Africa. Emberger noticed that olives and *Quercus ilex* (including *Q. rotundifolia* Lam.) can be used as primary criteria in certain parts of the Mediterranean, but they are not valid for defining the general ecological limits of the Mediterranean region. Both species are absent from vast areas of the Maghreb and *Q. ilex* is also absent throughout the Eastern Mediterranean region (Fig. 1), where it is somewhat replaced by *Q. calliprinos* Webb (considered by some authorities to be a part of the infra-specific variability of *Q. coccifera L.)*

Floristic methods can still give interesting results when used in particular regions. Thus, in North Africa the limit between the Mediterranean region and the Sahara can often be defined by the disappearance of *Stipa tenacissima* L. (Capot-Rey,

Gómez-Campo, C. (ed.), Plant conservation in the Mediterranean area.
© 1985, Dr W. Junk Publishers, Dordrecht. *ISBN 90 6193 523 7.*

Fig. 1. Map showning the area of distribution of *Quercus ilex* (incl. *Q. rotundifolia)* in the Mediterranean region. The broken line represents the limits of *Q. calliprinos* in the Eastern Mediterranean.

1952; Quézel, 1965). On the other hand, Le-Noble (1934) had previously studied the indicator role played by some species such as *Genista cinerea* (Vill.) DC., *Thymus vulgaris* L. or *Lavandula angustifolia* Miller, whose geographical limits seem to follow the real borders of the Mediterranean region.

These few examples show the limitations which are usually associated with floristic methods in the search for a definition of the Mediterranean region.

2.2 Criteria based on vegetation analysis

The development of modern methods of vegetation analysis has led to more practical and accurate methods to study the limits of the Mediterranean region. Flahault (1937) was a pioneer in proposing that not only *Q. ilex,* but the association to which it corresponds (precisely defined by Braun-Blanquet, 1915), should be considered. However, it quickly became apparent that neither the sclerophyllous forests nor the evergreen maquis and garrigues, with which they are generally associated, correspond to the actual extension of the biogeographical or ecological Mediterranean region. There are, for example, extensive regions covered with pines, deciduous forests or even with degraded forest that certainly cannot be excluded from the Mediterranean region if defined in this way. This is often the case for the Atlas, Taurus, Lebanon and Greek mountains as well as vast areas of the Maghreb and Anatolian highlands. There are also large areas where the theoretical forest climax was formerly constituted by such clearly Mediterranean elements as *Pinus halepensis* Miller or *P. nigra* Arnold, but these have been destroyed by man and replaced by a steppe-like vegetation whose appearance casts doubts on whether or not it belongs to the Mediterranean.

Thus the precise analysis of certain vegetation types clearly demonstrates that Mediterranean types of vegetation have often changed over relatively short periods of time either into European or Saharan types. There may also exist intermediate zones, often several hundred kilometres wide, where the eventual ascription and distribution of the biogeographical units largely depends upon ecological criteria, meso- and micro-climatic conditions or even historical events. Such is the case of the Saharan mountains (Quézel, 1965), where the remains of Mediterranean vegetation are still present after several millenia in disjunct localities that are more than 1.500 km distant from the rest of the Mediterranean-Maghreb zone.

By 1943 Emberger had already shown the serious imperfections of these floristic and phytosociological methods when they are used to establish the limits of the Mediterranean region. In his opinion, 'the only correct and non-subjective way to solve this problem is by trying to define the limits of the Mediterranean climate itself'.

2.3 Climatic criteria

Even though certain factors are unanimously accepted as defining the Mediterranean climate (such as summer drought), others have been subjected to differing interpretations (e.g. the distribution of rainfall, thermal factors, etc.).

De Martonne (1927) seems to have been the first to define the Mediterranean climate as a temperate climate of the subtropical zone, that can be differentiated by a cold temperature and humid season (average temperature for the coldest month >5°C) and by a hot and dry season (average temperature for the hottest month over 20°C) (see the definitions given by Gaussen, 1954; Walter & Lieth, 1960; Trewartha, 1954; and Peguy, 1970 on this subject). The results obtained by these authors are more or less similar for most countries bordering the west and central Mediterranean, but they are quite different for the Near and Middle East (Figs. 2 and 3).

It is also interesting to note that the criteria used by North American climatologists (Aschman, 1973) to define the Mediterranean climate of California are clearly more restrictive (annual rainfall between 275 and 900 mm with at least 65% of it occurring in winter, and average monthly temperature over 1°C). When this is applied to the Old World, the extent of the Mediterranean region is reduced (Fig. 4). It is evident that this view contradicts the observations derived from vegetation,

Fig. 2. Limits of the Mediterranean region as proposed by Gaussen (1954).

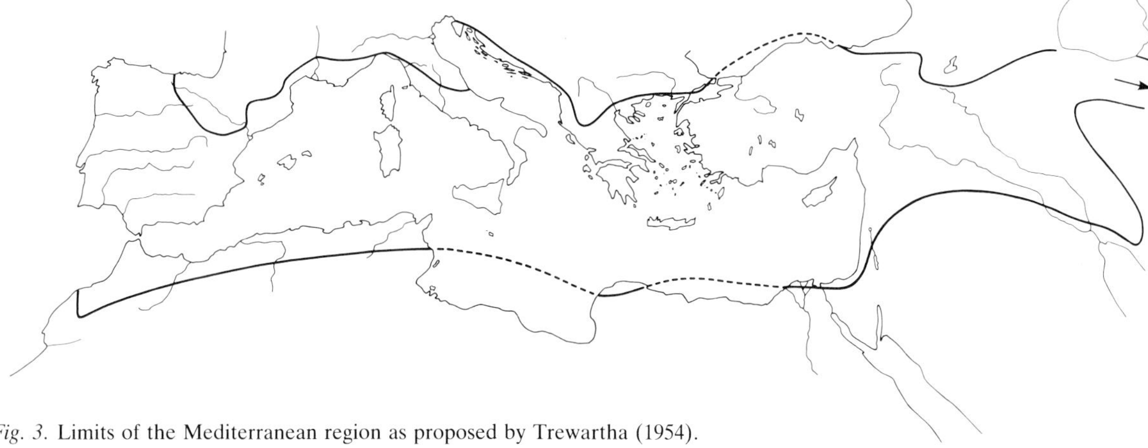

Fig. 3. Limits of the Mediterranean region as proposed by Trewartha (1954).

Fig. 4. Limits of Mediterranean region as proposed by Aschmann (1973).

since it leads, for example, to the exclusion of the Languedoc, a large part of Italy and the Dalmatian coast.

The interpretations of the Mediterranean climate by different authors may therefore differ widely and they may in turn diverge greatly from the often oversimplified conclusions of phytogeographers. Essentially for this reason it soon becomes apparent that a satisfactory answer to this problem can only be found by comparing and/or integrating climatic and ecological criteria. In fact, only one climatic criterion has been unanimously accepted by all workers, i.e. the presence of a dry, hot season during the summer when the vegetation needs more water than it receives. As indicated by Daget (1977a), the problem is one of water balance. It is nonetheless a very complex problem, whose mere estimation requires delicate equipment and experiments. That is why in order to find the most simple answer possible, bioclimatic criteria have been proposed by different authors.

2.4 Bioclimatic criteria

An excellent synthesis of this subject has been presented by Daget (1977a, b, 1980). Though there is no need to review it here in detail, it should be pointed out that the definitions given by Emberger (1930a, b) and Bagnouls & Gaussen (1953), are today used by most biologists and bioclimatologists working in the Mediterranean basin. Though largely empirical, these methods have very satisfactorily helped in estimating the variability within the Mediterranean climate, essentially as a function of summer drought and temperature.

In many English-speaking countries, especially in North America, other methods developed after the definition of Thorntwaite's (1948) coefficients are currently used. These coefficients were theoretically established as a function of the vegetation's hydric balance. They have been subjected to some criticism (Turc, 1961), and do not seem to be more accurate than those of Emberger or Gaussen.

A climate may be considered to be Mediterranean (Daget, 1977a) when it meets the following two criteria:

- summer is the driest season;
- there is a period of effective physiological drought.

This dry period can be evaluated without appreciable differences in the results by using the bioclimatic coefficients of the authors mentioned above. It should be noted that this definition does not take into consideration thermal criteria. Thus it does not agree with the opinions of Bagnouls & Gaussen (1953) for whom zones with an average monthly temperature lower than 0°C should not be included. Neither does it agree with the more restrictive viewpoints of the American authors referred to above.

This restriction explains the frequent confusion and the existence of doubtful classifications according to which the Mediterranean climate is limited to littoral bands with mild winters. This does not match any biogeographical reality and excludes wide zones where flora and vegetation are still indisputably Mediterranean.

On the other hand, if we admit that there are no limits to thermal minima when describing the Mediterranean bioclimates, which is quite evident in the Maghreb, Anatolia or the Near-East, we can then distinguish a whole scale of ecologically evident thermal variants in accordance to that factor (Emberger, 1939).

In a similar way, bioclimatologists who work on the Mediterranean bioclimate have defined several zones as a function of increasing humidities. Though atmospheric humidity is worthy of consideration, most indexes are based on annual rainfall. The terms most commonly used are: arid, semi-arid, sub-humid and humid (Emberger, 1930a, b; Thornthwaite, 1948).

These facts have recently permitted Daget (1977a) to define an isoclimatic Mediterranean area (Fig. 5) in correspondence with the whole of the northern tropical territories of the Old World. It is characterized by a Mediterranean bioclimate in agreement with the above cited definition.

It is to be noted that these bioclimatic conclusions are also in agreement with the positions that are now adopted by various biogeographers. Quézel (1978) has described a Mesogean sub-Empire within the Holarctic Empire where the Mediterranean, Irano-Turanian and Saharo-Arabian regions are grouped. The boundaries of this Mesogean sub-Empire approximately match those of the isoclimatic Mediterranean area.

3. General characteristics of the Mediterranean region

Defined and limited by the criteria discussed above, the Mediterranean region shows a number of general features which may not be completely diagnostic themselves, but certainly allow a better understanding of the region. The only common trait undoubtedly shown by these characters is heterogeneity.

3.1 Geological and geographical heterogeneity

With perhaps the most complicated geology in the world, the Mediterranean presents an extremely fragmented pattern that makes it appear a jigsaw puzzle. Marked reliefs, rugged massifs often frag-

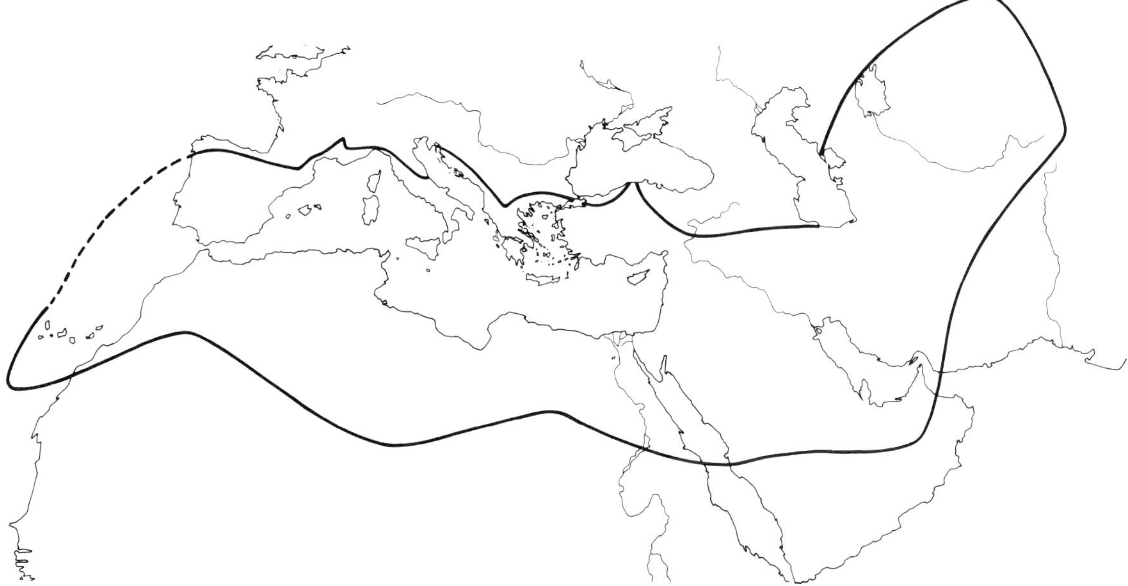

Fig. 5. Limits of the isoclimatic Mediterranean area (Daget, 1977).

mented with deep narrow valleys, highlands and enormous basins of deposition, jagged and winding coastlines with numerous islands, are just some of its outstanding geographical features. Geological maps show almost every type of substrata, though sedimentary rocks largely dominate. It is evident that al these are important factors in the diversification and characterization of the flora.

Such a remarkable variety of geomorphological structures due to either palaeogeological or palaeogeographical events or to present historical factors, has contributed to the number of ecological niches that are available to plant species.

3.2 Climatic and bioclimatic heterogeneity

The wide range of temperatures and of rainfall make the regions now bordering the Mediterranean into a curious blend of the extratropical world (Quézel, 1976). Annual rainfall may be as low as 100 mm in some predesertic zones and can climb to as high as 3000 mm on certain mountains. Average annual temperatures range from 5 to 18°C.

The role of continentality, atmospheric humidity and insolation, should also be considered. By combining these factors with annual rainfall and temperature, it is possible to identify bioclimatic units that correspond with a certain number of fundamental divisions of great ecological significance.

Emberger (1930a), has defined a series of étages and bioclimatic zones, which are classic in Mediterranean studies, based on his bioclimatic coefficient

$Q_2 = \frac{2000 \times P}{M^2 - m^2}$ (where P is the annual rainfall in mm, M is an average of the maxima of the hottest month, and m is an average of the minima in the coldest month). The values of P are obviously decisive in evaluating this coefficient, and some authors (Le Houerou, 1971; Stewart, 1975) have tried to use P instead of Q_2. This seems to be valid if used for relatively small areas.

Bioclimatic type	Q_2	P in mm (and for the cool thermic variant)
per-arid	<10	<100
arid	10 to 45	100 to 400
semi-arid	45 to 110	400 to 600
sub-humid	70 to 110	600 to 800
humid	110 to 150	800 to 1200
per-humid	>150	>1200

Each of these bioclimatic types might also be subdivided. It is also possible to distinguish a number of thermal variants defined as a function of m (*sensu* Emberger). Their generally admitted limits (Akman & Daget, 1971; Nahal, 1972; Daget, 1977) are as follows:

Variant	Values of m
very hot	m = 10°C
hot	m = between 7 and 10°C
temperate	m = between 3 and 7°C
cool	m = between 0 and 3°C
cold	m = between – 3 and 0°C
very cold	m = between – 7 and – 3°C
extremely cold	m = between –10 and – 7°C
icy	m = –10°C

The graphic combination of the bioclimatic zones and their variants leads to a diagram – the climagram (Emberger, 1933; Sauvage, 1963) – whose utilization is now classic in Mediterranean phytogeography.

The per-arid bioclimate corresponds to the Saharo-Arabian region, while the coldest variants (extremely cold to icy) correspond either to high mountains or to the Irano-Turanian region (Fig. 6).

3.3 Altitudinal étages

Following the works by Flahault (1901), Gaussen (1926), Schmid (1966), Ozenda (1975), Quézel (1974, 1976) and Quézel & Barbero (1981) an altitudinal zonation of species and vegetation types can be defined basically by thermal criteria. However, any strict comparison with the above-defined thermic variants, is very delicate (Achhal et al., 1979). The following étages are generally distinguished.

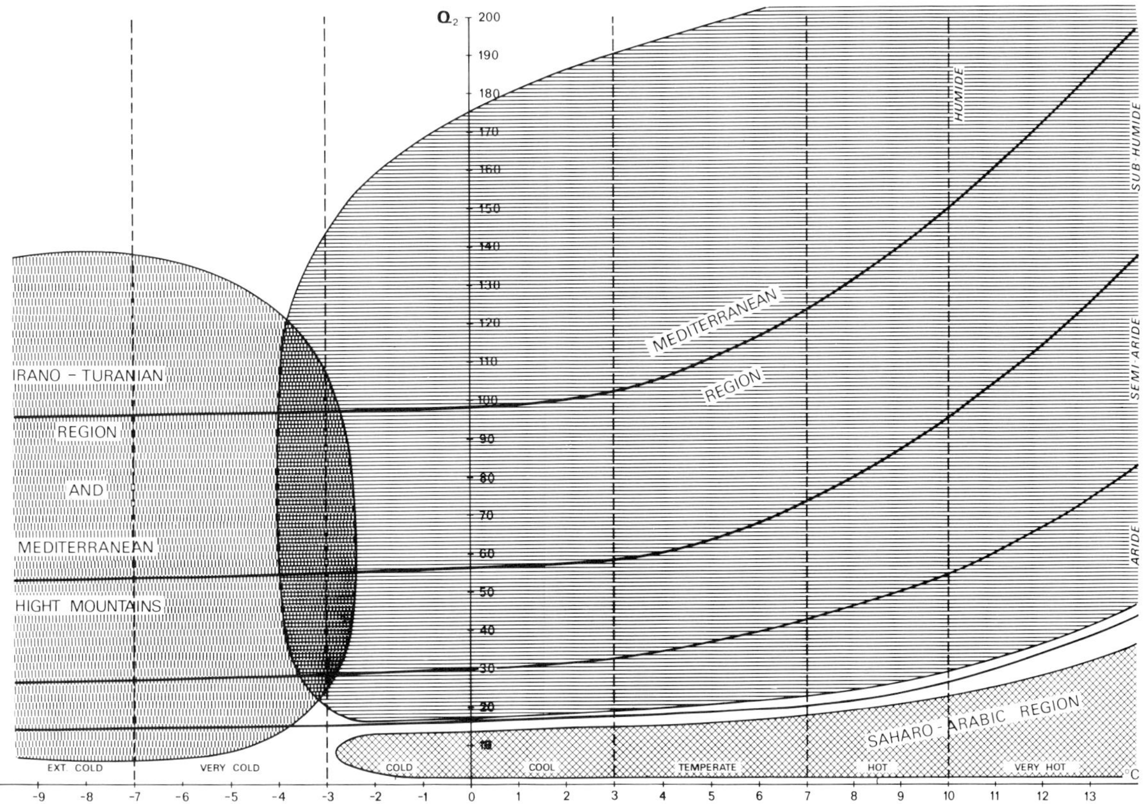

Fig. 6. Schematic projections of the Emberger's climagram of the Mediterranean, Saharo-Arabic and Irano-Turanian regions (including Mediterranean high mountains).

Infra-Mediterranean – which corresponds only to the Macaronesian zone of western Morocco, with *Argania spinosa* (L.) Skeels and *Acacia gummifera* Willd.

Thermo-Mediterranean – which is found all around the circum-Mediterranean area, and is formed by sclerophyllous stands of *Olea europaea* L., *Ceratonia siliqua* L., *Pistacia lentiscus* L., *Pinus halepensis* Miller, *P. brutia* Ten., and *Tetraclinis articulata* (Vahl) Masters.

Meso-Mediterranean – which is mostly formed by sclerophyllous forests of *Q. ilex* in the western and central Mediterranean, and of *Q. calliprinos* in the east.

Supra-Mediterranean – which is the main area of deciduous forests within the humid bioclimate. In sub-humid and even in semi-arid bioclimates as well as in the southern Mediterranean it is replaced by an Upper-Mediterranean level, dominated by sclerophyllous oaks.

Mountain-Mediterranean – which essentially corresponds to highland coniferous formations (*Cedrus, Pinus negra* and the Mediterranean firs).

Oro-Mediterranean – which is often made up by grazed grasslands or thorny xerophytic garrigues (Quézel, 1956).

Alti-Mediterranean – which is seldom developed except in the Atlas or in the Taurus Mountains, where scattered dwarf chamaephytes develop.

Though this succession is generally recognizable, the limits between the étages depend upon the latitude and other topographic and ecological criteria.

3.4 Surface area and a floristic inventory

Following the above interpretations, it is now possible to give an estimate of the surface area that is occupied by the Mediterranean region in the Old World, as well as an indication of its floristic richness.

3.4.1 Surface

Difficulties in estimating the surface area of the Mediterranean region arise from the lack of coincidence between its natural limits and political frontiers. Also, the attribution of certain territories as a part of the Anatolian highlands to the Mediterranean region remains doubtful. Therefore it would not be worthwhile attempting to give precise values. The following estimates are intended only to give a general idea.

The surface area of the Mediterranean region north of the tropics, in the Old World can be estimated as 2,300,000 km². Table 1 shows how this surface area is distributed among the different countries.

If we take into consideration the whole Mesogean sub-Empire this area would be tripled.

Defined as stated above, the following regions should be attached to the Mediterranean climatic and/or floristic region: (a) a significant portion of the Irano-Turanian region, especially the cool and cold climatic zones which occupy the south-western slopes of the Zagros Mountains and some regions in Afghanistan and Pakistan; and (b) the north-western region of Iran. They are characterized by a Mediterranean vegetation with sclerophyllous oaks and Mediterranean conifers. This can be clearly seen on the 1/5,000,000 bioclimatic and vegetational maps published by UNESCO.

Table 1. Surface area occupied by nineteen countries in the Mediterranean region.

Country	Surface area included in the Mediterranean region (in km²)	Approximate percentage of total area
Spain	400,000	17.3
Portugal	70,000	3.0
France	50,000	2.1
Italy	200,000	9.0
Yugoslavia	40,000	1.7
Albania	20,000	0.8
Greece	100,000	4.3
Turkey	480,000	20.8
U.S.S.R.	10,000	0.4
Cyprus	9.000	0.3
Syria	50,000	2.1
Lebanon	10,000	0.4
Israel	10,000	0.4
Jordan	10,000	0.4
Egypt	50,000	2.1
Libya	100,000	4.3
Tunisia	100,000	4.3
Algeria	300,000	13.3
Morocco	300,000	13.3

3.4.2 Floristic wealth and inventory

Again, if we agree with the limits given above, the floristic richness becomes very considerable, as has been often pointed out by various authors. Table 2 shows this very clearly.

The Mediterranean floristic richness appears more clear if we compare it to that of non-Mediterranean Europe. This covers an area of 9,000,000 km² and scarcely has 6,000 species (British Isles has for instance, 1,400 species in an area of 308,000 km²). But the richness of a flora is not to be measured by its total number of species but also by the number of endemics. The difficulty lies in the fact that the term 'endemism' has to refer to a defined area. The inventory given by Leon, Lucas and Synge in this volume for the endemics of each Mediterranean country, only provides an approximate idea of that diversity since species' distribution does not follow the political boundaries. For instance, 4,000 species grow in Mediterranean northern Africa (Quézel, 1978), and some more than 1,100 species are endemic while more than

Table 2. Mediterranean regions of the world. Estimates of surface area and number of species.

Region	Area 10^3 km²	Approximate number of species
Mediterranean region	2,300,000	25,000
California floristic province (Raven & Axelrod, 1978)	324,000	4,400
Australia (Specht et al., 1974)	7,716,000	15,000
Southern Africa (Goldblatt, 1978)	2,573,000	18,500

3,000 species should be considered as exclusively Mediterranean. It seems that out of the 25,000 above-mentioned species, more than half are endemic to the Mediterranean region as a whole.

It is at present almost impossible to supply a precise floristic inventory for the Mediterranean region. We do have various national ones available but they are often imperfect and the product of long and tedious work, and often open to criticism. The 'Med. Checklist' which is being prepared under the auspices of OPTIMA, will soon give us the first instructive estimate. Revision of the circum-Mediterranean flora will have to be carried out to reduce the numerous taxonomic uncertainties that now exist and the differences due to the different assessments of the systematic value of many taxa by different authors.

At the C.N.R.S. conference, in Montpellier (1975), which was devoted to the flora of the Mediterranean basin, a tentative inventory was presented of the data relating to our knowledge of the Mediterranean flora in different circum-Mediterranean countries. This has been completed and is now offered in Table 3.

This inventory was not primarily designed to cover the Mediterranean region, so we have tried to adjust it to the countries or territories involved within the Mediterranean world as defined above.

Even so, it is very risky to advance a precise number for the total flora. However, between 23 to 25 thousand species can be regarded as a reasonable approximation for the Mediterranean region *sensu stricto*. The Saharo-Arab region may add no more than 3,000 - 3,500 species (Quézel, 1978). For the Irano-Turanian region it is rather difficult to propose a number. However, the richness of its flora seems to be not far below that observed in the Mediterranean (probably 15 to 20 thousand species).

Table 3. Estimated floristic wealth of fourteen Mediterranean countries[α].

Country	Approximate number of species	Approximate number of species in the Mediterranean region
Portugal (Pinto da Silva, 1975)	3100	2500
Spain (Gaetano, 1975)	7500	6000
France	4500	3000
Italy (Moggi, 1975)	5500	3500
Jugoslavia	5000	2500
Albania	3000	2200
Greece (Greuter, Phitos & Runemark, 1975)	5500	4000
Turkey (Davis, 1975)	8000	5000
Cyprus (Osorio-Taffal & Seraphin, 1973)	1800	1800
Israel (Zohary, 1962)	2200	1500
Egypt (Boulos, 1975)	2100	1100
Libya (Boulos, 1975)	1600	1400
Algeria - Tunisia (Quézel, Bounaga, 1975)	3400	2800
Morocco (Sauvage, 1975)	4200	3800

α All references are from 'La Flore du Bassin Mediterranée, essais de systématique synthétique', Coll. Int. C.N.R.S. de Montpellier - Editions du C.N.R.S. Paris 1975, 576 pp.

4. The biogeographic significance of present Mediterranean flora

The problem of the biogeographical interpretation of the Mediterranean flora has been approached by various authors. The most recent are Walter & Straka (1970), Axelrod (1973), Axelrod & Raven (1978), Pignatti (1978), Quézel (1978), Quézel et al. (1980) and Quézel (1983).

An almost monolithic conception of the Mediterranean as a unit has given way to an interpretation which is much more precise and flexible. This has been the result of paleobotanical, palynological and even paleogeographical or paleoclimatical research in the past fifteen years. The flora is now viewed as a heterogenous entity associated with a region that is largely defined by climatic criteria.

In terms of the biogeographical significance of the flora, two major classes can be distinguished:

(a) of southern origin (tropical or subtropical);
(b) extratropical with an autochthonous or northern origin.

4.1 Elements of southern origin

Elements of southern origin play an important role in the present Mediterranean flora. This has been emphasized by various authors, particularly by Raven (1972, 1973) and Quézel et al. (1980). These elements actually belong to tropical taxa of undoubted African origin. However, they are not always easy to interpret and they are often associated with difficult biogeographic problems.

4.1.1 Elements showing pantropical disjunctions

The distribution of certain groups that are present today both in the Mediterranean region and in other distant territories, can only be explained as being derived fromsa situation prior to the fragmentation of the southern continental bloc. The European representatives of the families Gesneriaceae, Datiscaceae, Buxaceae, Compositae (Mutisieae) and the genera *Borderea* and *Dioscorea,* are allied to other South African or South-American taxa. The affinitites of *Tetraclinis* should be amongst the *Callitris* of Australia. The same occurs with *Aphyllanthes* and *Picconia* of the Canary Islands whose other representatives are in Australia or New Caledonia. *Naufraga,* from the Balearic Islands, seems to be related to *Schizeilema* of New Zealand.

There can be no doubt that these types of disjunctions date to the Lower Cretaceous.

A certain number of elements which are now simultaneously present in Asia, Africa, tropical America and even in Australia, should also be included in this group: Aquifoliaceae (*Ilex*), Aristolochiaceae (*Aristolochia*), Coriariaceae (*Coriaria*), Moraceae (*Ficus*), Myrtaceae (*Myrtus*), Styracaceae (*Styrax*), Ulmaceae (*Celtis*), Vitaceae (*Vitis*). The same applies to several taxa that, curiously, are meso-xerophilous or xerophilous: Celastraceae (*Maytenus*); Mimosoideae (*Acacia, Prosopis*), Rhamnaceae *(Zyziphus),* Santalaceae (*Osyris*), Tamaricaceae (*Tamarix*).

4.1.2 Elements showing north-tropical disjunctions

Another more recent and highly significant group is simultaneously present in the Mediterranean basin and in California (Axelrod, 1973; Axelrod & Raven, 1978; Quézel, 1978).

Their appearance can thus be regarded as being prior to the formation of the North Atlantic ocean: *Boerhavia, Cleome, Commicarpus, Fagonia, Lycium, Pistacia, Rhus, Smilax, Talinum, Trianthema* and *Vitex.* There are also some examples which appear in the Western Mediterranean, including the Canary Islands, and in Eastern North America (*Cneorum, Corema*).

4.1.3 Elements of palaeo-tropical origin.

Several distinct groups of taxa can be included here. An important number of genera of African (or even Madro-Tertiary) origin was already important in the circum-Mediterranean flora during the Oligocene and the Miocene, without reaching America (cf. Axelrod, 1973). Their establishment probably took place during these periods, i.e. after the widening of the North Atlantic Ocean. Many have subsisted in the Mediterranean region: genera such as *Asparagus, Capparis, Ceratonia, Chamaerops, Jasminum, Olea, Nerium* and *Phillyera* have originated from that stock.

Certain elements of the Canarian 'Laurisilva' have a similar significance, though some are also present in Northern America (*Clethra, Ocotea, Persaea*). Most of them, as *Catha, Myrica, Myrsine* and *Visnea,* were subsequently eliminated from the Mediterranean region and Europe during the ice ages.

The establishment of xerophilous taxa of African origin could not take place before the appearance of genuinely arid phases, which seemingly occurred at the end of the Miocene (particularly in the Messinian). For instance, *Acacia gummifera, Argania spinosa, Enneapogon, Gaillonia, Oropetium* (Gillet & Quézel, 1959). *Periploca, Tribulus, Trichodesma, Zygophyllum* and other elements are generally referred to an African desertic complex or 'randflora' (Aubreville, 1949; Monod, 1951; Quézel, 1958). Some genera or species which are now localized in the Canary Islands or in the Macaronesian sector of Morocco (as cactoid: *Euphorbia, Kalanchoe* or *Commelina*) as well as at the arid limits of the high Northern Africa or Saharan mountains (as *Hertzia, Pentzia*), have probably a similar history.

It is interesting to point out that certain tropical taxa with no endemic tendencies have recently migrated into the Mediterranean region. During their migration they probably took advantage of the Quaternary rainy phases (Quézel & Martínez, 1961). Some Andropogoneae (Quézel, 1958), some hygromesophytes in which zoochory plays a certain role (e.g. *Fimbristylis, Fuirena, Oldenlandia, Laurenbergia,* numerous Gramineae, *(Panicum, Echinochloa, Eragrostis,* etc.) and even *Acacia* or the Sahelian Capparidaceae all belong to this group.

In any case, the African north-south migrations constitute a very interesting problem whose importance has been recently demonstrated by Goldblatt (1978) and Quézel (1978). On the one hand, the relicts which grow on the African high mountains should not be forgotten. On the other, it is evident that these migrations could not have taken place without the presence of an orographic continuity, without important climatic variations, or better, without the concordance of these factors. The uplifting of African volcanic mountains that started at the beginning of the Miocene (Axelrod & Raven, 1978), has led to a consideration of the existence of periodical possibilities of exchange that has actually taken place in both directions. It also seems possible that the important climatic changes during the Messinian also helped to facilitate these contacts. However, the presence of pairs of generic vicariants between the Mediterranean region and the Cape region, suggests an ancient probably Miocenic disjunction: *Echium-Echiostachys, Iris-Moraea, Thymelaea-Passerina, Mercurialis-Seidelia, Buxus-Nothobuxus* (Burtt, 1971; Goldblatt, 1978), and also *Platycapnos-Discocapnos, Sarcocapnos-Cysticapnos.* The presence of the genus *Erica* in the Canary Islands where sclerophyllous oaks are absent, supports the same view.

4.2 Elements of extratropical origin

Extra-tropical (northern or autochtonous) elements, represent the greater part of the flora existing in the Mediterranean region. Their biogeographical analysis is not always easy, because of climatic perturbations in the Quaternary. However, at least three groups or elements can be identified:
(1) strictly Mediterranean,
(2) Mesogean,
(3) Holoarctic or Eurasiatic.

4.2.1 The strictly Mediterranean group

All the Tertiary-Mediterranean taxa and a great part of the flora of the Mediterranean mountains (the Oro-Mesogean flora) can be placed here.

The autochtonous Tertiary-Mediterranean flora is actually associated to a Mediterranean type of climate and it seems almost certainly to be ancient in spite of the absence of unequivocal evidence for the existence of this type of climate before the end of the Miocene, i.e. before the Messinian. Nevertheless, the presence in the Mediterranean climatic zones of North America of a considerable number of genera which are also present in the Mediterranean region of the Old World (Axelrod, 1973; Axelrod & Raven, 1978; Quézel, 1978), confirms a common origin that should be dated before the widening of the North Atlantic Ocean. Some examples are *Arbutus, Berberis, Helianthemum, Lavatera, Salvia* and several representatives of the genera *Cupressus, Pinus, Juniperus* (arborescent types) and the sclerophyllous oaks.

The development of this flora was favoured by the existence of continental zones or microplates, that appear to have been relatively stable during the Tertiary. These zones correspond to the present centres of Mediterranean endemism. Three of them are worth mentioning.

The Ibero-Mauritanian centre is the most remarkable (Quézel, 1978). At least 16 genera are localized in this centre as a whole, plus 25 which are endemic to north-west Africa and about 16 to the Iberian Peninsula; for example: *Boleum, Echinospartum, Euzomodendron, Ischaris, Ortegia, Petrocoptis, Rothmaleria, Securinega.* The importance of endemism among the Cruciferae (17 genera), or the Compositae (10 genera) is very great. This centre should be considered as the cradle of numerous genera such as: *Brachyapium, Cistus, Diplotaxis, Genista, Halimium, Helianthemum, Ionopsidium, Jasione, Linaria, Leucojum, Narcissus, Ptilotrichum, Retama, Teucrium, Ulex,* etc.

The Balkan centre is totally superposed with the ancient Apulian plate (Biju-Duval et al., 1977), and it plays a more discrete role. However, the following genera appear to be attached to it: *Degenia, Drypis, Edraianthus, Haberlea, Halacsya, Jankaea, Petromarula, Petteria,* etc. Emphasis should also be put on the rich representation of the following genera: *Alyssoides, Bornmuellera, Carum, Huetia, Pelaria, Sesleria, Silene,* and *Stachys.*

The Anatolian centre corresponds to the zone of contact between the Taurid and Arabic plates. Almost twenty genera can be fully associated with it. Among them are: *Cyprina, Dorystaechas, Gonocytisus, Michauxia, Microsciadium, Olymposciadium,* and *Thurya.* This region also represents a centre of diversity for various other genera such as *Aethionema, Alyssum, Asyneuma, Bolanthus, Ebenus, Phlomis, Petrorhagia, Ricotia, Rossularia* and *Verbascum.*

Other secondary centres such as the Tyrrhenian for *Naufraga, Morisia, Nanathea* and *Soleirolia* and the Cyrenaican for *Pachyctenium, Libyella* and *Euhespherida* are worth mentioning.

It is of interest to point out that among all these sets of taxa, no family is strictly endemic to the Mediterranean region. However, Globulariaceae, Resedaceae and Cistaceae are largely preferential to the area.

The uplift of the Alpine, Atlas and Himalayian ranges led to the development of a very important orophilous flora which was differentiated from autochthonous elements and benefited by the arrival of important new elements during the glacial periods. These phenomena became rather complicated by the presence of numerous centres of endemism, related to geographic and/or climatic isolation (Quézel, 1957-1978; Favarger, 1975).

A certain homogeneity is apparent in the flora of these mountains and it may be wondered whether it is original or whether it corresponds to multiple migrations, mainly from east to west. These became possible by the superposition of mountain-building phenomena and also through the Messinian climatic crises (Bocquet et al. 1978).

The existence of a primarily common alpine flora is evident through the presence of frequent segregation phenomena. These can be easily noted in genera as *Abies, Asperula, Berberis, Cedrus, Cotoneaster, Dracocephalum, Erigeron, Juniperus, Lonicera, Papaver, Quercus, Silene,* etc.

This oro-Mesogean preglacial flora was locally enriched by elements of Mediterranean, Irano-Turanian and even European origin, leading to the appearance of diversified groups with a high endemism ratio (i.e. 25% for the High Atlas) where the glaciations had a drastic effect. Alpine elements of European or even Boreal origin were also added.

Henceforth, these populations roughly followed the same path and the same vicissitudes as the whole Mediterranean region. It is only the high degree of local endemism, which has certainly arisen, since the ice ages, that needs to be emphasized.

4.2.2 Other Mesogean elements

The Irano-Turanian element. The development of a cold steppe flora is very important to the history of the Mesogean. The absence of African elements, seems to favour an eastern Mesogean development, which took place on the northern banks of the Tethys. The migration of some elements to north-western America is evident: *Artemisia, Astragalus, Ephedra.* Its development has been most probably favoured by dry and cold periods and perhaps the Messinian also played a role. The abundance of endemic genera also suggests an ancient origin.

On the other hand, its expansion in the Mediterranean region and also in Europe can easily by explained by the glacial phases of the Plio-Pleistocene, during which, full glacial and post-glacial periods with *Artemisia, Ephedra* and *Salso-*

la certainly correspond to important advances of this flora. Noticeable traces can be observed in the Northern Saharan and Maghrebi highlands (*Artemisia,* (part), *Astragalus,* (part), *Gymnocarpos, Noaea, Salsola* etc.) and even on the circum-Mediterranean high mountains.

The Saharo-Arab element. The vast deserts now existing in the south of the Mediterranean region did not appear before the Miocene-Pliocene boundary, and the Messinian crisis might have played an important, if not a decisive role. These observations are supported by geomorphological data (Butzer & Hansen, 1968; Coppens & Koeniguer, 1976; Rognon, 1967), as well as by palynological data (Maley, 1977).

As has been already stated by Quézel (1978), the present Saharo-Arab flora has differentiated from essentially xerophilous and biogeographically heterogeneous ancestors. However, Mediterranean elements are largely dominant (Boraginaceae, Caryophyllaceae, Cruciferae, Compositae, Gramineae). African elements *sensu-lato* also play a notable role (Asclepiadaceae, Aizoaceae, Capparidaceae, Mimosoideae). Endemism is generally at the species level in what seems to correspond to relatively recent evolutionary adaptations. However, a problem is posed by a few Saharo-Arabian genera, whose origin seems to be very old (*Anabasis, Cornulaca, Hammada, Halogeton, Traganum, Calligonum, Nitraria),* so that the existence of relations with Irano-Turanian flora should also be considered.

4.2.3 Holarctic or Eurasiatic elements

Four elements can be distinguished.

The mesothermic element. This contains a considerable number of genera whose present importance in the Mediterranean flora is variable. They usually characterize a temperate or even a hot-temperate climate. Two sub-elements can be distinguished: a Laurasian sub-element that is present in Northern America and in Eurasia with numerous arborescent elements such as Hammamelidaceae (*Liquidambar*), Hippocastanaceae (*Aesculus*), Juglandaceae (*Juglans, Pterocarya),* Platanaceae (*Platanus*), plus the genera *Carpinus, Cercis, Epimedium, Euonymus, Laurocerasus, Paeonia, Ostrya, Rhododendron* (macrophyllous types), *Rhamnus, Staphylea, Taxus* and *Viburnum,* and a more strictly Eurasian subunit that includes *Cotinus, Daphne, Fontanesia, Forsythia, Paliurus, Theligonium, Trachomitum, Wulfenia, Zelkova,* etc...

This element is of pre-Miocene origin, and it has played a very important role in the post-glacial floras of Southern Europe and even Northern Africa (Pons, 1964; Van Campo et al., 1965, 1968; Axelrod, 1973; Suc, 1978; Bessedik, 1979). However, it was widely scattered during the glacial periods and a large number of its former elements disappeared from these regions (*Carya, Engelhardtia, Gleditsia, Glyptostrobus, Pseudotsuga, Taxodium* and *Tsuga).* Many now occupy a very limited range, especially in the eastern Mediterranean: *Aesculus hippocastanum* L., *Rhazya orientalis* A.DC., *Forsythia europaea* Degen & Bold., *Liquidamber orientalis* Mill., *Pterocarya fraxinifolia* (Poiret) Spach, *Zelkova abelicea* (Lam.) Boiss., etc.

A number of these taxa are now localized in the Hyrcanian or the Euxinian zones because of their prevailing climatic conditions.

The microthermic element. This corresponds to a group of genera with a distribution within the Laurasian area and also present in the Mediterranean, where it is primarily found in the mountains or in hygrophilous surroundings. It has narrow ecological requirements in respect of thermic factors. Its origin is certainly ancient and prior to the formation of the Northern Atlantic Ocean. Most of these genera were already important in the preglacial flora all around the Mediterranean and even in Africa (Arambourg, et al., 1953), but during the Quaternary climatic fluctuations, their range accordingly varied. Many herbaceous taxa and the following three genera can be considered to belong to this element: *Acer, Betula, Corylus, Fagus, Fraxinus, Pinus (*cf. *sylvestris*), *Quercus* p.p., *Tilia* and *Ulmus.*

Generic endemism is absent but specific endemism is quite high, particularly as a result of geographic disjunction. Nevertheless, the presence of relatively high endemic levels in sections suggests that at least some of these elements have existed, mainly in the mountains, since the Pliocene, as in the case of the *Gentiana* of the section *Pseudotricha* (*G. tornezyana* Litard. & Maire) or of *Draba* section *Helicodraba* and *Acrodraba* (*D. hederifolia* Coss. and *D. oreadum* Maire) in the Atlas.

The Sarmatic element. Many biogeographers recognize a unit of Sarmatic flora (Braun-Blanquet, 1923), or Pontic flora (Walter & Straka, 1970). This corresponds to taxa that escaped the Mediterranean climate and became attached to a cold, continental steppe climate. They seem to derive from Eurasian and chiefly Southern Siberian and Arabo-Caspian ancestors. Their role in the Mediterranean region is very small. However, genera such as *Stipa, Dasypyrum, Eremopyron, Asperugo,* and partly *Seseli, Trinia, Agropyron, Festuca, Aster,* etc., can be referred to this element. Particular mention should be made concerning some Chenopodiaceae and Amaranthaceae, for which the presence of disjunct populations, particularly in Spain (*Krascheninnikovia, Kalidium*), or of endemic species (*Kochia saxicola* Guss., *Corispermum leptoterum* Iljin, *Polycnemon fontanesii* Dur. & Moq., *Aster* of the group *Amellus,* Quézel, 1957) may suggest relatively ancient adaptations.

The Arctic-Alpine element. Taxa that are usually referred to this group play a rather modest role in the Mediterranean region, where they grow on the high mountains. About twenty species can be assigned to it in the High Atlas Mountains (Quézel, 1956). In the Taurus Mountains there are no more than 15 species. Their establishment probably dates back to the glacial phases. An example may be *Drosera rotundifolia* L. which reached northern Lebanon.

The presence of some 20 species of this group in Corsica (Contandriopoulos, 1962; Contandriopoulos & Gamisans, 1974) raises the problem of the date of their arrival on the island, since isolation, according to most authors, dates back to the middle Miocene. It is possible to envisage a more recent arrival through the Cyrino-Toscan bridge. This interpretation does not take into consideration the remarkable absence from the montane flora of Corsica of numerous boreal-alpine elements (*Androsace, Gentiana, Pedicularis, Primula, Oxytropis, Salix* etc.) that have reached most other Mediterranean mountains.

The low level of endemism among these elements could be either explained by a great genetic stability (Contandriopoulos, 1962) or by a recent establishment. The recently discovered endemics *Luzula italica* Parl. and *Erigeron paolii* Gamisans (group *uniflorus*) should be mentioned.

It is therefore becoming more and more obvious that the present Mediterranean flora exhibits a very heterogeneous biogeographical value, corresponding to the extreme paleogeographic and paleoclimatic complexity of the region. In spite of important impoverishment during glacial ages, it possesses a notable floristic richness today. This could be explained both by the high ratio of endemism in some privileged regions and by the extremely variable bioclimatical types that are now present, and have probably existed, at least locally, since the Pliocene.

It should be emphasized that in the light of recent progress in paleogeography as well as in paleoclimatology and in palynology, a certain number of satisfactory models have been proposed to explain how the main biogeographic elements have been established. However, detailed taxonomic research into the whole Mediterranean flora will allow us to develop further and correct existing interpretations, especially with regard to the biogeographic value of many taxa.

References

Achhal, A., Akabli, A., Barbero, M. Benabid, A., M' Hirit, A., Peyre, C., Quézel, P. & Rivas-Martinez, S. (1979). A propos de la valeur bioclimatique et dynamique de quelques essences forestières au Maroc. Ecol. Medit. 5: 221-249.

Akman, Y. & Daget, Ph. (1971). Quelques aspects synoptiques des climats de la Turquie. Bull. Soc. Congr. Geogr. 5: 3.

Arambourg, C., Arènes, J. & Depape, G. (1953). Contribution a l'étude des flores fossiles d'Afrique du Nord. Arch. Mus. Hist. Nat. 2: 1-81.

Ashmann, H. (1973). Distribution and peculiarity of Mediterranean ecosystems. In: Di Castri F., & Mooney H.A. (Eds.), Mediterranean-type Ecosystems: Origin and Structure. Ecological Studies No. 7, pp. 225-283. New York: Springer Verlag.

Aubreville, A. (1949). Climats, forêts et désertification de l'Afrique tropicale. Paris: Soc. Edit. Geogr. Marit. et Colonn.

Axelrod, .D.I. (1973). History of the Mediterranean ecosystem in California. In: Di Castri F., & Mooney H.A., (Eds.), Mediterranean-type Ecosystems: Origin and Structure. Ecological Studies No. 7, pp. 225-283. New York: Springer Verlag.

Axelrod, D.I. & Raven, P. (1978). Late cretaceous and tertiary history of Africa. In: Werger, M.J.A. (Eds.), Biogeography and Ecology of Southern Africa, pp. 77-130. The Hague: Jung.

Bagnouls, F. & Gaussen. H. (1953). Saison sèche et indice xérotermique. Bull. Soc. Hist. Nat. Toulouse 83: 193-239.

Bessedik, M. (1979). La flore pollinique d'un niveau gypseux de la série de Portel (Aude): Oligocène termina. Rapport D.E.A. Université des Sciences et Techniques du Languedoc, Montpellier.

Biju-Duval, B. & Montadert, L. (1977). Structural history of the Mediterranean Basin. Proc. Int. Symp. Split (Tecnip., (Eds.), Paris.

Birot, P. & Gabert, P. (1964). La Méditerranée et le Proche-Orient., Vol. 1. Orbis, Press Universitaire. Paris: France.

Bocquet, G., Widler, B. & Kiefer, H. (1978). The Messinian model: A new outlook for the floristics and systematics of the Mediterranean area. Candollea 33: 269-287.

Braun-Blanquet, J. (1915). Les Cévennes méridionales. Soc.Gen. d'Impr., Geneve.

Braun-Blanquet, J. (1923). L'origine et le developpement des flores dans le Massif Central de France. Paris: L'homme ed.

Burtt, B.L. (1971). From the south: An African view of the floras of Western Asia. In: Davis, P. (Eds.), Plant Life in South-West Africa, pp. 134-149. Aberdeen: Harper & Hedge.

Butzer, K.W. & Hansen, C.L. (1968). Desert and River in Nubia. Madison: University Wisconsin Press.

Capot-Rey, R. (1952). Les limites du Sahara français. Trav. Inst. Rech. Sah. Alger 8: 23-48.

C.N.R.S. (1975). La Flore du Bassin Méditerranéen, essais de systematique synthétique. Coll Int. C.N.R.S. de Montpellier. Paris: Editions du C.N.R.S.

Contandriopoulos, J. (1962). Recherches sur la flore endémique de la Corse et sur ses origines Ann. Fac. Sc. Marseille 32: 1-352.

Contandriopoulos, J. & Gamisans, J. (1974). A propos de l'element arcticoalpin de la flore corse. Bull Soc. Bot. Fr. 121; 175-204.

Coppens, Y. & Koeniguer, J.C. (1978). Signification climatique des paléoflores ligneuses du Tchad Bull. Soc. Geol. Fr. 18: 1009-1015.

Daget, Ph. (1977a). Le bioclimat Méditerranéen, caractères géneraux, modes de caractérisation. Vegetatio 34: 1-20.

Daget, Ph. (1977b). Le bioclimat Méditerranéen: analyse des formes climatiques par le système d'Emberger. Vegetatio 34: 87-104.

Daget, Ph. (1980). Un élément actuel de la caracterisation du monde Méditerranéen: le climat. Naturalia Monspelliensia 237: 101-126.

De Martonne E. (1927). Traité de géographie physique, Vol. 1. Paris: A. Colin.

Drude, O. (1884). Die Florenreiche der Erde. Petermanns Mitteilung, Ergänzungsheft, Leipzig.

Durand, E. & Flahault, Ch. (1886). Les limits de la région Méditerranéenne en France. Bull. Soc. Bot. Fr. 33: 23-34.

Emberger, L. (1930a). Sur une formule climatique applicable en géographie botanique. C.R. Acad. Sc. 191: 389-390.

Emberger, L. (1930b). La végétation de la région Méditerranéenne. Essai d'une classification des groupements végétaux. Rev. Gén. Bot. 42: 641-662.

Emberger, L. (1933). Nouvelle contribution à l'étude de la classification des groupements végétaux. Rev. Gen. Bot. 45: 473-486.

Emberger, L. (1939). Apercu général sur la végétation du Maroc. Veröff. Geobot. Inst. Rübel, Zürich 14: 40-137.

Emberger, L. (1943). Les limites de l'aire de végétation Méditerranéenne en France. Bull. Soc. Hist. Nat. Toulouse 78: 159-180.

Favarger, C. (1975). Données caryosystématiques concernant la flore des pays Méditerranéens: La Flore du bassin Méditerranéen Montpellier: pp. 145-158. Coll. Int. C.N.R.S.

Flahault, C. (1901). Introduction. In: Coste, H., Flore de France. Paris: Lechevalier ed.

Flahault, C. (1937). La distribution géographique des végétaux dans la région Méditerrannéenne française. Encyclop. Biologique, Vol 18. Paris: Lechevalier.

Gaussen, H. (1926). Végetation de la moitié orientale des Pyrénées. Paris: Lechevalier.

Gaussen, H. (1954). Théorie et classification des climats et microclimats. VII Congr. Int. Bot. Paris 7: 125-130.

Gillet, H. & Quézel, P. (1959). Le genre *Oropetium* en Afrique française. Jour. Agr. Trop. et Bot. Appl. 1-3: 37-58.

Goldblatt, P. (1978). An analysis of the flora of Southern Africa: Its characteristics, relationships and origins. Ann. Missouri Bot. Gard. 65: 369-436.

Hedberg, O. (1965). Afroalpine flora elements. Webbia 19: 519-529.

Le Houerou, H.N. (1971). An assessment of the primary and secondary production of the arid grazing lands ecosystems in North Africa. Symp. of Ecophysiol., Leningrad.

Le-Noble, F. (1934). Sur la définition de la région Méditerranéenne en géographie botanique et ses limites dans le S.E. de la France. Bull. Soc. Bot. France 81: 88-96.

Maley, J. (1977). Analyses polliniques et paléoclimatologie des douze derniers millénaires du bassin du Tchad (Afrique Centrale) In recherches françaises sur le Quaternaire hors de France. Com. Nat. INQUA, Xe Congrés Birmingham. Suppl. Bull. AFEQ 1(50): 187-197.

Monod, Th. (1951). Biologie des régions arides. In: Basses écol. regen. veg. zones arides pp. 33-44. Paris: U.I.S.B. Publ. UNESCO.

Nahal, I. (1972). Contribution à l'étude des bioclimats et de la végétation naturelle de Turquie. Hannon, Beyrouth 7: 116-129.

Ozenda, P. (1975). Sur les étages de végétation dans les mon-

tagnes du bassin Méditerranéen. Doc. Cart. Ecol. 16: 1-32.
Peguy, Ch. (1970). Précis de climatologie. Paris: Masson.
Pignatti, S. (1978). Evolutionary trends in the Mediterranean flora and vegetation. Vegetatio 37: 175-185.
Pons, A. (1964). Contribution palynologique à rétude de la flore et de la végétation pliocènes de la region rhodanienne. Ann. Soc. Nat. Bot. Paris. 5: 499-722.
Quézel, P. (1956). Contribution à l'étude des forêts de chênes à feuillages caduques d'Algerie. Mém. Soc. Hist. Nat. Afrique du Nord 1: 1-57.
Quézel, P. (1957). Peuplement végétal des hautes montagnes de l'Afrique du Nord. Paris: Lechevalier.
Quézel, P. (1958). Mission botanique au Tibesti. Inst. Rech. Sahar. Alger. Memoire n. 4
Quézel, P. (1965). La végétation du Sahara. Stuttgart: Fischer Verlag.
Quézel, P. (1974). Les fôrets du pourtour Mediterranéen. Notes Techn. MAB-UNESCO 2: 9-34.
Quézel, P. (1976). Les fôrets du pourtour Méditerranéen: écologie, conservation et aménagement. Notes Techn. MAB-UNESCO 2: 9-33.
Quézel, P. (1978). Analysis of the flora of Méditerranean and Saharan Africa. Ann. Missouri Bot. Garden 65: 479-534.
Quézel, P. (1983). Flore et végétation de l'Afrique du Nord, leur signification en fonction de l'origine, de l'evolution et des migrations des flores et structures de végétation passées. Bothalia 14: 411-416.
Quézel, P. & Barbero M. (1981). Definition and characterization of Mediterranean type ecosystems. Ecologia Mediterranea 8: 15-29.
Quézel, P., Gamisans, J. & Gruber, M. (1980). Biogeographie et mise en place des flores Méditerranéennes. Naturalia Monspeliensia 237: 41-51.
Quézel, P. & Martínez, C. (1961). Le dernier pluvial au Sahara Central Libyca 6-7: 211-227.
Raven, P. (1972). Plant species disjunction: A summary. Ann. Missouri Bot. Garden 59: 234-246.
Raven, P. (1973). The evolution of Mediterranean flora. In: Di Castri, F. & Mooney, H.A. (Eds.), Mediterranean-type Ecosystems: Origin and Structure. Ecological Studies nr. 7, pp. 213-224. New York: Springer Verlag.
Raven, P. & Axelrod, D.I. (1978). Origin and relationships of the Californian flora. Univ. Calif. Public. Bot. 72: 1-34.
Rognon, P. (1967). Le massif de l'Atakor et ses bordures (Sahara Central). Edit. du C.N.R.S., Sér. Géol. 9: 1-559.
Sauvage, Ch., (1963). Etages bioclimatiques. Notice et carte au 1/2.000.000. Atlas du Maroc Sect. II, pl. 6b, Comité Géographie Maroc, Rabat.
Schmidt, E. (1966). Die Vegetationsgürtel der iberischen barbarischen Gebirge Ver. Geobot. Inst. Rübel, Zürich 31: 124-163.
Specht, R.L., Roe, .EM. & Broughton, V.H. (1974). Conservation of major plant communities in Australia and Papua. Austral. J. Bot. suppl. sér 7: 1-667.
Stewart, Ph. (1975). Un nouveau climagramme pour l'Algérie. Bull. Soc. Hist. Nat. Afrique du Nord 59: 23-36.
Suc, J.P. (1978). L'étude palynologique du Pliocène du Sud de la France, méthode d'approche et résultats. Ann. Mines Belgique 6: 120-126.
Trewartha, C. (1954). An Introduction to Climate. New York: McGraw-Hill.
Thornthwaite, C.W. (1948). An approach toward a rational classification of climate. Geogr. Rev. 38: 55-94.
Turc, L. (1961). Evaluation des besoins en eau d'irrigation, évapotranspiration potentielle. Ann. Agron. 12: 13-49.
UNESCO (1962). Carte bioclimatique de la région Méditerranéenne. Paris: UNESCO Edit.
Van Campo, M., Cohen, J., Guinet, P. & Rognon, P. (1965). Contribution à l'étude du peuplement végétal quaternaire des montagnes sahariennes. Pollen et spores. 7: 361-371.
Van Campo, M., Guinet, Ph. & Cohen, J. (1968). Fossil pollen from late Tertiary and middle Pleistocene deposits of the Kurkur oasis. In: Butzer, K.W. & Hansen, C.L. (Eds.), Desert and River in Nubia, pp. 515-520. Madison: University of Wisconsin Press.
Walter, H. & Lieth, H. (1960). Klimadiagram Weltatlas. Jena: Fischer Verlag.
Walter, H. & Straka, H. (1970). Arealkunde. Stuttgart: Verlag Eugen Ulmer.

Faculté des Sciences at Techniques de Saint-Jérome
Rue Henri Poincaré
F-13397 Marseille Cedex 13, France

CHAPTER 3

The history of the flora and vegetation and past and present human disturbance in the Mediterranean region

A. PONS and P. QUÉZEL

1. The history of the Mediterranean flora and vegetation

The constantly high level of heterogeneity that can be seen in any aspect studied of the Mediterranean region has long suggested to researchers a complex and eventful history. Our knowledge of the history of the region's flora and vegetation has begun to take objective shape only during the past few years, largely due to data from the relatively new field of pollen analysis. This technique has permitted us to update the results of many older palaeobotanical studies of macrofossils, which were concerned primarily with the past history of taxa. Pollen analysis has also made it possible to obtain some information on the history of past vegetation communities on both a local and a regional scale.

The geological sciences too have contributed valuable and complementary paleogeographic and paleoclimatic data, the application of which has generally inspired as much prudence amongst paleobotanists as it has inspired boldness amongst biologists who have less familiarity with the past.

Despite recent progress in all these disciplines, paleobotanists are not yet in a position to formulate a detailed, wide-ranging and continuous historical scheme for the taxa or the communities which now characterize, or have characterized, the vegetation of the Mediterranean. Thus, only an outline scheme of this history can be presented.

One fact dominates this scheme, the existence of 'Mediterranean' vegetational assemblages can be supported on the basis of currently available data only for those periods from the Middle Pliocene onwards. However, the more ancient history of the taxa themselves (and less frequently their fluctuations in the vegetation), has been the subject of numerous papers (Suc, 1980; Pons & Suc, 1980; Medus & Pons, 1980). Three important methodological considerations should be taken into account if we are to understand the situation properly:

Firstly, macroscopic plant remains allow relatively precise comparison between extant taxa and those of past periods (Older Tertiary or Upper Cretaceous for the Angiosperms, and Lower Cretaceous for the Gymnosperms). However, the natural processes which lead to the preservation and fossilization of bulky organs – wood, leaves, fruits or seeds – only occur under exceptional conditions. Therefore, deposits are usually rare, scattered in location and variable in age. Macrofossils essentially reflect their immediate depositional environment and the subsequent changes which have occurred in the fossilization media. Thus they do not convey a clear image of the nature of the prevailing vegetation.

Secondly, macropaleobotanical research, including that of early research workers, is a valuable source of floristic knowledge for periods before the Miocene (that is, until approximately 20 million years ago). Our knowledge of vegetation as revealed by pollen analysis commences with the beginning of the Pliocene for Europe, the begin-

Gómez-Campo, C. (ed.), Plant conservation in the Mediterranean area.
© 1985, Dr W. Junk Publishers, Dordrecht. *ISBN 90 6193 523 7.*

ning of the Upper Pleistocene for the Middle East and, with the exception of two small floras (Arambourg et al., 1953; Van Campo E., 1978), the last 6,000–8,000 years for North Africa. Therefore, a gap exists between the Miocene and more recent periods covered by pollen analysis.

Lastly, in fossiliferous sediments pollen grains vastly outnumber macroscopic remains. The amount of pollen produced varies widely between species, but is largely constant within each. Pollen grains, widely dispersed around the parent plants, are endowed with an exine or outer wall of sporopollenin which is extraordinarily resistant to most chemical and physical degradation. Pollen is commonly found preserved in a wide range of sediments, especially those that are waterlogged or acidic, and its presence gives an approximate picture of ancient vegetational assemblages through the relative frequency of different pollen-types. If the sediment was laid down gradually, as in a lake bed or in a peat bog, the pollen analyst can sample a specially taken core at regularly spaced intervals to assess what changes occurred in the types of pollen deposited. At each level, relative proportions of the individual pollen types are recorded; later, the data from all the levels are compiled to form a pollen diagram. By joining up on the diagram the proportional values of different levels, each pollen type acquires its own 'curve' which, upon analysis, can provide valuable information on past vegetation (Fig. 1). Although pollen morphology generally permits identification of the parent plant to genus level, as we go further into the past the number of pollen types that can be correlated with present living taxa is reduced. Before the Miocene, for instance, only a few genera of extant trees can be identified (*Acer, Engelhardtia, Rhizophora,* etc.). Beyond the Eocene, morphological identification is possible, but based only on purely paleontological classification and nomenclature.

The history of the flora and vegetation of the Mediterranean region can be conveniently divided into two parts; one part can be seen as the history of the ancestors of the present flora prior to the middle of the Pliocene (about 3 million years B.P.), while the other part deals with the more recent history of the vegetation, in which 'Mediterranean' assemblages play a dominant part. For the past 7,000 years or so, the vegetation has been strongly affected by man's action and this complex period should perhaps form a separate area of study.

Reference periods and their duration are shown in Figure 2. Some paleogeographic details concerning these periods can be usefully summarized as follows:

a) The Iberian sub-continent has provided an almost constant link with the African continental block (Biju-Duval et al. 1977), since at least the beginning of the Cretaceous. For much of that time, it represented the only emergent land amongst the countries bordering the present Mediterranean Sea. The eastern link between Arabia and the Middle East does not appear to be older than the Eocene.

b) Although the widening of the north Atlantic Ocean began during the Cretaceous (70 million years B.P.*), the North American block and the Laurasian and North African blocks were not fully separated before the Middle Eocene (50 million years B.P.), whereas South America and Tropical Africa had already been separated for twice that length of time.

c) The installation of the continental masses that surround the Mediterranean Sea at the present day began at the end of the Jurassic period. They have drifted northwards constantly; 20 degrees since the Cretaceous and 12 since the Eocene (Hughes, 1973).

d) The Mesogean zone, formed by closed and presumably shallow basins during the Oligocene and the Miocene, was strongly affected by a final 'salinity crisis' (Ryan & Hsu, 1973; Busson, 1979) which might have arisen from a very peculiar system of relations with the Atlantic Ocean. This crisis is often called the 'Messinian crisis' after the Messinian period, which occurred at the end of the Miocene.

e) Alpine orogenesis or mountain building continued during the Pliocene and has even occurred during the Pleistocene (Fourniquet, 1977).

* B.P. = before present, the present being defined as 1950 for radiocarbon dating.

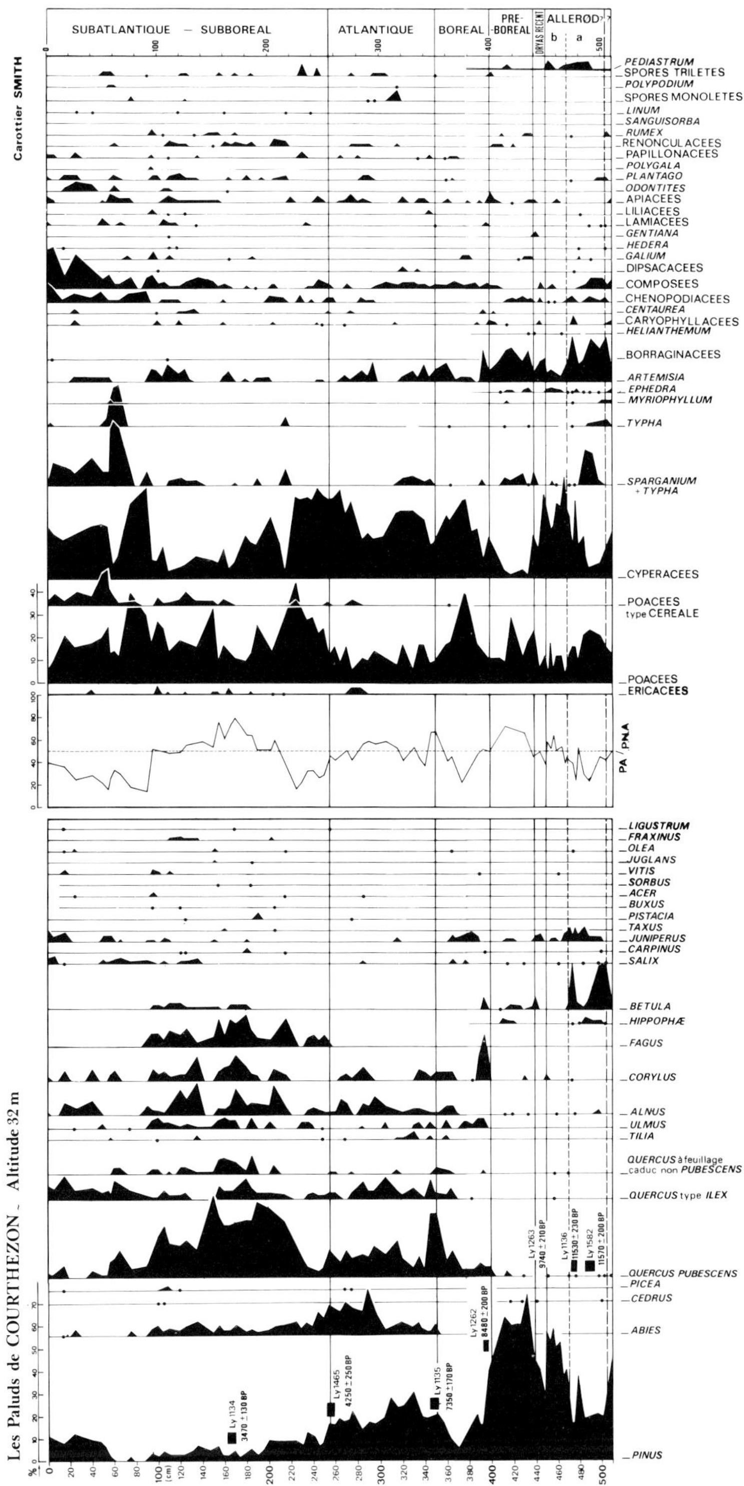

Fig. 1 An example of a pollen diagram from the French Mediterranean region. It clearly shows the important role of *Pinus* at the beginning of the Postglacial and also the long history of man's influence. However, a valid reconstruction of the history of vegetation in a given region can only be based upon several diagrams that cover a long time-span and are from sites in different ecosystems.

1.1 A regional history: Cretaceous to mid-Pliocene

The earliest remains, mostly woody, that can be identified as plant taxa occurring at present in the Mediterranean region, have been found in various deposits from the Lower Cretaceous in Northern France and Belgium (Depape, 1963). These remains have all been identified as Gymnosperms and belong almost exclusively to the genus *Cedrus*. This genus was widely distributed in the Northern Hemisphere during the Jurassic and Cretaceous periods (Fergusson, 1967) but became less impor-

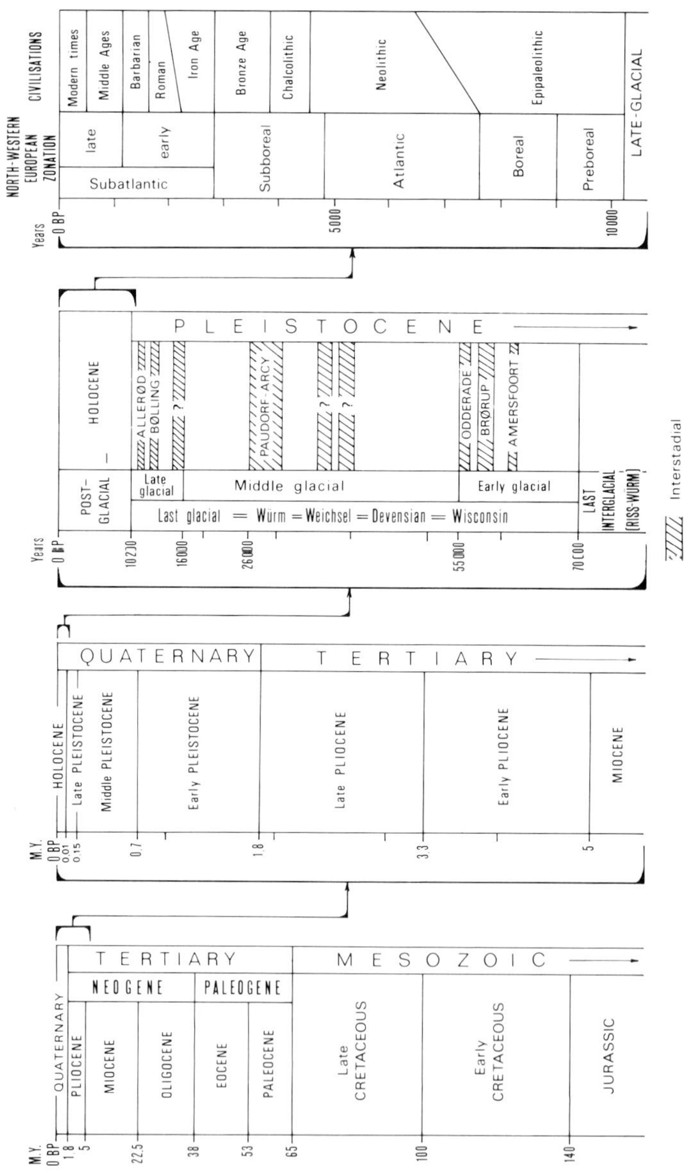

Fig. 2. A diagram showing the different geological and climatic periods mentioned in this chapter and their duration.

tant during the Pleistocene. Fossil remains of *Cedrus* cannot therefore be considered to represent any exclusively 'Mediterranean ' feature.

The famous floras of Provence from the Upper Cretaceous contain only one element related to present Mediterranean genera, *Laurus preatavia* Sap., from Bagnols-sur-Cèze (Saporta, 1890). Other macrofloras of this epoch are similarly poor in Mediterranean elements (Depape, 1959). Amongst many imprecise and debatable identifications (Saporta, 1890), there are some suggestions that the predecessors of the genus *Quercus* and of the families Rhamnaceae, Leguminosae and Liliaceae were present. Meanwhile, in Westphalia, the first *Nerium* appears as a variant which is different from both *N. oleander* L., from the Mediterranean, and *N. odorum* (Solander) Aiton, from eastern India, Iran and Japan (Saporta, 1888).

Pollen analysis reveals a number of pollen types which are limited to the north-western border of the Mesogean, including definite *Classopollis* (Pflug) Pocock and *Jansonius,* representing the Chierolepidaceae family during the Lower Cretaceous, and definite *Normapolles* during the Upper Cretaceous (Medus & Pons, 1980). Pollen analysis also shows a distinct pre-Tertiary flora in the African 'Austral' plate (as opposed to the 'Boreal' north Mesogean or European plate).

In the Paleocene, the older part of the Paleogene, only some late and poor florulas are known in Belgium and Northern France (Depape, 1925) where unconfirmed *Myrtus, Laurus, Ziziphus* and some archaic *Dryophyllum* and *Pradionopsis,* representing Fagaceae, should be noted.

A very rich flora from the Early Eocene is known from Sézanne. Dépt. Marne, France. Almost one half of the plant remains belong to genera that show no precise correspondence with living ones (Saporta, 1868), and thus cannot be confidently regarded as predecessors of present day Mediterranean taxa. More recent Eocene macrofloras show a higher percentage of identifiable remains that can be referred to living genera but the ancestors of present Mediterranean taxa are not evident there either. These Eocene floras are widely considered to be 'Indo-Malaysian' tropical assemblages. The rich macroflora of the London-Paris basin confirms this well (Reid & Chandler, 1933; Fritel, 1908 to 1924). *Cupressinoxylon* and *Juniperoxylon* (Grambast, 1955), as well as *Laurus, Ficus* and *Quercus,* are of uncertain affinities. Only *Callitris brongiartii* Endl. appears to represent a putative ancestor of the now existing *Tetraclinis articulata* (Vahl) Masters from North Africa, also S.E. Spain and Malta, (Saporta, 1888; the distinction between *Callitris* and *Tetraclinis* had not at that time been established).

The foliar flora of the Early Oligocene has revealed only *Pistacia lentiscus* L. var. *oligocenica* Mars. (Marion, 1872) from Ronzon Dépt. (Haute-Loire). Other macrofloras formerly assigned to the same period in South-eastern France in fact correspond to the Upper Oligocene and even to the Oligocene-Miocene boundary (Anglada et al., 1978). They contain a low but constant quantity of fossil remains surely attributable to more or less direct ancestors of present Mediterranean species. At Aix-en-Provence, for example, of 89 recorded genera, 21 are represented in Provence today; and of 231 species, 17 can be referred to Mediterranean species. The 17 species are *Cheilanthes fragrans* L., *Tetraclinis articulata* Masters, *Juniperus* cf. *phoenicea* L., *Pinus*-type *pinaster* Aiton, *Pinus nigra* subsp. *salzmani* Dun., *Ostrya carpinifolia* Scop., *Paliurus aculeatus* L., *Smilax aspera* L., *Quercus*-type *ilex-coccifera, Nerium oleander* L., *Osyris alba* L., *Ceratonia siliqua* L., *Cercis siliquastrum* L., *Rhus coriaria* L., *Pistacia lentiscus* L., P. *saportae* Burnat and *Ficus carica* L. (Saporta, 1872).

An exact equivalent of these floras can be found in the macroflora from the Early Oligocene of Hungary (Andreansky, 1963). In all these Oligocene floras, remains are found that can be attributed to rather xeromorphic plants. Some of them can be considered to be the ancestors of South African or Mediterranean plants, while others are similar to the Australian genera *Lomatites, Banksites, Knightites* in Proteaceae, and *Leptospermites* in Myrtaceae (Saporta, 1872). Besides some aquatic European plants which occur today, the most frequent species are those related to Asiatic tropical taxa, followed by a group of plants related to

American taxa. Within the latter group, those related to northern and to tropical taxa are roughly balanced. The remainder is a small group that can be referred to taxa that are irregularly distributed at the present day.

In the Portel series from the end of the Oligocene (Languedoc, France), a detailed analysis has only revealed Juglandaceae and Hammamelidaceae pollen, an abundance of *Olea*-type pollen (19%), *Pistacia, Nerium, Rhamnus* and a number of Mimosoideae *(Prosopis)* which now exist in Algeria and Palestine (Bessedik, 1979).

The disappearance of southern xeromorphic elements occurred at the beginning of the Miocene; the proportion of Mediterranean ancestors was maintained and some new elements appeared. These trends can be demonstrated in the flora of Armisan, Dépt. Aude (Saporta, 1866), which includes *Pinus halepensis* Mill., *Myrtus communis* L., *Pistacia atlantica* Desf. and *Rhamnus buxifolius* Duh. Holoarctic forest trees that had been previously restricted to northeastern Europe and northern Asia, such as *Ulmus minor* Mill. and *Acer campestre* L., became more and more abundant, together with the ancestors of other non-tropical American elements such as various members of the Pinaceae, Taxodiaceae and Fagaceae.

Evidence for the Mediterranean 'salinity crisis' of the late Miocene, is only known from pollen assemblages from sea drilling sites. However, the identification of the pollen is somewhat doubtful (Bertolani-Marchetti & Cita, 1975). A few continental deposits from the late Miocene at Coirons in Ardèche (Grangeon, 1958; Naud & Suc, 1975), and in northeastern Spain (Van Campo, 1976), show only the most frequent 'Mediterranean' taxa of the Oligocene such as *Pistacia, Phillyrea, Olea* and *Quercus*-type *ilex*.

Mediterranean taxa were well represented in the Pliocene pollen flora, which was very rich, widely diversified and to some degree dominated by Gymnosperms. Among those found are *Cathaya, Picea, Pinus,* Cupressaceae and Taxodiaceae, Betulaceae (*Alnus*), Ericaceae, Fagaceae (*Quercus*), Hammamelidaceae, Juglandaceae (*Carya* and *Engelhardtia*) and Ulmaceae. At Rio Maior (Portugal) the Mediterranean taxa present are Cupressaceae, *Cistus, Helianthemum, Myrtus,* Rhamnaceae, *Vitis* and abundant Umbelliferae (Diniz, 1967, 1969). In Languedoc-Rousillon, France, and in Cataluña, Spain Cupressaceae, *Pistacia, Rhus* cf. *Cotinus, Helianthemum, Knautia, Scabiosa, Quercus*-type *ilex, Asphodelus, Olea, Phillyrea, Rhamnus* and *Vitis* have been reported (Pons, 1964; Planchais, 1974; Suc, 1976, 1980 a, b). The pollen of Cupressaceae, *Olea* and *Phillyrea* reaches high percentages within short, repeated periods (Suc, 1980b).

Therefore, a considerable number of plants identified as direct ancestors of extant Mediterranean species seem to have appeared for the first time during the Oligocene. Paleobotanists are not yet in a position to decide whether this event was the direct result of progressive, gradual evolution or whether major climatic changes were involved as well. The first hypothesis is supported by the fact that, at this time, an increase in the proportion of identifiable remains has been observed that bears no relation to distribution of the present descendants. The second hypothesis takes into account the fact that the Antarctic ice cap was then forming and that important tectonic movements had marked the beginning of the Oligocene (Cavelier, 1976). The appearance of xeromorphic plants which are directly related to taxa that now live in countries far from the Mediterranean, such as South Africa and Australia, also suggests climatic change (Andreanski, 1963) and supports the second hypothesis.

1.2 A regional history: Mid-Pliocene to Holocene

By the end of the early Pliocene, many 'exotic' taxa had become very rare (Hammamelidaceae, Juglandaceae) or had vanished (Taxodiaceae) and relatively poor floras are observed in the fossil record (Ichkeul lake in Tunisia: Van Campo, 1978; Languedoc-Rousillon coast: Suc, 1980b). At the same time, Mediterranean taxa increased substantially. That these changes were synchronous is shown by the stratigraphy of mammal remains, especially those of rodents, found in pollen-bearing sediments. These changes can only be interpreted as the result of a progressive climatic modification, mainly through decreasing rainfall, particularly during the warmest season (Suc, 1980b). The role played in the pollen spectra by xerophyllous taxa (Cupressaceae, *Pinus, Quercus*-type *ilex, Olea, Phillyrea, Cistus, Helianthemum, Rhus* cf. *Cotinus, Rhamnus)* suggests that Mediterranean vegetation communities arose for the first time on low hillsides with dry calcareous soils.

During the Late Pliocene and the Early Pleistocene, pollen diagrams from the northwestern Mediterranean regions show marked fluctuations between two main types of vegetation:

a) Steppe communities dominated by *Artemisia*, Amaranthaceae, Chenopodiaceae and *Ephedra* in northwestern Spain (Villaroya: Remi, 1958) or open communities with *Pinus*, Cupressaceae and similar herbs at low altitudes in Cataluña and Languedoc (Suc, 1980a).There may also have existed well-developed Mediterranean communities (with 'Cupressaceae, Quercus-type *ilex, Olea,* and *Plantago'* or *'Olea, Phillyrea* and *Pistacia'* even *Pistacia, Ceratonia* and *Olea')* on the plains and coasts of Languedoc (Suc, 1980a), in Campania (Camerata) (Bazile et al., 1980) and in Southern Italy (Ricciardi, 1965; Bertoldi, 1977; Bertolano-Marchetti et al., 1978).

b) Forest communities rich in deciduous species that are no longer present in the region, (*Carya, Pterocarya, Parrotia persica)*, with abundant gymnosperms at higher altitudes.

This alternation reflects climatic oscillations that were undoubtedly related to warming and cooling periods known to have affected the northern hemisphere at this time. Each cold period resulted in a decrease in rainfall that manifested itself regionally in dry summers (Suc, 1980b).

The period after the Early Pleiocene, the last million years, is characterized by thermic oscillations, corresponding to the fluctuations of the ice cap in the northern hemisphere. These oscillations had a dominant rhythm of about 100,000 years. Traditionally, the cold periods have been called 'glacials', whereas the temperate periods have been called 'interglacials'. Today, scientists studying the Pleistocene distinguish between interglacials, periods with a climatic optimum sufficiently warm to have allowed the development of a temperate flora similar to today's, and interstadials, periods which were either too short of too cold to have permitted the development of temperate deciduous forest. These terms are most useful in Northern Europe, where the proximity of the ice-sheets would have accentuated thermal oscillations. Further to the south, the differences between interglacial and interstadials were not likely to have been as marked, due to the amelioration of the climate.

Consequently, a forest vegetation that developed during temperate periods was alternating with a herbaceous steppe vegetation during colder periods. It is difficult to determine the exact nature of the herbaceous assemblages that appeared in the Mediterranean region during the cold stages. The rare occurence of *Cistus* and *Phlomis* cf. *fruticosa* L. pollen grains suggests increased aridity rather than a fall in temperature, but our inability to separate certain important and/or large pollen groups, that characterize recent periods, into subgroups or to distinguish them to species level such as *'Artemisia', 'Juniperus', 'Chenopodiaceae'*, 'Compositae' and 'Gramineae', prevents the elucidation of the details of the real affinities of the vegetation*.

In the Middle Pleistocene, when most of the 'exotic' taxa were vanishing, available data is restricted to sequences from Tenaghi-Philipon (Greece) and Padul (Sierra Nevada, Spain). The only published pollen diagram for the first site (Wijmstra, 1971) shows fluctuations of the type

* It should be noted that there is a complex relationship between the general climatic oscillations of the Plio-Pleistocene and the important changes that occurred in the climate of the Sahara (Maley, 1980; Rognon, 1976). During two periods with higher humidity, a Mediterranean flora and vegetation developed in the Hoggar Massif at a favourable altitude, where remnants still survive, though much reduced (Quézel, 1965). In the Upper Pliocene, an eastern Mediterranean montane flora prevailed (Rognon, 1967). During the Holocene, an unquestionably Mediterranean pollen spectrum with a predominance of Cruciferae, Anacardiaceae (*Pistacia:* 10%), Ericaceae (12%), Oleaceae and Umbelliferae has been found in a well dated autochthonous organic deposit (Pons & Quezel, 1958). This occurred within a period immediately preceding the climatic deterioration which led to the present desert conditions in this zone.

already known from the Early Pleistocene. No particular indication of Mediterranean taxa is evident. At Padul, evidence of a Mediterranean-type of vegetation can be observed only during short periods that correspond to sudden changes in pollen curves.

During the early stages of an interglacial period just preceding the 'Ris I' Glacial, evidence for a warm and humid open forest can be seen. In this phase, pollen of Oleaceae and *Cistus* reach values of approximately 5% (Florschutz et al. 1971). Both at Padul and at Tenaghi Philippon, the curve for Ericaceae closely follows that of a group of species considered to be Mediterranean. However, because of the absence of an accurate scheme of identification for the Ericaceae, a difficult pollen type, the interpretation of this information remains problematical. In the upper part of a 'Riss Interstadial', the presence of *Quercus*-type *ilex* L. and Oleaceae pollen suggests the existence of a humid open forest variant of the present *Quercus ilex* zone (Florschulz et al., 1971).

The history of Mediterranean-type vegetation in the late Pleistocene is still far from clear. Evidence of the last interglacial is scanty and little known in the Mediterranean region (Frenzel, 1980). In the diagram from St-Paul-lez-Durance (Dépt. Bouches-du-Rhône) there are no pollen curves of truly Mediterranean taxa. However, the presence of *Pinus halepensis* has been revealed in the surroundings of St-Paul. This is the only clear indication of the presence of Mediterranean species, as *Quercus*-type *ilex* pollen is rare and only one *Myrtus* pollen grain, a few of *Pistacia* and *Cistus*, and three grains of *Rhamnus* (probably *alaternus* L.) have been recorded. Thus, during the Riss-Würm, the Mediterranean vegetation had not yet reached its full differentiation and extension. *Pinus* is a prolific producer of pollen: thus, if the over-representation of pine is taken into account, Mediterranean deciduous oak wood may have been locally common (Baulieu, 1972). In Macedonia (Wijmstra, 1969) or in Padul (Florschulz et al. 1971), *Quercus ilex, Pistacia* and Oleaceae are present in very low percentages, so that the Mediterranean character of the forest (deciduous oak woodland with a very rich understory) is unclear. Towards the transition into the next glacial period these percentages increase and indicate more open vegetation.

Numerous plant remains of taxa that are now important constituents of Mediterranean vegetation (*Phillyrea angustifolia* L. and *P. latifolia* L., *Quercus ilex* L., *Pinus nigra* Arnold subsp. *salzmanii* Dun. etc.) or are present in them (*Buxus*) have been reported from various tufa deposits in Southern France. Though some older works date them back to the last interglacial, their stratigraphical significance remains uncertain (Pons, 1969; Bazile et al., 1980).

Stronger evidence for a Mediterranean type of vegetation is available from the first part of the last glacial period between 70,000 and 55,000 B.P. Pollen grains of Mediterranean taxa, whether shrubs or sclerophyllous forests *(Arbutus, Phillyrea, Quercus*-type *ilex, Pistacia*) reach particularly high levels (above 50%) within a formation that seems to correspond to an early period of the last glaciation in Sardinia (Bertolani-Marchetti, 1964).

The diagram of Padul shows that the same pollen types were regularly present, at least during the last glacial period when the climate was more or less temperate. But only at the beginning and at the end of this interstadial (correlated with the classical interstadial of Brørup) do these types reach substantial percentages, during brief periods of climatic variations, where spectra reflect maquis-type community structures (Florschulz et al., 1971). Percentages observed at Tenaghi Phippon for the same species are higher, but only during short intervals, ones that correspond to the end of an interstadial (which is thought to be older than the Brørup) and another at the beginning of a more recent interstadial (Wijmstra, 1969).

At the height of the last glacial (between 55,000 and approximately 160,000 B.P.), no evidence for Mediterranean type of vegetation has been found on the western side of the Mediterranean, even though charcoals from prehistoric sites have been examined and identified in southeastern France by Vernet (1972) and Bazile & Bazile (1978). Pollen analysis of infilled archeological sites in caves have been published from this region, but the methodology involved makes it difficult to assess their significance (Coûteaux, 1977; Bottema, 1975).

At the middle altitude site of Calabre (Gruger, 1977), a diagram some 37,000 years old, and probably reflecting an interstadial, shows the presence of a local beech-fir forest in which very low percentages of *Quercus*-type *ilex, Olea* and *Ostrya* pollen are present. Therefore, Mediterranean forest assemblages might have existed during temperate periods at low altitudes.

Pollen analysis from sites near the coast of Provence (Triat, 1979) that indicate the beginning of postglacial reafforestation, confirm that even during the most severe glacial periods, refuges existed in which mesothermic plants of low altitudes survived. Likely refugia are the lower part of the Rhône valley and the neighbouring coastal paleothalwegs, which are today submerged. Triat (1978) suggests that Mediterranean taxa could have also survived on abrupt slopes and on refuges with favourable aspect, but among the sparse research done in this area, there is no solid evidence for the survival of true Mediterranean vegetation on the northern shores of the Mediterranean during the most arid and cold stages of the last glaciation (about 18,000 B.P.). Only the survival of small areas of extramediterranean forest at middle altitudes in the southern European mountains is well established (Beug, 1975a; Gruger, 1977).

By the end of the last glacial (Between 16,000 and 10,000 B.P.) rare pollen grains of *Oleaceae, Quercus*-type *ilex, Pistacia* and *Phillyrea* can be found in Provence and in Padul, but due to the problems of long-distance pollen transport, they cannot be taken as firm evidence for the existance of Mediterranean vegetation at these sites. Periods of ameliorating climate during the Late-Glacial period are characterized by high pollen frequencies of *Juniperus,* suggesting a shrubby vegetation. However, the impossibility of an accurate specific determination of the pollen makes it impossible to describe the vegetation in further detail. This is also true for *Pinus* which, especially in Provence, characterized the second period of improved climate, called the Allerød (11,900 - 10,800 B.P.); in these pine-groves a southern variant of *Pinus sylvestris* L. dominated, but *P. halepensis* Mill. may have also played an important part (Triat, 1978).

The data from the eastern Mediterranean region for both the above-mentioned periods are complex*. Near the present Mediterranean coasts (Tenaghi Philippon: Wijmstra, 1969; Xinias in Greece: Bottema, unpubl.; at Ghab Lake in northwestern Syria; Niklewski & Van Zeist, 1970), the period before 40,000 B.P. is marked by a high percentage (more than 10%) of tree pollen, with *Quercus*-type *coccifera* L., *Cedrus, Pistacia* cf. *atlantica* Desf., *Carpinus orientalis-Ostrya, Juniperus* and even *Olea.* In Syria, the presence of Mediterranean-type vegetation on a thick layer of sediments is indicated. Many erratic and somewhat enigmatic fluctuations in arboreal pollen frequencies occur between 40,000 and 30,000 B.P., whereas between 30,000 and 16,000 B.P. three clear stages appear with a high proportion of arboreal pollen. But, again in Syria, a new increase is noted in Mediterranean tree pollen during the two periods before 22,000 B.P. Within these periods, the percentage of tree pollen is always much lower in the more continental western Iran (Van Zeist & Bottema, 1977). There, pollen curves are sometimes discontinuous and show only weak variations.

In the eastern Mediterranean region, the Late Glacial period between 16,000 and 11,000 B.P. seems to have been particularly unsuitable for tree growth, and steppes, with Chenopodiaceae and *Artemisia,* prevailed.

In the early part of the Postglacial period human impact on vegetation was still negligible. Little evidence for Mediterranean vegetation assemblages appears in the northwestern Mediterranean regions, but the situation is quite different on the southeastern border.

Juniperus pollen is still found at higher levels in Provence at the beginning of the Postglacial period and these levels are maintained, while *Pinus* declines due to the migration of the Mediterranean deciduous oak woodland into the area. *Juniperus* only disappears itself with the establishment of a modest but continious curve of *Quercus ilex* L.

* Long-sequence pollen diagrams from Ulah Lake (Horowitz, 1968) cannot be taken into account here because they are less detailed than other diagrams from that region.

This juniper decline takes place about 7,500 B.P., somewhat earlier in the North of the Alpilles and slightly later in the South. It can be interpreted as a reflection of the establishment of *Quercus ilex* groves on the steeper slopes where *Juniperus* and other steppe elements had previously grown (Triat, 1978).

Emphasis should be placed on the fact that the establishment of deciduous oak forests and their dominance in the early Postglacial was the general pattern throughout the Mediterranean basin, not only at higher altitudes (Corsica and North Africa: Reille, 1975 and 1977a; Iran: Van Zeist & Bottema, 1977; Turkey: Van Zeist et al., 1975) but also at middle altitudes (Spain: Menéndez-Amor & Florschulz, 1961-1964) and on the plains (central Italy: Frank, 1969; Dalmatia: Beug, 1967; Greece: Wijmstra, 1969; and Bottema, 1974; Western Syria: Niklewski & Van Zeist, 1970, and Israel: Horowitz, 1971).

Diagrams from two coastal sites are characterized by a brief and sudden rise in pollen of Mediterranean taxa during the Holocene: At Fos-sur-Mer, this episode (with *Phillyrea* cf. *angustifolia* L., *Pistacia* cf. *lentiscus* L. and *Juniperus*) occurs during the Atlantic period (about 6,000 B.P.) and is interpreted as evidence for a dune complex that was first formed near the site and was then destroyed by the postglacial marine transgression (Triat, 1975). At Malo-Jezero (Beug, 1960) on Mliet Island, 60 km northeast of Dubrovnic, on the Dalmatian coast, a similar but slightly older (between 7,600 and 6,700 B.P.) episode is characterized by high percentages of *Phillyrea* and *Juniperus* pollen only. It might be attributed to a well-developed forest phase corresponding to a prominent climatic fluctuation. But it may also be the indirect effect of the postglacial marine transgression, that brought a belt of thermophilous Mediterranean and rocky shore vegetation into the vicinity of this island site before invading more continental areas. That this may not be due to a climatic fluctuation, but rather to more local causes, is shown in a diagram from the neighbouring valley of Neretva (Brande, 1973) where the phenomenon is much less distinct.

In Corsica, Reille (1975) showed that *Erica arborea* L. has played a dominant role since the beginning of the Atlantic period, both in Cape Corse and at middle altitudes. The arid and compacted nature of the soil substrate which is found in Corsica may have accounted for the reduced competitive success of many species, and explains the exceptional role played by *Erica arborea* L. The existence of maquis before the development of *Quercus ilex* assemblages is therefore completely natural on this island.

In the southeastern regions, vegetation of a Mediterranean type already was a component of the first Holocene forest which appeared towards 11,000 B.P. Assemblages of *Quercus coccifera* L., *Pistacia* and *Olea* in the plains of nortwestern Syria, were replaced at higher altitudes by *Carpinus orientalis-Ostrya,* deciduous oaks and *Cedrus* (Niklewski & Van Zeist, 1970). At the same time, *Juniperus* and *Quercus* – that was probably of a Mediterranean type – were present in southeastern Turkey. In western Iran (Van Zeist & Bottema, 1977) pollen spectra indicate steppe, with occasional scattered *Pistacia* and *Prunus*. The steppe becomes subsequently richer in trees towards 6,000 B.P., and finally gives rise to an oak forest similar to the present forest of the Zagros Mountains. Generally, an increase in humidity accompanies any rapid extension of forest but the replacement of *Cedrus* by *Pinus nigra* Arnold noted in Anatolia towards 6,000 B.P. is ambiguous as far as humidity is concerned.

A new and inexorably increasing influence – that of man – begins to affect the development of the flora and vegetation during the second part of the Postglacial, at about 7,000 B.P.

Before proceeding, let us summarize the most essential points of the above accounts:

1) Families and even genera which today include Mediterranean species, were already present in the region during the Eocene. However, it is only by the end of the Oligocene that they were present as a consistent group within the essentially tropical flora which was characterized by the presence of xeromorphous elements.
2) The first true Mediterranean vegetation com-

munities arose towards the middle of the Pliocene, probably in relation to the establishment of an annual summer drought.

3) During the Late Pliocene and Pleistocene, Mediterranean forest communities were important only in open types of vegetation during short periods of climatic change. This pattern is more pronounced for sclerophyllous shrub formations.
4) The exact nature of the herbaceous formations that appeared during the cold glacial periods of the Pleistocene remains an unresolved problem.
5) The continued existance of true shrub and arboreal Mediterranean vegetation during the dry and cold periods of the Pleistocene cannot be properly demonstrated in the northern and western parts of the Mediterranean, but refugia undoubtedly existed which ensured the survival of these Mediterranean taxa. In the Middle East, refugia probably occurred near present coastal zones.
6) Some juniper scrub and pine forests of the early Postglacial in the northwestern region were probably of a Mediterranean type. On the southeastern side, Mediterranean tree formations directly contributed to the appearance of widespread forest. In any case, deciduous oak forest was prevalent before human disturbance affected the area. In contrast, evergreen stands only occupied sites on unfavourable soils.
7) This account does not cover the southern Mediterranean regions because of a scarcity of available data.

2 The history of human disturbance

In the assessmant of the history of human influence on the vegetation, the earliest prehistoric disturbance is essentially studied through the fossil record, especially by pollen analysis in conjunction with archaeology. By contrast, the information concerning vegetational changes over the last two or three centuries is based largely on written records, together with modern ecological observations.

2.1 Human influence in prehistoric times

Early man was a hunter and a gatherer, and had relatively little influence on natural vegetation. Human influence began to be more perceptible with the development of fairly permanent agricultural and pastoral activities. This profoundly and directly damaged natural vegetation, as forests were cleared to provide pasture and to free fertile soils for cultivation. Also, the regular production of food led to man's continuous population expansion.

Since it exhibits geographical variation (Reille et al., 1980), human influence can best be described through an inventory of the main phases through which it has successively passed. However, it should be emphasised that much recent information concerning human influence in the Mediterranean region is derived from pollen analysis and cannot be readily and unambiguously interpreted. Morphological distinctions that are desirable for separating taxa are not possible under the light microscope, i.e. certain wild grasses produce large pollen grains that are undistinguisable from those of cereals. Similarly, the differentiation between pollen grains from wild and cultivated olive trees is impossible; indeed, the pollen of *Olea* is similar to that of *Phillyrea.* In addition, many of the sites studied are mountainous, due to a long history of destruction of lowland sites. These higher-altitude sites may be marginal in relation to the Mediterranean climatic zone as a whole.

Human influence on the vegetation of the Mediterranean region can be defined as having a very ancient origin in paleolithic localized disturbance, an intensification of clearance by the end of the Neolithic, with widespread agricultural activity following the introduction of cultivated plants by the beginning of historic times and a further intensification of land-use during recent centuries.

Agriculture and livestock breeding began more than 10,000 years ago in the Near East (Van Zeist, 1980), but because of the location of pollen sites the impact of these activities on the vegetation is not well known. The same difficulties apply to the extension of agriculture into Crete and mainland Greece by 8,000–7,500 B.P. However, there is some archaeological evidence for the diffusion of agriculture into the western Mediterranean region (Italy and southern France) on the evidence of 'cardial' ceramic pottery (decorated with imprints of *Cardium* shells) by 7,500 B.P. At Courthézon, in the lower Rhône plain (Triat, 1978) an horizon of black charcoal dated to 7,350 ± 170 B.P corresponds to a decrease in the percentage of deciduous oak pollen, accompanied by high percentages of pollen of Labiatae and Leguminosae, of which many species were weeds of cereal cultivation in prehistoric agriculture. The presence of *Plantago* (several species are resistant to trampling), various Compositae, and other herbs, is also noted.

Similar phenomena are known from the Alps bordering Provence, where cereal-type pollen appears by the same period (Beaulieu, 1977). This confirms the existence of land clearance at the beginning of the Neolithic period even in regions that were poorly accessible (Jalut, 1976). Temporary decreases in *Quercus pubescens* Willd. pollen percentages are noted in Provence during earlier periods, towards 8,300 B.P. The simultaneous increases in *Plantago,* Ericaceae and other heliophilous or light-demanding species, unaccompanied by any indications of agriculture, suggest land clearance for grazing of animals in a period prior to the Neolithic.

Cereal cultivation was widespread at low altitudes in southern France at the end of the Atlantic period* and had spread to middle altitudes towards 5,000 B.P. with the Chassean civilization (Triat, 1978). Many other mediterranean pollen diagrams (from the Rif to the Western Taurus, through the southern Alps, Corsica and southeastern Greece) indicate land clearance, grazing and cultivation, even for sites at higher altitudes. Stronger effects of anthropogenic origin are seen after the beginning of the Sub-Boreal (4,500 B.P.) (Fig. 1) perhaps as a result of the general population expansion during the Chalcolithic (Reille et al., 1980).

* The classic periods of the Postglacial are not used here in any climatic sense which might be conveyed by their names, but simply as reference periods: Atlantic 7500–4700 B.P., Sub-Boreal 4700–2800 B.P. and Sub-Atlantic after 2800 B.P. (see Fig. 2).

An expansion of pastures and cultivated lands took place between the beginning of historic times and the Roman period (by 3,200 B.P. in northwestern Greece, 2,800 B.P. in Provence, 2,700 B.P. on the Dalmatian coasts and in western Iran and 2,500 B.P. in the Corsican mountains). This process became progressively more intensified until the Middle Ages and was accompanied by an increase in the planting and cultivation of trees such as *Castanea, Olea, Juglans* and *Platanus,* shown by the sudden increase in the frequency of these pollen types.

In more recent periods, sudden changes in vegetation, reflecting historical trends, can often be observed in pollen diagrams. Events such as the decline of cultivation at the end of the Roman Empire, a general increase in human activities at the time of the Arab invasion of northern Africa, a sudden rise of agriculture during the Middle Ages in the Provence Alps and in northwestern Greece, farming fluctuations mirroring economic crises in Southern France, and an acceleration of deforestation in Corsica as a consequence of its colonization by the Republic of Genova, can be cited as examples. Pollen analysis often allows us to locate the extinctions of natural vegetation communities that have taken place in recent times.

What other changes in the vegetation have been brought about by human action, besides the appearance of cultivated and ruderal or nitrophilous plants correlated with forest clearance?

The most general consequence has been the decline or even total disappearance of deciduous oak forest. The progressive deterioration of *Quercus pubescens* forests was to the advantage of the sclerophyllous oaks, and thus led to the increase of *Quercus ilex* and *Q. coccifera* in Provence (Triat, 1979) and in Istria (Beug, 1977) and to the increase of *Q. ilex* and *Buxus sempervirens* L. in Languedoc (Vernet, 1972). In the Rif of Morocco, *Q. ilex, Q. rotundifolia* Lam. and *Q. suber* Willd. forests have spread at the expense of *Q. canariensis* Willd. and *Q. pyrenaica* Willd. (Reille, 1977b). In Kroumiria the almost total disappearance of *Q. canariensis* forest at middle altitudes allowed the spread of *Q. suber* (Reille et al., 1980).

In Corsica, the consequences of the total destruction of the deciduous oak forest, in which *Q. petraea* (Mattuschka) Lieblein and *Q. pubescens* Willd. had been dominant, and more complex. The lower altitudes are now occupied by *Q. ilex* and the upper altitudes by *Pinus nigra* subsp. *laricio* (Poiret) Maire. Some accompanying understorey species that belong to the *Querceto-Fagetea* demonstrate the common origin of these two forests. In Dalmatia (Beug, 1975a), the clearance of the deciduous oak forest resulted in an increase of *Carpinus orientalis* and *Ostrya,* This was followed by the appearance of maquis; a decline of broad-leaved elements finally gave rise to a *Pistacia-Phillyrea-Juniperus* maquis.

A similar decline can be shown for other important forest populations of the Mediterranean mountains. Examples are the *Cedrus atlantica* (Endl.) Carriére forests in the Rif and High Atlas mountains (Reille, 1977b), *Fagus sylvatica* L. and *Abies alba* Miller forests in Corsica (Reille, l.c.) and in the mountains surrounding Provence (Triat, l.c.), and mixed forests of *Pinus nigra* Arnold and *Cedrus libani* Richard in Anatolia (Van Zeist et al., 1975.

An expansion of *Pinus* during historical times has also been a consequence of human action and this is particularly evident in pollen diagrams. This spread can be seen for *Pinus pinaster* Aiton forests in the High Atlas (Reille, 1976) or in Corsica (Reille, 1975), *Pinus halepensis* Miller in Languedoc (Vernet, 1972), Provence (Triat, 1978) and Dalmatia (Beug, 1967; Brande, 1973) and *Pinus nigra* Arnold in Anatolia (Van Zeist et al., 1975).

The extension of dwarf-shrub or herbaceous formations is often demonstrated in the pollen record, as the spread of thorny xerophytes in the High Atlas Mountains and in Corsica (Reille et al., 1980). These alpine assemblages were established as climax vegetation by the end of the late-glacial period and have since migrated into areas that had been dominated by forest from the latter part of the Sub-Atlantic.

A problem is posed by the Mediterranean maquis and garrigues, whose major taxa (Labiatae, Leguminosae) are weak pollen producers or produce pollen grains which do not allow the necessary distinctions to be made either morphologically (*Quercus ilex* and *Q. coccifera* cannot be distinguished from each other) or phytogeographically (Gramineae, Compositae, Ericaceae, *Helianthemum, Rhamnus* or *Juniperus* cannot be dissociated). In some rare cases, however, pollen analysis allows us to identify certain elements of a community with confidence. In northwestern Greece, the occurrence of *Loranthus* pollen allows us to propose *Juniperus oxycedrus* L. as an element of recent assemblages (Bottema, 1974). In Dalmatia (Beug, 1975b) and in the Rif (Reille, 1977) shrubby formations can be shown to be the final result of human disturbance. However, only the study of macroremains (Vernet, 1973; Pons, 1981) reveals the constant increase in the percentages of shrubby taxa during the recent past which has been so heavily dominated by man. The large percentages of *Pinus* pollen in the upper, modern part of the diagrams are largely responsible for the difficulties in discerning sclerophyllous formations in the spectra due to 'flooding' or 'masking' effects. In this respect, it would be interesting to consider the phytosociological relationships that exist between *Pinus* and many shrub communities.

At middle altitudes in the northern parts of the Mediterranean, garrigue and maquis are certainly the result of more or less continuous human disturbance. In coastal regions, however, there is evidence for shrubby sclerophyllous communities (*Pistacia* cf. *lentiscus,* with *Phillyrea cf. angustifolia* and *Juniperus*) prior to human influence (Reille et al., 1980). This indicates the natural origin of these assemblages at low altitudes.

In the eastern part of the Mediterranean, the anthropogenic origin of maquis and garrigue is evident only for sites near to the coast (Van Zeist et al., 1975), although in western Iran and northwestern Syria, the difficulty in separating deciduous and evergreen *Quercus* pollen makes it difficult to interpret variations in other elements, so that little information is available.

2.2 Human influence in modern times

During the period for which precise written documentation exists, man's activities have been controlled by political and economic conditions in a manner that appears to have been much more important than it had been previously.

Cultivation and farming became common throughout the regions bordering the Western Mediterranean under the stimulus of the Americas-based boom in precious metals in the 16th century. During the period 1780–1850, increased exploitation of natural vegetation took place as a consequence of the Industrial Revolution and the establishment of railways (Guillerm & Trabaud, 1980). Land suitable for cultivation or grazing was made available for use through extensive clearance of natural vegetation (Kunholtz-Lordat, 1938, 1958; Sigaut, 1975; Seigue, 1972).

Before the intensive use of coal, forest timber was required to fuel local industries (forges, pottery kilns, glassworks etc.) as well as to provide firewood for centres of population. From the beginning of the 18th century, areas increasingly distant from workshops, factories and towns were cleared in response to this need for energy (Blondel, 1941).

Forests, therefore, needed protection because they were the only available sources of energy, although at the same time, regional and local self-sufficiency made arable and pastoral agriculture equally necessary. Therefore, a certain level of equilibrium, illustrated by the classical notion 'silva-saltus-ager', was achieved (Kunholtz-Lordat, 1946). Human influence on the Mediterranean flora and vegetation had not yet produced permanent changes on a catastrophic scale.

3. The present influence of man on the Mediterranean environment

Widespread changes occurred towards the end of the 19th century which were to have serious consequences for the vegetation of the Mediterranean region. Traditional methods of working the land were abandoned, the human population underwent an explosion, and new paterns of modern society came into being. The considerable increase in population and the progressive mechanization of farming led to a radical transformation of a great part of the circum-Mediterranean area in less than a century, with great damage to both vegetation and flora. This transformation is not only a product of man's immediate effect on the environment (Le Houerou, 1981; Pignatti, 1983) but also of indirect action by intensive grazing, fires, the use of herbicides and pesticides and by pollution.

The particular problems concerning the conservation of the flora in circum-Mediterranean countries are considered in other chapters and only general aspects of this subject will be covered here. Some of the changes directly caused by modern man on the Mediterranean environment will be presented first.

Olive and Carob trees of the coastal fringes of Syria and Lebanon have almost vanished due to unregulated urbanization of the coastline (Abi Saleh et al., 1976). In the Maghreb, the continuing search for arable soils is often the cause for the disappearace of these thermomediterranean sclerophyllous formations which now only survive as isolated trees or in tiny populations at marabout sites*.

The extensive exploitation of sclerophyllous oak forests for charcoal and firewood is mainly responsible for their degradation into coppices. The ageing of the surviving trees and continuing exploitation has led to a degradation resulting in maquis and garrigue which in turn have been exposed to overgrazing and erosion. The example provided by the Mogod Mountains in northern Tunisia is highly significant. *Quercus suber* forests have been entirely destroyed and replaced by a landscape of garrigue or degraded grassland that is now scarred by numerous erosion gullies. In Morocco, tree cover in the Mamora forest appears to have been reduced by half during the last 20 years. In Greece for Anatolia it is almost impossible to find native forests of sclerophyllous oaks in good condition (Barbero & Quézel, 1976).

In spite of their vigour, *Pinus halepensis* and *P. brutia* Ten. communities have also been affected by human disturbance. In fact, conifers were the first to be heavily affected in the semi-arid zones of the Maghreb. The magnificent populations of *P. halepensis* that once grew in the highlands of the

* Burial place of a Moslem monk worshipped as a saint.

Saharan Atlas Muntains of Algeria, and in the Tunisian interior, were locally cleared during the War of Indepence and later degraded through uncontrolled cutting.

Deciduous oak forests have also suffered, especially in the Near East. The forests of 'Zen oaks' (*Querius canariensis* Willd.) in the Mahgreb have been protected for a long time and they were often in locations that were difficult of access. But as these forests always occupy deep soils, farming expansion has led to their complee destruction in the Near East and in Anatolia. Oak forests of the *Quercus aegilops* group in Western Anatolia and *Q. pseudocerris* in Southern Anatolia have been reduced to isolated stands, deliberately preserved by local communities, or in marabouts. Excessive grazing is directly accountable for the total deterioration of *Quercus pubescens* subsp. *anatolica* O. Schwarz forests over vast areas in the semi-arid zones of the Anatolian highlands; these areas are now covered by steppe vegetation or scrub that rarely exceeds 1 m in height (Akman, Barbero & Quézel 1978).

The future of mountain conifer forests is very uncertain and locally very problematic. They often include particularly fine stands of trees.

The black pine of Mauritania is no longer represented in Morocoo save for a few tiny populations hardly covering 20 hectares; in Algeria a few hundred individuals have been left as relics following fire.

The decline of the cedar forests at high altitudes in the eastern Atlas Mountains of Morocco and the absence of any seedling regeneration, under existing condition, is very spectacular. In the region of Anemzi-Agoudim, thousands of hectares of forest were transformed within 30 years into a waste-land. In Algeria, the cedars of Bou-Thaleb and those of Chelia have undoubtedly diminished by half within the same period. North African fir forests have also suffered greatly in the last decades. The populations of *Abies pinsapo* subsp. *maroccana* (Trabut) Emb. & Maire in Jebel Tazzout (Rif) were partially burned a few years ago. Populations of Numidian firs in the Babor were thinned by uncontrolled felling during the past few years in spite of the creation there of a National Park.

Fortunately, the situation is less grave in Anatolia and in the Near East (Akman, Barbero & Quézel, 1978; Chalabi, 1980; Abi Saleh et al., 1976). The forests have been over-exploited during the last century but, particularly in Turkey, a very effective national forestry service now ensures their protection and is even striving towards their re-establishment. However, natural regeneration is still in question and the deleterious effects of excessive grazing are felt locally. It should also be pointed out that during World War II the magnificent fir forests of Khamoua'h (Chouchani, Khouzani & Quézel, 1974) in Northern Lebanon were entirely destroyed in order to supply the necessary timber for railway sleepers.

A very interesting example amongst non-arboreal ecosystems is the progressive disappearance of vast areas of steppe vegetation in the Maghreb and the Near East. Clearing and the cultivation of cereals became possible through the development and utilization of agricultural machinery. In the past 20 years, thousands of hectares that had a very unique floristic assemblage have disappeared in Anatolia alone, without any measures being taken to protect or safeguard them. Even though no attempt has been made to record the species present, it is probable that a few dozen species have disappeared or are on their way to extinction in these territories.

Drainage and infilling projects to allow the cultivation of former marsh, estuarine or other environments have been carried out for the last century in all of the Mediterranean countries, and have led to the disappearance of many floristically rich ecosystems. This is the case for Pontin Marsh in Italy or Rharb Marsh and La Calle Lake in North Africa, where more than one hundred species have disappeared in less than a century. Similarly, the remarkable habitat of ephemeral *Isoetes* marshes has become very rare around the Mediterranean and many associated species have already vanished forever.

On the coasts of the Mediterranean crowd pressure brought by tourism, urban development, holiday residences and the intensive trampling of sand dunes and seaside rocks are ruining an environment that is equally rich in rare species and endemics.

Indirect human influence on natural environments around the Mediterranean already plays a notable role and these pressures will undoubtedly increase in the years to come. Many authors have studied the effect of grazing on the flora and vegetation of the Mediterranean region (Le Houerou, 1972). Grazing is, of course, as old as the civilizations that have occupied the Mediterranean basin. However, after only a century or even a few decades the former balance has been disturbed to the detriment of various vegetation communities.

Especially in the Maghreb and the Near East, an uncontrolled increase in sheep and goat numbers, together with irrational land use methods, have often led to intense degradation of pastures and to considerable modification of steppes and forests. These changes favour the rapid multiplication of noxious, unpalatable weeds that hinder forest regeneration. This is widely responsible for the expansion of secondary steppe vegetation and can even accelerate desertification (Le Floch & Floret, 1972; Le Houereou, 1968, 1969).

The reverse problem can be detected in some countries on the northern shores of the Mediterranean. Here, reduction in herds of grazing livestock has become deleterious to vegetation and a progressive invasion by scrub has led to a decrease in species diversity and to an increased fire risk.

The incidence of serious fires in forests or in scrubland is increasing in the Mediterranean region (Seigue, 1972; Chautrand, 1972; Le Houerou, 1973). Fire can be considered as a natural ecological factor, but now it seems clear that the considerable increase of burnt-over areas is directly related to more frequent human presence, especially by tourists. The influence of fire on plant species and vegetation is well known and need not be discussed here. It results in reduced diversity and aids the spread of several common species that are well adapted to fire.

The general use of herbicides and pesticides in agriculture over the past few years has been the cause of radical changes in the status of weeds of cultivation and ruderal species that occur in man-made environments. We are now witnessing the progressive disappearance of a number of plant species that were very common some 30 years ago and can now hardly be found. As pointed out by several authors (Aymonin, 1965, 1976; Guillerm & Trabaud, 1980) this problem is a serious one, and protective measures should be taken to ensure the survival of various weed and ruderal species, particularly those associated with cereal fields.

The effects of pollution on vegetation are very localized, but certainly worth mentioning here. These effects are chiefly seen in aquatic ecosystems during processes of eutrophication which usually lead to radical changes, including the disappearance of various aquatic vegetation communities and the replacement of unusual species by common, invasive ones.

References

Abi-Saleh, B., Barbero, M. & Quézel, P. (1976). Les series forestières de végétation au Liban, essai d'interprétation schématique. Bull. Soc. Bot. Fr. 123: 541–560.

Akman, Y., Barbero, M. & Quézel, P. (1978). Contribution à l'étude de la végétation forestière d'Anatolie Méditerranéenne. Phytocoenologia 5: 189–276.

Andreansky, G. (1963). Das Trockenelement in der alterter-tiären Flora Mitteleuropas auf Grund paläobotanischer Forschung in Ungarn. Vegetatio 11: 95–111.

Anglada, R., Arnaud, M. Babinot, J.F., Brunel, P., Catzigras, F., Colomb, E. & Jacquet, D. (1978). L'extension de la transgression aquitanienne dans les Bouches-du-Rhône: les calcaires du Pied d'Autry (Allauch) et d'Eguilles. C.R. Acad. Sc. Paris. sér. D. 286: 1563–1566.

Arambourg, C., Arènes, J., & Depape, G. (1953). Contribution à l'étude des flores fossiles d'Afrique du Nord. Arch. Mus. Nat. Hist. Nat. sér. 7, 2: 1–81.

Aymonin, G.C. (1965). Origines présumées et disparition progressive des adventices messicoles calcicoles; pp. 1–11. C.R. 2nd Colloque Mauvaises Herbes, Grignon.

Aymonin, G.C. (1976). La baisse de la diversité spécifique dans la flore des terres cultivées, pp. 195-202. C.R. 5me Colloque Int. Ecol. Biol. Mauvaises Herbes, Dijon.

Barbero, M. & Quézel, P. (1976). Les groupements forestiers de Grèce centroméridionale. Ecol. Medit. 2: 1–86.

Bazile, E. & Bazile, F. (1978). Evolution des climats au Würm récent en Languedoc méditerranéen d'après l'analyse anthracologique essentiellement. Geobios. 11: 933–935.

Bazile, E., Suc, J.P. & Vernet, J.L. (1980). L'Histoire floristique des régions méditerranéennes depuis le début du Pliocène. Naturalia Monspeliensia 237: 33–40.

Beaulieu, J.L. (1972). Analyses polliniques des tourbes éémiennes de St. Paul-lez-Durance (Bouches-du-Rhône). Bull. Assoc. Franc. pour l'Etude du Quatern. 3: 195–205.

Beaulieu, J.L. (1977). Contribution pollenanalytique à l'histoire tardiglaciaire et holocène de la végétation des Alpes méridionales françaises. Thèse des Sciences. Faculté des Sciences et Techniques St. Jérôme, Marseille.

Bertolani-Marchetti, D. (1964). Richerche palinilogiche in sedimento torbosi a Porto Conte, presso Alghero (Sardegna). Arch. Bot. e Biogeogr. Ital. 9: 402–406.

Bertolani-Marchetti, D., Accorsi, C.A., & Bandini-Mazzanti, M. (1978). Primi dati palinilogici sulla serie marine pliopleistocenica di Vrica presso Crotone (Calabria). Giornale Bot. Ital., Rome, pp. 112, 296.

Bertolani-Marchetti, D. & Cita, M.B. (1975). Studi sul Pliocene e sugli strati di passagio dal Miocene al Pliocene (Palynological investigations on the late Messinian sediments records at D.S.D.P. Site 132 [Thyrrhenian basin] and their bearing on the deep basin desiccation model). Riv. Paleontol. Ital. 81: 281-308.

Bertoldi, R. (1977). Studio palinologico della serie di Le Castella (Calabria). Atti. Acad. Naz. dei Lincei, Rome 62: 547-555.

Bessedik, M. (1979). La flore pollinique d'un niveau gypseux de la série de Portel (Aude): Oligocène termina. Rapport D.E.A. Université des Sciences et Techniques du Languedoc, Montpellier.

Beug, H.J. (1960). Beiträge zur postglazialen Floren- und Vegetationsgeschichte in Süddalmatien: Der See 'Malo Jezero' auf Mljet. Flora 150: 600–631.

Beug, H.J. (1967). On the forest history of the Dalmatian coast. Review of Palaeobot. and Palinol. 2: 271-279.

Beug, H.J. (1975a). Changes of climate and vegetation belts in the mountains of Mediterranean Europe during the Holocene Biuletyn Geol. 19: 101-110.

Beug, H.J. (1975b). Man as a factor in the vegetational history of the Balkan Peninsula. In: Jordanoc, D. et al. (Eds.) Problems of the Balkan Flora and Vegetation, pp. 72-78. University of Sofia.

Beug, H.J. (1977). Vegetationsgeschichtliche Untersuchung im Küstenreich von Istiren. Flora 166: 357-381.

Biju-Duval, B., Dercourt, J. & Le Pichon (1977). From the Tethys ocean to the Mediterranean seas: A plate tectonic model of the evolution of the Western alpine system. In: Biju-Duval & Montadert (Eds.) Basins, pp. 143-164. Split (Yugosl.) Technip. Paris.

Blondel, R. (1941). La végétation forestière de la région de Saint-Paul près de Montpellier. Mem. Soc. Vaudoise Sc. Nat. 46: 307-382.

Bottema, S. (1974). Late quaternary vegetation history of Northwestern Greece. Thèse. University Groningen.

Bottema, S. (1975). The interpretation of pollen spectra from prehistoric settlements (with special attention to Liguliflorae). Palaeohistoria 17: 17-35.

Brande, A. (1973). Untersuchungen zur postglazialen Vegetationsgeschichte im Gebiet der Neretva-Niederung (Dalmatien, Herzegowina). Flora 162: 1–44.

Busson, G. (1979). Le 'géant salifère' messinien du domaine Méditerranéen: interpretation génetique et complications paléographiques. Ann. Géol. Pays Hellén, Tome hors série Fasc. 1, pp. 227-238.

Cavelier, C. (1976). La limite Eocène-Oligocène en Europe occidentale. Thèse ès Sciences, Paris. Résumé in Bull. Inf. Géol. Bassin. Paris 13: 65–84.

Chalabi, M. (1980). Analyse phytosociologique, phytoécologique, dendrométrique et dendroclimatologique des forèts de *Quercus cerris* subsp. *pseudocerris* et contribution à l'étude taxonomique du genre *Quercus* en Syrie. Thèse ès Sciences, Faculté des Sciences et Techniques St. Jérôme, Marseille.

Chautrand, L. (1972). Les incendies de forêt en Provence-Côte d'Azur. Bull. Tech. Inform. 268: 405–414.

Chouchani, B., Khoueani, A. & Quézel, P. (1974). A propos de quelques groupements forestiers du Liban. Biol. Ecol. Médit., Marseille 1: 63–77.

Couteaux, M. (1977). A propos de l'interpretation des analyses polliniques de sédiments mineraux, principalement archéologiques. In: Aproche écologique de l'homme fossile. Suppl. Bull. de l'Assoc. Franc. pour l'Etude du Quatern 1: 259-277.

Depape, G. (1925). La flore des grès landéniens du Nord de la France. Ann. Soc. Géol. Nord France 50: 10–48.

Depape, G. (1959). Les flores fossiles du Crétace superieur en France. C.R. Congrès Soc. Sav. Dijon, pp. 61-94.

Depape, G. (1963). Les flores infracrétaciques de la France et des pays voisins. Colloque Intern. 'Crétacé Inferieur' Orleans, pp. 349-371. Paris: Bureau de Rech. Géol. et Minières.

Diniz, F. (1967). Une flore tertiaire de caractère méditerranéen au Portugal. Rev. Palaeobot. Palynol. 5: 262-268.

Diniz, F. (1969). Ombellifères pliocènes de Rio Maior (Portugal). Naturalia Monspeliensia 20: 77-88.

Fergusson, D.K. (1967). On the phytogeography of Coniferales in the European Cenozoic. Paleogeogr., Palaeoclimatol. Palaeoecol. 3: 73-110.

Florschutz, F., Menéndez-Amor, J. & Wijmstra, T.A. (1971). Palynology of a thick quaternary succession in southern Spain. Palaeogeogr., Palaeoclimat., Palaeoecol. 6: 67-85.

Fourniquet, J. (1977). Mise en évidence de mouvements actuels, verticaux, dans le Sud-Est de la France par comparaison de nivellements succesifs. G.R. SOMM. Soc. Géol. France 5: 266–268.

Frank, A.H.E. (1969). Pollen stratigraphy of the lake of Vico (central Italy). Palaeogeogr., Paleoclimatol., Palaeoecol. 6: 67-85.

Frenzel, H.B. (1980). Quelques remarques sur la situation de la stratigraphie de Pléistocène dans la partie méridionale d'Allemagne. In: Chaline J. (Eds.), Problems de stratigraphie quatenaire en France et dans les pays limitrophes. Suppl. Bull. de l'Assoc. Franc. pour l'Etude du Quatern 2: 141-147.

Fritel, P.H. (1908). Contribution à l'étude des flores éocenes du Bassin de Paris. C.R. Congr. Soc. Sav., Sciences, pp. 315-328.

Fritel, P.H. (1909). Révision de la flore fossile des grès yprésiens du Bassin de Paris. Journ. Bot. 22: 101-112.

Fritel, P.H. (1910a). Etude sur les végétaux fossiles de l'étage sparnacien du Bassin de Paris. Men. Soc. Géol. France 40: 1-37.

Fritel, P.H. (1910b). Observation sur la flore fossile des grès thanétiens de Vervins et une Revision des espéces qui la composent. Bull. Soc. Géol. France 10: 191-709.

Fritel, P.H. (1923). Flore sparnacienne des 'Gres des lignites' des environs de Laon et de Soissons (Aisne). Bull. Mus. Nat. Hist. Nat. 29: 189-193.

Fritel, P.H. (1924). Suite et addition à la révision de la flore cuisienne des grés de Belleu. Bull. Soc. Géol. France 24: 150-175.

Grambast, L. (1955). Un *Juniperoxylon* particulier dans l'Eocène inferieur du Bassin de Paris. Arch. Mus. Nat. Hist. Nat. 7: 3-24.

Grangeon, P. (1958). Contribution à l'étude de la paléontologie végétale du massif du Coiron (Ardèche, S.E. du massif central français). Mém. Soc. Hist. Nat. d'Auvergne 6: 1-300.

Gruger, E. (1977). Pollenanalytische Untersuchung zur würmzeitlichen Vegetationsgeschichte von Kalabrien (Süd-italien). Flora 166: 475-489.

Guillerm, J.L. & Trabaud L. (1980). Les interventions récents de l'homme sur la végetation au nord de la Mediterranée et plus particulièrement le Sud de la France. Naturalia monspeliensia 237: 157-171.

Horowitz, A. (1968). Upper Pleistocene-Holocene climate and vegetation of the Northern Jordan Valley (Israel); Thesis University Jerusalem.

Horowitz A. (1971). Climatic and vegetational developments in Northeastern Israel during Upper Pleistocene-Holocene times. Pollen et Spores 13: 255-278.

Hughes, N.F. (1973). Mesozoic and Tertiary distributions and problems of land-plant evolution. Special Papers in Paleontology 12: 188-198.

Jalut, G. (1976). Les débuts de l'agriculture en France, les défrichements. In: La Prehistoire Française, pp. 180-185. Paris: C.N.R.S.

Kunholtz-Lordat, G. (1938). La terre incendiée. Nimes: La Maison Carrée.

Kunholtz-Lordat, C. (1946). La silva, le saltus et l'ager de Garrigue. Ann. Ecol. Nat. Agric., Montpellier 26: 1-82.

Kunholtz-Lordat, G. (1958). L'écran vert. Mém. Mus. Nat. Hist. Nat. 9: 1-276.

Le Floch, E. & Floret, C. (1972). Désertification et ressources pastorales de la Tunisie présaharienne. Coll. désert. Gabés, Tunisie, pp. 1-11. Mimeograph.

Le Houerou, H.N. (1968). La désertization du Sahara septentrional et des steppes limitrophes. Ann. Alger. Geogr. 3: 2-27.

Le Houerou, H.N. (1969). La végetation de la Tunisie steppique. Tunis: I.N.R.A.T.

Le Houerou, H.N. (1972). An assesment of the primary and secondary production of the arid grazing lands ecosystems of North Africa. Symp. of Ecophysiol. Leningrad: Nauka.

Le Houerou, H.N. (1973). Fire and végétation in the Mediterranean Basin. Proc. Ann. Tall. Timbers Fire Ecol. Conf. 13: 237.

Le Houerou, H.N. (1981). Impact of man and his animals on Mediterranean Vegetation. In: Di Castri et al. (Eds.,) Mediterranean-type Shrublands, pp. 479-517. Amsterdam: Elsevier.

Maley, J. (1980). Les changements climatiques de la fin du Tertiaire en Afrique, leurs conséquences sur l'apparition du sahara et de sa végétation. In: The Sahara and the Nile, , pp. 63-86. Rotterdam: A.A. Balkema.

Marion, A.F. (1872). Plantes fossiles des calcaires marneux de Ronzon. Ann. Sc. Nat. Bot. 14: 326-364.

Medus, J. & Pons A. (1980). Les prédécesseurs des végétaux méditerranéens actuels jusq'au début du Miocène. Naturalia Monspeliensia 237: 11-20.

Menéndez-Amor, J. & Florschutz, F. (1961). Contribucíon al conocimiento de la historia de la vegetación en España durante el Cuaternario. Estudios Geológicos 17: 83-99.

Menéndez-Amor, J. & Florschutz, F. (1963). Sur les éléments steppiques dans la végétation quaternaire de l'Espagne. Bol. Real. Soc. Esp. Hist. Nat. 61: 121-123.

Menéndez-Amor, J. & Florschutz (1964). Results of the preliminary palynological investigation of samples from a 50 m. boring in southern Spain. Bol. Real Soc. Esp. Hist. Nat. 62: 251-255.

Naud, G. & Suc, J.P. (1975). Contribution à l'étude paléofloristique de Coirons (Ardèche): premières analyses polliniques dans les alluvions sous-basaltiques et interbasaltiques de Mirabel (Miocène supérieur). Bull. Soc. Géol. France 17: 818-825.

Niklewski, J. & Van Zeist, W. (1970). A late quaternary pollen diagram from Northwestern Syria. Acta Bot. Neerland. 19: 737-754.

Pignatti, S. (1983). Human impact on the Mediterranean basin. In Holzner, W. et al., Man's Impact on Vegetation, pp. 151-162. The Hague: Junk.

Planchais, N. (1974). La palynoflore du Tertiaire final à Sète (Hérault). C.R. Acad. Sc. Paris 278: 855-858.

Pons, A. (1964). Contribution palynologique à l'étude de la flore et de la végétation pliocènes de la region rhodanienne. Ann. Sc. Nat., Bot., Paris 12: 499-722.

Pons, A. (1969). Les macroflores quaternaires de France. In: Etudes Françaises sur le Quaternaire, pp. 85-93. VIIIe Congress INQUA, Paris.

Pons, A. (1981). The history of the Mediterranean shrublands. In: Di Castri et al. (Eds.), Mediterranean-type Shrublands, pp. 131-138. Amsterdam: Elsevier.

Pons A. & Quézel, P. (1958). Premières remarques sur l'étude palynologique d'un guano fossile du Hoggar. C.R. Acad. Sc. Paris 224: 2290-2292.

Pons, A. & Suc, J.P. (1980). Les témoignages des structures actuelles de végétation méditerranéennes durant le passé anterieur à l'action de l'homme. Naturalia Monspeliensia 237: 69-78.

Quézel, P. (1965). La végétation du Sahara. Stuttgart: Fischer Verlag.

Reid, E.M. & Chandler, M.E.J. (1933). The flora of the London Clay. British Museum (Natural History) London.

Reille, M. (1975). Contribution pollenanalytique à l'histoire tardiglaciaire et holocène de la végétation de la montagne corse. Thèse des Sciences, Fac. Sciences et Techn. St-Jerôme, Marseille.

Reille, M. (1976). Analyse pollinique de sédiments postglaciaires dans le Moyen Atlas et le Haut Atlas Marocains: premiers resultats. Ecologia Mediterranea 2: 153-170.

Reille, M. (1977a). Quelques aspects de l'activité humaine en Corse durant le Subatlantique et ses conséquences sur la végétation. Bull. de l'Assoc. Franc. pour l'Etude du Quatern. 47: 329-342.

Reille, M. (1977b). Contribution pollenanalytique à l'histoire holocène de la végétation des montagnes du Rif (Maroc Septentrional). In: Recherches françaises sur le Quaternaire Suppl. Bull. de l'Assoc. Franc. pour l'Etude du Quatern. 1: 53-76.

Reille, M., Triat, H. & Vernett, J.L. (1980). Les témoignages des structures actuelles de végétation méditerranéenne durant le passé contemporain de l'action de l'homme. Naturalia Monspeliensia 237: 79-87.

Remy, H. (1958). Zur Flora und Fauna der Villafranca-Schichten von Villaroya, Prov. Logrono, Spanien. Eiszeitalter und Gegenwart, Ohringen 9: 83-103.

Ricciardi, E. (1965). Analisi polliniche di una serie stratigraphica dei sementi lacustri del Pleistocene inferiore nel baccino di Leonessa (Rieti, Italia centrale). Giorn. Bot. Ital. 72: 62-82.

Rognon, P. (1967). Le massif de l'Atakor et ses bordures (Sahara Central). C.N.R.S., Sect. Géol., Paris 9; 1-559.

Rognon, P. (1976). Essai d'interpretation des variations climatiques au Sahara depuis 40.000 ans. Rev. Géogr. Phys. Géol. Dyn. 18: 251-282.

Ryan, W.B. & Hsu, K.J. (1973). Initial Report of the Deep Sea Drilling Project. Washington: U.S. Govern. Printing Office.

Saporta, G. (1866). Etudes sur la végétation du Sud-Est de la France a l'époque tertiaire. Paris: Masson.

Saporta, G. (1868). Prodrome d'une flore fossile des travertins anciens de Sézanne. Mém. Soc. Géol. France 8: 289-436.

Saporta, G. (1872). Etudes sur la végétation du Sus-Est de la France à l'époque tertiaire. Suppl. I: Révision de la flore des gypses d'Aix. Paris: Masson.

Saporta, G. (1888). Origine paléontologique des arbres cultivés ou utilisés par l'homme. Paris: Baillière.

Saporta, G. (1890). Revue des travaux de paléontologie végétale. Rev. Gén. Bot. 2: 182-190.

Seigue, A. (1972). Les incendies de la forêt méditerranéenne: Historique, essai prospectif. Bull. Tech. Inform. 268: 415-423.

Sigaut, L. (1975). L'agriculture et le feu. Paris: Mouton.

Suc, J.P. (1976). Quelques taxons-guides dans l'étude paléoclimatique du Pliocène et du Pleistocène inférieur du Languedoc (France). Rev. Micropal. 18: 246-255.

Suc, J.P. (1980a). Apercu sur la végétation et le climat des regions méditerranéennes d'Europe occidentale au Pliocène et au Pleistocène inférieur d'apres l'analyse pollinique. Mem. Mus. Nat. Hist. Nat. 27: 177-181.

Suc, J.P. (1980b). Contribution à la connaissance du Pliocène et du Pleistocène inferieur des régions méditerranéennes d'Europe occidentale par l'analyse palynologique des dépôts du Languedoc-Roussillon (sud de la France) et de la Catalogne (nord-est de l'Espagne). Thèse dès Sciences, Univ. des Sciences et Techniques du Languedoc, Montpellier.

Triat, H. (1975). Les survivantes glaciaires de Fos-sur-Mer (Bouches-du-Rhône). In: Les données historiques et l'étude de la flore méditerranéenne. Paris: C.N.R.S.

Triat, H. (1978). Contribution pollenanalytique à l'histoire tardi- et postglaciaire de la végétation de la basse vallée du Rhône. Thèse ès Sciences, Facult. des Sciences et Techn. St-Jerôme, Marseille.

Triat, H. (1979). Histoire de la fôret provençale depuis 15 000 ans d'après l'analyse pollinique. Forêts Medit. 1: 19-24.

Van Campo, E. (1976). La flore sporopollinique du gisement miocène de Venta del Moro (Espagne). Univ. Scienc. et Techn. du Languedoc, Montpellier.

Van Campo, E. (1978). Paléoflores et paléoclimate néogenes au nord-est de la Tunisie. Thèse spec. Univ. Scienc. et Techn. du Languedoc, Montpellier.

Van Zeist, W. (1980). Aperçu sur la diffusion des végétaux cultivés dans la région méditerranéenne. Naturalia Monspeliensia n° spécial: 129-145.

Van Zeist, W. & Bottema, S. (1977). The pleniglacial, lateglacial and early postglacial vegetation of Zeribar and their present-day counterparts. Palaeohistoria 19: 19-85.

Van Zeist, W., Woldring, H. & Stapert, D. (1975). Late Quaternary vegetation and climate of Southwestern Turkey. Palaeohistoria 17: 55-143.

Vernet, J.L. (1972). Contribution à l'étude de la végétation du sud-est de la France au Quaternaire. Etude de macroflores de charbon de bois principalement. Thèse ès Sciences. Univ. du Languedoc, Montpellier.

Vernet, J.L. (1973). Etude sur l'histoire de la végétation du Sud-Est de la France au Quaternaire d'après les charbons de bois principalement. Paléobiologie continentale 4: 1-93.

Wijmstra, T.A. (1969). Palynology of the first 30 m of a 120 m deep section in Northern Greece. Acta Bot. Neerl. 18: 511-527.

Wijmstra, T.A. (1971). The floral record of the late Cenozoic of Europe. In: Van der Hammen, E. et al. (Eds.), The Late Cenozoic Ages, pp. 391-424. New Haven and London: Yale University Press.

Faculté des Sciences et Techniques de Saint-Jérôme,
Rue Henri Poincaré
F-13397 Marseille Cedex 13, France

Part II

Conservation problems in different sub-areas

Case histories

CHAPTER 4

The Iberian Peninsula

C. GÓMEZ-CAMPO and J. MALATO-BELIZ

1. Endemism in the Iberian Peninsula

According to data extracted from *Flora Europaea* (Tutin et al., 1964-80), the flora of the Iberian Peninsula includes 15 endemic genera and 986 endemic species plus 272 endemic subspecies. These figures exclude the rich floras of the politically dependent Atlantic and Mediterranean islands, but include Pyrenean endemics when these growat both sides of the French-Spanish border. They also include all those taxa (18% approximately) which are not assigned a number in *Flora Europaea* due to uncertain taxonomic status. If the subspecific level is considered, the distribution of endemics in the peninsular countries is as follows:

Present in Portugal only	114 species + subspecies
Present in Spain only	684 species + subspecies
Present in both Portugal and Spain	228 species + subspecies
Pyrenean, also present in French territory	195 species + subspecies
Northern range from Portugal to the French side of the Pyrenees	37 species + subspecies
Total	1258 species + subspecies

That these figures are at odds with those given by the Council of Europe (Anonymous, 1983: 497 endemic species for Spain and 60 for Portugal) stems from their exclusion of the subspecific level and of taxonomically doubtful species, while including some taxa that did not appear in *Flora Europaea*.

A more precise assignment of these taxa to fifty natural geographic subdivisions of the Iberian Peninsula gives a better idea of their distribution and indicates which regions are the richest with respect to their content of endemic species and subspecies. This distribution is summarized in Figure 1; some more details can be seen in Gómez-Campo et al. (1984). The great abundance of endemics in the Baetic mountain range, extending from the Strait of Gibraltar to Cabo San Antonio, Alicante, is very noteworthy. The few square kilometres of the Sierra Nevada, with 177 endemics present - 66 of them exclusive to that area - probably represents the highest concentration of plant diversity in continental Europe. Other areas of interest are the Pyrenees, the Serra da Estrela in central Portugal, and, to a lesser extent, the Picos de Europa in northwestern Spain, the Gredos mountains in central Spain, the mountains of Teruel and Barcelona in eastern and northeastern Spain and the Serra de Monchique in southern Portugal.

It should be remarked that the detailed study of the geographic distribution of plants provides a highly useful and objective criterion for the evaluation of areas for subsequent land use planning.

By the same procedure it can also be demonstrated that at least a half of the Iberian endemics are

Gómez-Campo, C. (ed.), Plant conservation in the Mediterranean area.
© 1985, Dr W. Junk Publishers, Dordrecht. *ISBN 90 6193 523 7.*

local, often 'narrow' endemics, and as such are restricted to only one of the above-mentioned fifty subregions. This is very important with regard to their conservation. Though not all of these taxa are threatened, the numbers speak eloquently of the magnitude of plant conservation problems in the Iberian Peninsula.

Taxonomically, a half of the Iberian endemics belong to only 27 genera, while the other 50% are divided among 286 additional genera (Table 1). These figures suggest that the endemic flora of the Iberian Peninsula is to a large extent composed of recently evolved groups rather than by old relict ones. The relatively low number of taxa which are endemic at the generic level is in agreement with this view. This point should be taken into account when formulating conservation policies.

Another important component of Iberian plant diversity is the so-called Ibero-African group of plants, i.e. species from northwestern Africa whose sole occurence on the European continent is in the Iberian Peninsula, largely in the southern and southeastern regions. The number of Ibero-African plants is estimated to be not much lower than that of true endemics. Though they should obviously have a secondary importance from the con-

Table 1. Genera with the highest representation in the endemic flora of the Iberian Peninsula[α], indicating the number of endemic species and subspecies they contain.

	Number of endemic species	Number of additional endemic subspecies	Total, to subspecies level
Centaurea (Compositae)	38	40	78
Hieracium (Compositae)	45	1	46
Linaria (Scrophulariaceae)	38	5	43
Saxifraga (Saxifraceae)	32	8	40
Armeria (Plumbaginaceae)	26	5	31
Sideritis (Labiatae)	20	7	27
Limonium (Plumbaginaceae)	19	6	25
Narcissus (Amaryllidaceae)	18	6	24
Galium (Rubiaceae)	23	0	23
Campanula (Campanulaceae)	16	6	22
Teucrium (Labiatae)	19	3	22
Thymus (Labiatae)	20	2	22
Dianthus (Caryophyllaceae)	12	9	21
Festuca (Gramineae)	18	3	21
Antirrhinum (Scropulariaceae)	16	3	19
Arenaria (Caryophyllaceae)	15	3	18
Jasione (Campanulaceae)	9	8	17
Silene (Caryophyllaceae)	16	1	17
Genista (Leguminosae)	12	3	15
Senecio (Compositae)	14	1	15
Carduus (Compositae)	10	5	15
Ranunculus (Ranunculaceae)	10	4	14
Avenula (Gramineae)	11	2	13
Biscutella (Cruciferae)	12	0	12
Euphorbia (Euphorbiaceae)	12	0	12
Iberis (Cruciferae)	6	6	12
Thymelaea (Thymelaeaceae)	11	1	12
Total endemics in the above 27 genera	498	138	636
Total endemics in 286 other genera	488	134	622
Grand total	986	272	1258

[α] Data extracted from Tutin et al. (1964–1980), 'Flora Europaea', Cambridge University Press.

servational point of view, sometimes they give rise to interesting plant assemblages that should not be neglected. A well-studied example is *Tetraclinis articulata* Vahl (Rigual & Esteve 1952; Templado, 1974). It forms extensive, often dense stands in Morocco and Algeria, and Algeria, and exists in some localities of Tunisia, Lybia and Malta. The only remaining populations on the European continent are located near Cartagena, southeastern Spain, in a very restricted area that has suffered from at least 3,000 years of disturbance, largely in the form of mining operations. Only a few hundred individual trees are taller than 1,5 m but the presence of young saplings suggests recent regeneration. Only a few dozen reach the normal height of 7-8 m.

2. The significance and conservation problems of a very rich flora.

The ecological point of view is a major consideration in conservation. Though much further work is needed to confirm the data we give here, it can be estimated that at least 60% of the Iberian endemics grow in alpine or montane habitats; the map in Figure 1 is pertinent in this respect for those who know the geography of the Iberian Peninsula. Many endemics are rock-dwelling or grow in stony situation. Perhaps another 20% are of coastal habit, growing in maritime rock, dune, or litoral salt-tolerant steppe vegetation. The remaining 20% are largely continental, distributed on plains and plateaus, either in steppes or humid zones. Of the steppe species, an important group are those adapted to gypsum-deposit soils. These are abundant in the east of the Iberian Peninsula, although rare in other parts of Europe. Few, if any, of the Iberian endemics seem to be adapted to a climax type of vegetation. Most are adapted to intermediate successional stages, and some, such as *Guiraoa arvensis* (Gómez-Campo, 1980) are even weedy.

The taxonomic status of many endemics are often imperfectly established. This reflects the general state of knowledge regarding the Iberian and Mediterranean flora, a field which obviously needs

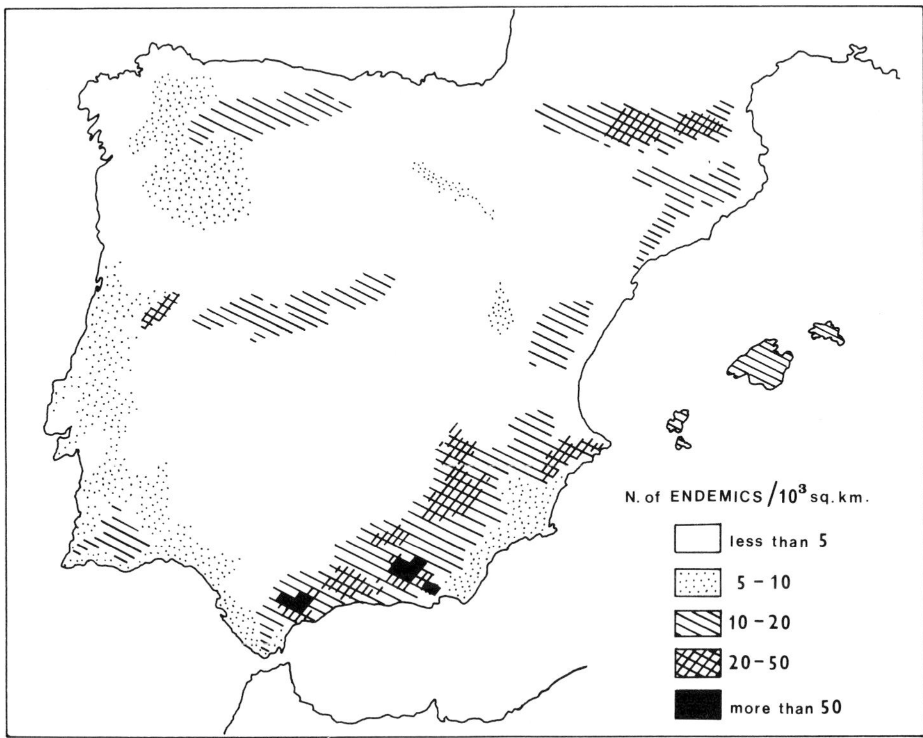

Fig. 1. Map showing the comparative density of endemism in different regions of the Iberian Peninsula and Balearic islands.

much further investigation. Each year, new species or subspecies are described, and most of these are endemic. Indeed, there are at least 150 recently described endemic taxa that were not included in *Flora Europaea* and thus have not been computed in the figures given above. Certainly, several means exist whereby these numbers could decrease, as for instance when some supposedly endemic taxon is discovered in North Africa or elsewhere, or when a taxon is disqualified from its present taxonomic status. However, an overall tendency towards further increases in the list of endemics is to be expected.

In the past, botanical publications concerning Iberian endemics have seldom mentioned their conservation. Some notable exceptions are the papers by Bolós (1962), Rivas-Goday (1959) and Tavares (1939, 1959) which are pioneering efforts concerning different aspects of conservation. The first of these gives an account of some reintroduction experiments of the now extinct species *Lysimachia minoricensis* Rodr. into its former locality on Menorca Island. In the second paper, a list is given of thirty-eight species to be protected, including some non-endemic ones. The works of Tavares refers to several aspects of conservation concerning Portuguese endemics. More recently, references to conservation in the botanical literature have still been somewhat scarce, but with a tendency to increase. A few examples are Vasconcellos (1970), Rivas-Martinez (1971), Malato-Beliz (1975), Prentice (1976) and Gómez-Campo (1977, 1980, 1981a).

An account of the general significance of Iberian endemics and more particularly of the paleoendemic component has been given by Montserrat & Villar (1972). Their origins and the relationship to soil factors is commented upon by Rivas-Martinez (1972). The causes of the floristic richness of the Iberian Peninsula seem to be mainly paleohistoric. Several events that had a deleterious effect on the flora and vegetation of other Mediterranean countries, such as increasing seasonal variation, the Messinian salinity crisis, the glaciations of northern Europe and the desertification of the Sahara (Bramwell & Richardson, 1973) did not affect the Iberian Peninsula as severely due to the buffering effect of the Atlantic Ocean upon temperature and humidity. A deep gulf which occupied the present Guadalquivir Valley (which was only completely infilled by sediments during historical times) may have played an important role in this amelioration of the local climate. In addition, the mountain ranges are predominantly oriented from east to west: a favorable range of habitats permitted plants to survive changing conditions by slowly moving from one altitude to another. Thus, while many extinctions were taking place in other areas, recent diversification has occurred with many plant groups of the Iberian Peninsula.

The Iberian Peninsula has experienced the same presures from man's disturbance that are commonplace for the rest of the Mediterranean region. The most heavily affected areas belong to the *Oleo-Ceratonion* coastal zones where the climax vegetation is now almost completely destroyed. Since the Phoenicians, Greeks and Punics visited these coasts in antiquity, there has been a long history of over-use and exploitation. The valley of the Guadalquivir River is roughly included in the *Oleo-Ceratonion* area and it too has supported intensive agriculture since ancient times. As for the central plateaus, deforestation of the northern half seems to have preceded by a few centuries that of the southern half. The excessive priiileges granted to shepherds in Medieval and Early Modern times which permitted them unlimited grazing access were detrimental to the vegetation, and during the 15th, 16th and 17th centuries when both peninsular countries were major naval powers, much forest was cleared to prvide timber for shipbuilding.

Today, public awareness of the need for plant conservation in Spain is still poor and protective laws are either scarce or absent. As a very significant portion of the endemic flora occurs in the mountains, any type of development taking place above 1200–1500 m in altitude should be regarded with discomfort.

A very rich flora, containing high numbers of endemics that are still often poorly known, poses a real challenge for any organieation dedicated to its conservation. It seems obvious that some preliminary priorities should be established soon. For example, selection of those taxa with a limited geographic distribution or with narrow habitat requirements has been the first step in a national programme to elaborate red data sheets in Spain (see Gómez-Campo, 1981b). Though work is now in progress to determine the degree of urgency with which every case should be treated later, difficulties of taxonomy or the study of the geographic distribution make it slow. The following are a number of cases which may serve as examples of the problems and situations that might arise in future conservation efforts (Fig. 2).

3. Some case histories of rare and threatened species in the Iberian Peninsula

3.1 Coronopus navasii *Pau (Cruciferae), a very restricted hardy endemic (Fig. 3)*

In the province of Almería there is a small circular pond, located at 1800 m in altitude in the Sierra de Gádor. Its diameter is 16 m in summer, but the autumn rainfall increases its size to some 70 m (Fig. 4). During most winters the surface water freezes intermittently sometimes becoming covered by snow. The water level recedes again by spring and early summer.

Incredible as it might seem, a species of perennial plant lives in between the two water levels. *Coronopus navasii* Pau (1922) has a stout woody root 3-5 cm in diameter that descends vertically to reach the ground water-table, usually two or three meters below. The small aerial part consists of several creeping herbaceous branches with pinnatisect leaves and tiny white flowers. Flowering occurs in August. The plants are intensively trampled by the flocks of sheep that come daily to water at the pond. Far from being harmful, this seems to be a mechanism used by the species for the dispersal of its fruits. The aerial part dies after fructification, leaving buds to survive the severe winter conditions of cold temperatures and immersion, to sprout again the following season.

The species lives, on its own within the above-

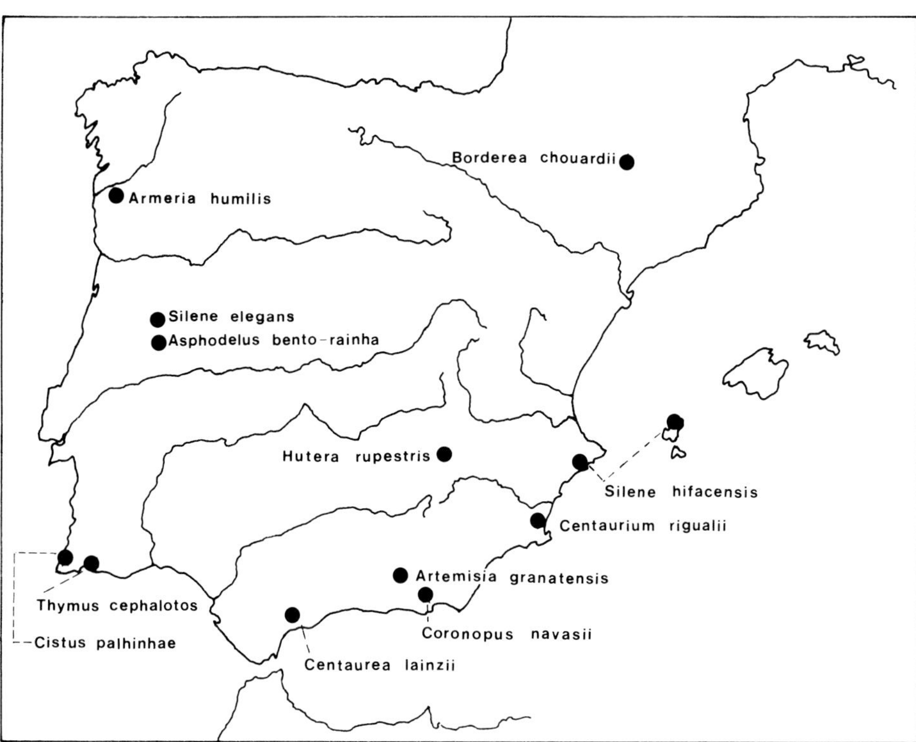

Fig. 2. Map showing the geographic positions of the endemic species whose conservation status is commented upon in this article.

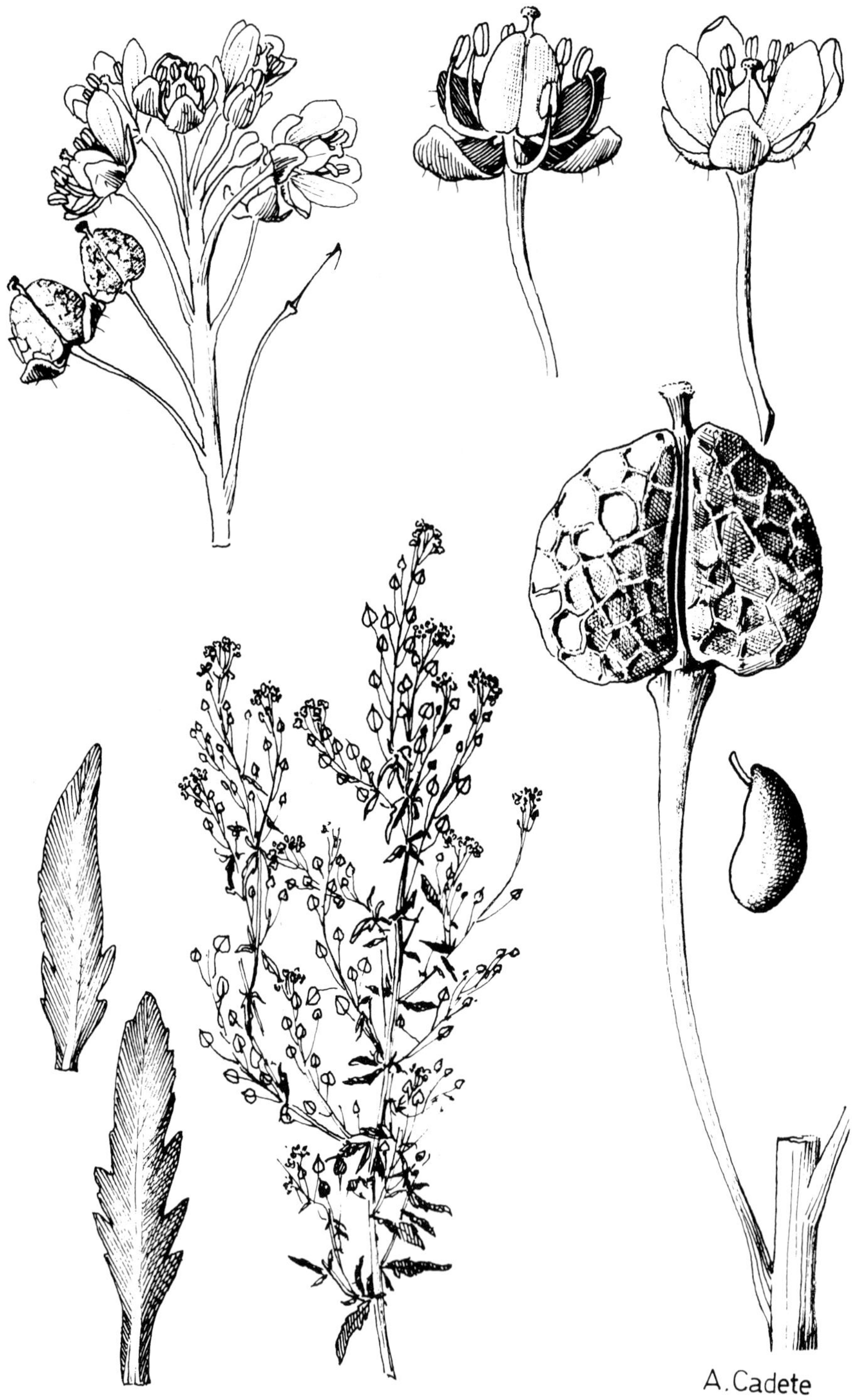

Fig. 3. Coronopus navasii Pau.

Fig. 4. The habitat of *Coronopus navasii* Pau in the Sierra de Gádor, southwestern Spain.

mentioned 3,750 square metre area, and maintains a population of about 14,000 individuals, with an approximate density of four individuals per square metre. In actual fact, the density is higher near the waterline of the pond and lower at the periphery of the area, in correspondence with the vertical distance to the ground water-table. In a second pond seven km distant, the conditions are similar and there exists a second population. The plant seems to be unknown from any other locality. If we take into account the total extent of the area in which the plant is found - less than one hectare - or the strict ecological conditions under which the species thrives, the total population of about 30,000 seems to be a fairly large one.

Though it has been suggested that *Coronopus navasii* may be cospecific with the North African *C. violaceus* (Munby) O. Kuntze (Ball, 1964), the shape of the fruit and other characters suggest that these two are very different. In fact these two species are as distinct to each other as to any other *Coronopus* species *(C. squamatus* (Forsk.) Asch., *C. didymus* (L.) Sm., *C. lepidioides*, (Coss. & Dur.) O. Kuntze, etc.) whose geographic distribution is much wider.

From a conservational point of view, the tiny size of the area in which this plant is distributed coincides with a high degree of ecological specialization. It is evident that if the ponds are ever drained or transformed in any way, the plant is in danger of becoming extinct. It is unlikely that total desiccation of the ponds will occur at the present time because of the important role they play in providing water for the livestock of the surrounding area. But any well-meaning modernization of the watering system, such as the construction of a well with a pump and some drinking troughs, would be fatal for *Coronopus navasii.*

Since a prosperous agricultural system with irrigation and greenhouses is established at the foot of the Sierra de Gádor, a disruption of the present delicate balance of the whole Sierra will always remain a menacing possibility. Damming, flooding or water pumping from lower levels could eventually affect the ponds around which this *Coronopus* lives.

3.2 Centaurea lainzii *Fernández-Casas (Compositae), a sexually sterile triploid (Fig. 5)*

This handsome species was described in 1975 from the Sierra de Estepona, Málaga Province, in Southern Spain. It grows at altitudes between 500 and 900 m above sea level and forms small but compact groups of leafy rosettes in a few localities of the Sierra. It is a sterile triploid clearly distinct from other related *Centaurea* species (such as *C. clementei* Boiss, or *C. prolongoi* Boiss) which also grow in that area. It has large flower heads, 35 mm in diameter, at the end of unbranched stalks approximately one meter high.

Abundant seeds were collected by the first author in the summer of 1979, but when they were cleaned and airblown, the sample was reduced to a small number of malformed seeds that were unable to germinate. By the end of that year it was reported (Fernández-Casas & Fernández-Morales, 1979) that this species is a natural triploid, a condition that renders it completely sterile.

Reproduction is, therefore, strictly vegetative. The roots are thick and well developed. The union between root and stem is fragile as it apparently develops an abscission zone which breaks easily when the plant is manipulated. New shoot primordia are then produced by the rootstock, and new leaf rosettes soon appear. Fernández-Casas & Fernández-Morales have demonstrated that small root portions when cultured are also able to produce leaf rosettes.

The method of dispersal of root propagules is not known. The known localities are a maximum of eight kilometres apart, but each usually consists of no more than a few square metres and less than two hundred rosettes. It is possible, however, that more intensive exploration of the Sierra would yield additional localities. In fact (Asensi & Diaz-Garretas, 1977) *Centaurea lainzii* seems to be expanding its area along the embankments of a newly-built mountain road.

As a taxon that does not reproduce sexually, the plant should be viewed as a short-term evolutionary adaptation; this might well be correlated to the higher intrinsic vulnerability to destructive factors, in connection to its lower genetic variability. If the plant eventually achieves duplication of its chromosome number and becomes sexually fertile, it might become a successful and even an ecologically aggressive species. This duplication might be done experimentally, but, in that case, there are ethical reasons whereby the resulting plant should never escape the garden or laboratory boundaries.

The wild boar (*Sus scropha*) is now becoming increasingly abundant in the Iberian Peninsula, and this could pose a real threat for *Centaurea lainzii* due to their habit of rooting about in search for edible subterranean plant parts. But perhaps this could also provide a method for the dispersal of plant propagules. Development of housing for the tourist trade is very actively taking place only twelve kilometres away and this could pose another series of threats.

Positive action might consist of artificially expanding the number of existing stands of *Centaurea lainzii* by introducing the plant into similar sites on the same Sierra. This could certainly reduce the possibility of extinction.

3.3 Silene hifacensis *Rouy ex Willk. (Caryophyllaceae) an overcollected species*

This species was described from the Peñón de Ifach (or Hifac), Alicante Province in southwestern Spain (Fig. 6). The Peñón is a huge scenic rock that once served as a landmark and lighthouse for ancient navigators. *Silene hifacensis* grew on vertical calcareous cliffs on the northern side of the rock and its numbers were never very abundant, perhaps no more than one or two hundred individuals.

The Peñón was a classic collecting place for visiting botanists who were often eager to collect this particular species. The inaccessibility of the spots where this *Silene* grew seemed not to inconvenience the collectors because rock climbing techniques were used when necessary. The habit of collecting one specimen for the collector himself and also additional ones for exchange, was probably fatal to the population of *Silene hifacensis* because after 1925 it has never been found again in that locality.

Fortunately, the plant is not yet completely ex-

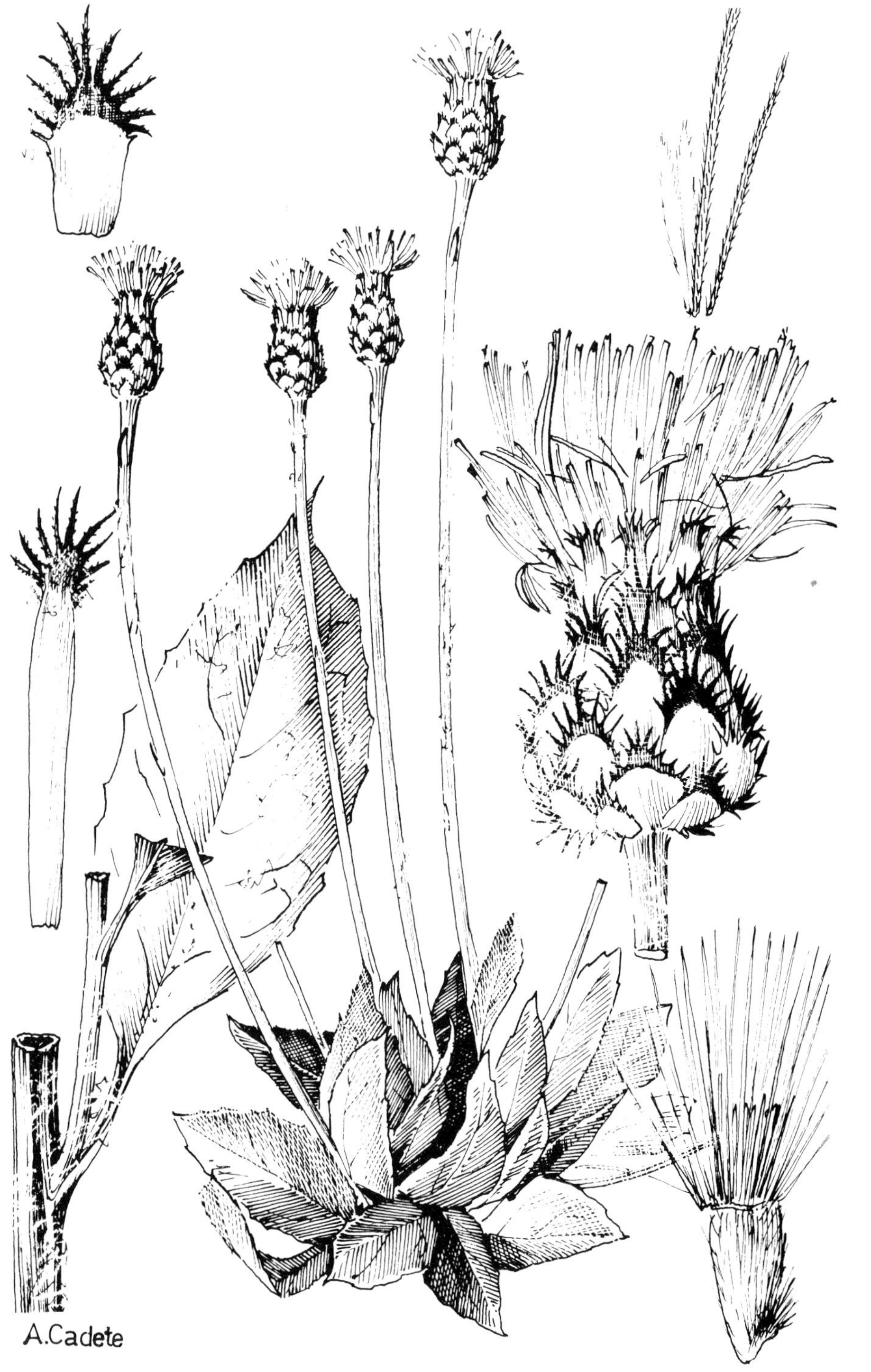

Fig. 5. Centaurea lainzii Fdez. Casas

Fig. 6. Peñón de Ifach, Alicante Province, the former locality of *Silene hifacensis* Rouy ex Willk.

tinct because at least two or three other localities with similar habitats exist on the neighboring island of Ibiza. There, the species is not much more abundant that it was formerly in Ifach. Luckily, botanists are now increasingly conservation-minded, so that the same overcollection is not likely to be repeated. Many are known to have visited these sites in the past few years with no apparent harmful effects on the populations. Mr. Jean Y. Lesouef, who visited the island in 1974, was able to collect some mature seeds which are now stored for long-term preservation.

S. hifacensis Rouy et Willk. has affinities with *S. italica* L., a taxon which is widely distributed in southern and central Europe. However, *S. hifacensis* is more robust and can be easily distinguished by its longer carpophore. Both belong to Section *Siphonomorpha* Otth. (Chater & Walters, 1964).

An important point remains to be solved, that is, whether or not the Ibizan plant is exactly the same as that formerly found on Ifach. But even if there were some varietal differences, it is believed that given the present situation, the reintroduction of living material from Ibiza to Ifach – already carried out on an experimental scale by Sáinz-Ollero & Hernández-Bermejo (1979) – would be a suitable way to restore *Silene hifacensis* to its *locus classicus*.

As stated repeatedly in this book, overcollection may be one of the major direct threats for many members of the endemic Mediterranean flora. *Silene hifacensis* demonstrates that its interest to botanists was sufficient to ensure its extinction on Ifach.

3.4 Centaurium rigualii *Esteve Chueca (Gentianaceae) – Found and then lost*

The history of this taxon cannot be simpler, but we consider it worth mentioning because it exemplifies a case that could easily be repeated. *Centaurium rigualii* was described in 1965 from Los Alcázares, Murcia Province in southwestern Spain, where only a small number of individuals was seen. Any attempt to find it since that time has failed. Its discoverer Prof. Esteve-Chueca tried to find it in the wild, and Dr. Rigual, to whom the species is dedicated tried also (personal communications) as well as the first author of this article, who searched for it for several hours. None of these efforts was successful.

In our opinion, this alerts botanists to the advisability of taking some viable seed for long-term preservation each time a new species is discovered. First contact with a new species is through one of its populations, and the botanist tends to assume that such a population is not unique but that others are present either nearby or in other areas. This may not be so. Furthermore, today, unique populations of newfound plants may be a comparatively common situation, since undiscovered taxa are usually so because of their very rarity.

The closest relative of *C. rigualii* may be *C. linariifolium* (Lam.) G. Beck. (Melderis, 1972) but the first can be easily distinguished by its annual habit and the absence of a persistent leaf-rosette.

3.5 The genus Hutera *(Cruciferae) in the Sierra Morena*

A group of four species is considered together here because they represent an interesting case of ongoing evolution with challenging conservational implications. Only one of these species is really endangered (*H. rupestris* Porta; Lucas & Synge, 1978) but the other three contain several tiny local populations that are phenotypically intermediate and whose scientific and educational value warrants their protection. The whole group reminds us that we perhaps oversimplify the problem of plant conservation when we refer to it as the mere conservation of individual species.

The group grows in a 200 km long area in the Sierra Morena range that extends from East to West between the Guadiana and the Guadalquivir Valleys. Found on the Western side, *Hutera hispida* (Cav.) Gómez-Campo is the same species which grows in other Iberian mountains. Eastwards in the main Sierra Morena range is *H. longirostra* (Boiss.) Gómez-Campo, while in the some small foothills slightly to the north, *H. leptocarpa* Gonzáles Albo can be found. *H. rupestris* Porta, very local in the Sierra de Alcaraz, is the only one growing on a calcareous substrate. Geography is closely correlated to morphology, thus giving rise to a clinal series that, in this case, is Y-shaped. While the siliquas of *H. hispida* are narrow and erecto-patent with a comparatively short beak, *H. longirostra* has pendent long beaked fruits. The evolutionary branch *H. leptocarpa-H. rupestris* is characterized by increasingly shortened fruits with thickened beaks.

H. rupestris (Fig. 7) only occurs at the junction of two karstic canyons where it grows on inaccessible north-facing cliffs. The total area where the plant grows is less than 0,2 square km even if one includes some rocks two kilometres to the south of the main population where the plant also grows. The total population was estimated in 1980 to be between 500 and 1,000 flowering plants plus perhaps twice as many non-flowering rosettes. This species went undiscovered until 1891, when Porta and Rouy found the plant independently and described it as a new genus. Its affinities to the other members of the group have been studied in detail by Gómez-Campo (1977). It seems clear that *H. rupestris* is not a Tertiary relict as had been stated before, but rather a recent invader of the Eastern limestones adjacent to the Sierra Morean. *H. rupestris* seeds were formerly used to make a locally appreciated mustard (González-Albo, 1934) but fortunately this use seems to be now forgotten. Present threats are from collectors and from building developments in the vicinity. In its present habitat, *H. rupestris* has the competition of *Sisymbrium arundanum* Boiss. which occupies the same ecological niche and might be more aggressive. Though there is not direct evidence that *Sisymbrium* is displacing *Hutera,* such a possibility should be monitored.

It is doubtful whether *H. leptocarpa* should be labelled 'rare' or 'vulnerable' as a species, but certainly many of its populations are endangered. Apart from the main group in the Sierra de Alhambra, all other populations are morphollogically different to each other, ranging from phenotypes similar to *H. hispida* to others resembling *H. rupestris.* A newly - found population near Moral de Calatrava (Peinado-Lorca, 1980) might even be an acid-loving ecotype of *H. rupestris* and, in any case, represents a discontinuity in the clinal series

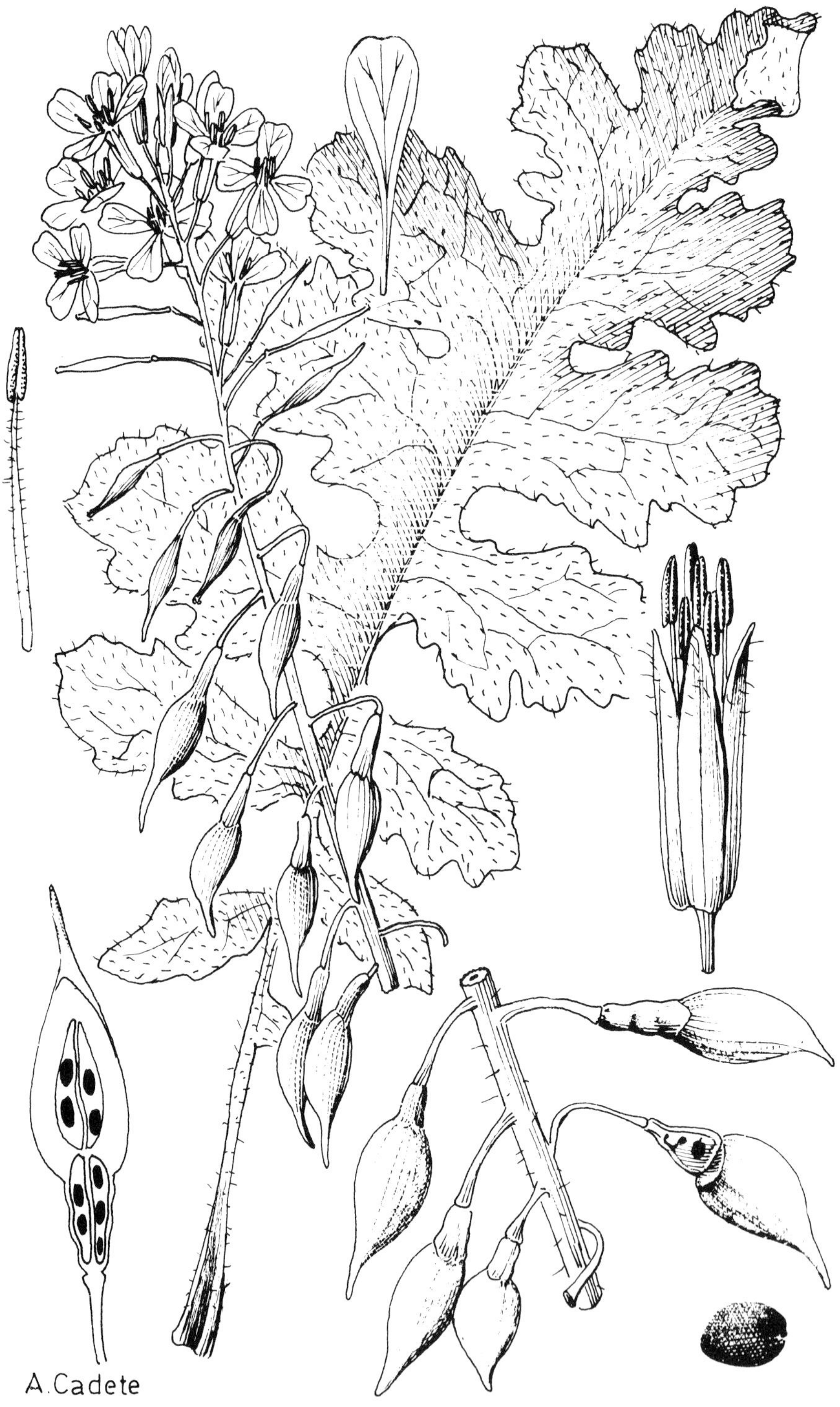

Fig. 7. Hutera rupestris Porta

cited above. In turn, *H. hispida* includes some populations that grade into either *H. leptocarpa* or *H. longirostra.*

Only a three-hour drive from Madrid, the whole complex has enormous educational possibilities for students of plant evolution, who could be helped to rediscover by themselves the pheno-geographic pattern, to perceive evolution taking place, and to discuss its possible significance. Also, there is a vast scope for further specialized evolutionary studies in cariology, genetics, phytochemistry, etc. Although everybody would agree that such a complex should be preserved in its entirety, the difficulty lies in how to achieve this. Full conservation of every subspecific taxon and every population would be an almost impossible goal. Clinal series and evolutionary situations like the one described are rather abundant in the Mediterranean region. Many are likely to have been destroyed before they had been discovered.

In summary, *Hutera* seems to be a young, rapidly evolving, western European genus (Heywood, 1964, partly under *Rhynchosinapis* Hayek); Leadley & Heywood (in prep.), which achieves maximum diversity in the Iberian Peninsula. It can be taken as a representative example for dozens of other genera containing a great diversity of neoendemic taxa.

3.6 Artemisia granatensis *Boiss. (Compositae) and four dozen narrow endemic companions in the Sierra Nevada (Fig. 8).*

The 'manzanilla real' *(Artemisia granatensis* Boiss.) is a well-known example of a species that has been brought to the brink of extinction by overcollection. An infusion made with its leaves became so popular in Granada that within a few decades the species had practically dissapeared from the heights of the Sierra Nevada (Lucas & Synge, 1978). Many other endemic species and subspecies which share the same general area and habit are now under threat. The preservation of 250 square km above an altitude of 1,800 m should constitute the first priority for plant conservation within the Iberian Peninsula. At least 177 Iberian endemic species and subspecies can be found there, of which 66 are endemic to the Sierra Nevada itself; those recognized as species by Flora Europaea (49) are listed in Table 2. In the approximately sixteen sq km above 3,000 m altitude, some 75% of the flora is endemic.

The landscape of the Sierra Nevada (Fig. 9) is apparently barren and denuded in summer when seen from afar, but careful inspection of the soil reveals a multiplicity of small plants that occupy the spaces amongst the bright micaceous Paleozoic schists (Prieto, 1975). Many od these plant species are small perennials adapted to survive several months of heavy snowfall. It is no wonder that Boissier (1839–45) was able to describe a large number of new species during a single trip to the area in 1837. A belt of Mesozoic limestone surrounds the Sierra at an average altitude of 2,000 m and makes an important contribution to the floristic richness of the area.

Spinose shrublets such as *Ptilotrichum spinosum* (L.) Boiss., *Vella spinosa* Boiss., *Bupleurum spinosum* Gouan and *Arenaria pungens* Clemente ex Lag. are abundant and their habit may represent adaptation to discourage grazing by the wild goat *Capra pyrenaica.* The goats were hunted to near extinction and their role taken over by sheep flocks from the lowland muicipalities. Wild goats have been protected over the past few decades, so that numbers have risen to over 2,000. The sheep flocks continue to graze here, and excessive grazing pressure is now affecting the rich and unique vegetation of this zone.

Worst of all, the urban development that accompanies tourism, aggressively climbs the slopes which face the city of Granada. At about 2,200 m a very important ski resort has been in operation for several years, and all around it, up to the summit of the Picacho de Veleta (3,392 m), the landscape is being drastically changed (Prieto, 1973) by housing, construction and cabin installations. Wrongly sited tracks have given rise to accelerated erosion. In one location at 2,600 m altitude, the observer can count two paradors, a meteorological station, a watertank, a ski-lift terminal station, a chapel, an astronomical observatory, a herzian tower, among others, to the extent that almost every visible small summit is occupied by man-made installations. Lit-

Fig. 8. Artemisia granatensis Boissier

Fig. 9. The Mulhacén at 3,460 m is the highest point in the Iberian Peninsula. Sixty-six plant taxa are endemic to the area which immediately surrounds it.

Table 2. Plant species endemic to the Sierra Nevada, Southern Spain.

Aconitum nevadense Uechtr.	*Leontodon microcephalus* DC.
Antirrhinum rupestre Boiss. & Reut.	*Linaria glacialis* (Boiss. ex DC) Boiss. & Reut.
Arenaria nevadensis Boiss. & Reut.	*Linaria nevadensis* (Boiss.) Boiss. & Reut.
Artemisia granatensis Boiss.	*Linaria salzmanii* Boiss.
Avenula levis (Hackel) J. Holub	*Lotus glareosus* Boiss. & Reut.
Campanula willkommii Witasek	*Lotus granadensis* Zertova
Carex camposii Boiss. & Reut.	*Pimpinella procumbens* (Boiss.) Wolff.
Centaurea monticola Boiss.	*Pinguicula nevadensis* (Lindb.) Casper.
Chaenorhinum glareosum (Boiss.) Willk.	*Plantago nivalis* Boiss.
Delphinium nevadense G. Kunze	*Potentilla nevadensis* Boiss.
Erigeron frigidus Boiss ex DC	*Potentilla reuteri* Boiss.
Erigeron major (Boiss.) Viehr.	*Ptilotrichum purpureum* (Lag. & Rodr.) Boiss.
Erodium astragaloides Boiss. & Reut.	*Ranunculus acetosellifolius* Boiss.
Erodium boissieri Cosson	*Reseda complicata* Bory
Erodium rupicola Boiss.	*Rothmaleria granatensis* (Boiss ex DC) Font Quer.
Euphorbia nevadensis Boiss. & Reut.	*Santolina elegans* Boiss & DC
Evax nevadensis Boiss.	*Saxifraga nevadensis* Boiss.
Festuca clementei Boiss.	*Sempervivum nevadense* Wale
Festuca frigida (Hackel) K. Richt.	*Senecio elodes* Boiss & DC
Galium rosellum (Boiss.) Boiss. & Reut.	*Senecio nevadensis* Boiss. & Reut.
Hieracium argyrocomum (Fries) Zahn	*Tanacetum funkii* Schultz Bip. ex Willk.
Holcus caespitosus Boiss.	*Taraxacum nevadense* H. Lindb. fil.
Lactuca singularis Wilmott	*Trisetum glaciale* (Bory) Boiss.
Laserpitium longiradium Boiss.	*Viola crassiuscula* Bory.
Leontondon boryi Boiss. ex DC	

ter made up of objects of every kind is commonplace. One small lake, Laguna de las Yeguas, a *locus classicus* for some of the *'nevadensis'* endemics, has been deeply transformed and now supplies water to the tourist complex. Some peat deposits are being exploited, as well as entire stands of *Genista baetica* Spach and *Juniperus communis* subsp. *nana* Syme that are burned as fuel in Granada ovens.

Fortunately, they are still many square kilometres in a near virgin state, affected only by the above-mentioned sheep-grazing problem. If the boundaries of the tourist and ski areas were quickly delimited and as much area as possible effectively protected to the east of Veleta peak and around the Mulhacen (3,460 m) and the Alcazaba (3,336 m), a satisfactory result could still be obtained. Ideally, the protected area should include a part of the calcareous soils at about 2,000 m altitude. There are overwhelming reasons whereby the Sierra Nevada should have been declared the site of a National Park or Biological Reserve many years ago. Though plans for this are now under way, the slowness with which they are being put into practice contrasts sharply with the speed at which development is transforming the area.

3.7 Asphodelus bento-rainhae *P. Silva (Liliaceae),* Silene elegans *Link. ex Brot. (Caryophyllaceae) and other endemics from the Serra de Gardunha and Serra da Estrela*

Serra da Estrela and Serra da Gardunha are close to each other, which justifies the joint consideration of their endemics here. Serra da Estrela includes the highest peak in Portugal (Torre, at 1,992 m) and is formed by huge granitic masses of striking appearance. The area contains a high concentration of Portuguese endemics. The following species seem to be exclusive to the Serra da Estrela and adjacent mountains:

Angelica angelicastrum (Hoffm. & Link) Cout.
Asphodelus bento-rainhae P. Silva
Centaurea rothmalerana (J. Arènes) Dostál
Festuca henriquesii Hackel
Senecio cespitosus Brot.
Silene elegans Link ex Brot.
Teucrium salviastrum Schreber

Silene elegans Link ex Brot (Fig. 10) is probably the most locally restricted. It lives above 1,700 m altitude in a small area, at the base of granite rocks or in rock crevices. It was once confused with *S. ciliata* Pourret which is restricted to other South European mountains. The Portuguese species can be distinguished by its smaller size and the presence of only one or two flowers per stem. *S. elegans* has closer affinities to *S. legionensis* Lag. which is endemic to northern Spain and northwestern Portugal but can be recognized by its flowering stems which arise terminally from the centre of the leaf rosettes, the anastomosed calyx nerves, the shorter carpophore, and the fewer number of flowers (1 or 2) (Sampaio, 1947; Franco, 1971).

The species possibly is of preglacial origin and belongs to the primitive stock of Mediterranean mountain vegetation. For several decades, the main threat to the plant was the voracious appetite of the goats that were once frequent in the area. Fortunately it survived, and now though no special protective measures are taken, it is covered by the general protection granted to the area when the Serra da Estrela became a Natural Park a few years ago. A small population within a restricted area will always be at risk from collectors; also the park is an important tourist and ski resort. The problems confronting other Serra da Estrela endemics are similar, but most of them have a comparative wider distribution within the protected territory.

Thirty-five km to the south, in the Serra da Gardunha, *Asphodelus bento-rainhae* P. Silva (Fig. 11) might exemplify a plant in a similar situation but without the benefit of any protective measures. The species was described in 1965 and differs from *A. ramosus* L. by its fusiform tubercles, smaller perianth and smaller miter-shaped capsules. It lives on granitic soils. The plant's two means of reproduction – sexual by abundant seeds and vegetative by tubercles – make the species to appear to be resistant to depredation. But as the known area is very small and situated in the midst of agricultural lands, accidental or intended fires and the use of herbicides might be among the possible threats to this species. Exact delimitation of the area of distribution and the establishment of appropiate protective measures would be desirable.

Fig. 10. Silene elegans Brot.

Fig. 11. Asphodelus bento-rainha P. Silva

3.8 *Armeria humilis (Link) Schultes (Plumbaginaceae) from the northwest of Portugal*

This small orophytic species lives in Minho Province, northwestern Portugal. Its habitat consists of stony pasturelands and crevices in granitic rocks in the Serras of Arga, Amarela and Gerês, between 800 and 1,400 m altitude. To a certain degree, the plant is similar to *Armeria trachyphylla* Lange, an endemic of the eastern and southeastern mountains of Spain. But the latter species lives on calcareous soils, while the Portugese species is strictly silicicolous. Morphologically, *A. humilis* exhibits mucronulate external leaves and reduced involucre bracteoles, while *A. trachyphylla* presents muticous leaves and well-developed bracteoles (Bernis, 1949).

The type subspecies of the species is exclusive to the two Serras Amarela and Gerês. In Arga, and parts of Amarela, a subspecies - *Armeria humilis* subsp, *odorata* (Samp.) P. Silva - is found. This is characterized by several, mostly dimensional differences.

Armeria humilis is not rare at higher altitudes of the Serras mentioned. There, its stability is due to its preference for the crevices of almost inaccessible rocks and its perennial habit. Though it has not yet been the object of special protective measures, the conservation of this plant is helped by the protective rules which apply in the Parque Nacional da Peneda-Gerès. The park covers the area in which this species occurs, with the exception of the Serra of Arga.

3.9 *Some representative endemics from the region of Cabo Sao Vincente:* Cistus palhinhae *Ingram Cistaceae),* Thymus cephalotos *L. and (Labiatae)* Linaria lamarckii *Rouy (Scrophulariaceae)*

In the very southwest of Europe, the land the Arabs named 'al-gharb' (the west, the sunset) now known as the Algarve of southern Portugal, a remarkable group of coastal endemics exists. The neighbouring Serra de Monchique is also rich in local endemics. We will mention some examples from the coastal group.

Cistus palhinae Ingram (Fig. 12) is relatively closely related to the common 'Xara' or 'Esteva' (*Cistus ladanifer* L.) but while the latter often reaches 2 m in height, has linear lanceolate leaves and has ovaries with ten locules, *Cistus palhinhae* is usually much shorter (up to 50 cm in heigh) bears oblanceolate or spathulate leaves, and has ovaries with six locules only (Burtt, 1951). These differences, together with the fact that the habitats are dissimilar, were not at first appreciated and the plant was ascribed to *C. ladanifer*.

Cistus palhinhae plants live on calcareous rocky platforms near the sea, along a short length of the coast between Odeceixe and Sagres. Together with *Juniperus phoenicea* L., *Armeria pungens* (Link) Hoffmans et Link, *Teucrium polium* subsp. *vincentinum* (Rouy) D. Wood and *Thymus camphoratus* Hoffmanns, it forms an association of open spreading small shrubs which are strongly beaten by the prevailing sea winds.

Thymus cephalotos L. (Fig. 13) a member of sect. Pseudothymbra is a relative of *T. longiflorus* Boiss., and extends along the area named 'Berrocal' between Cabo Sao Vincente and the vicinity of Tavira. It is a dwarf shrub which lives on calcareous, often stony soils, either in open zones associated with other plants of similar size, or occupying cleared areas within more robust shrubbery or evergreen forest (Braun-Blanquet, Silva & Rozeira, 1964; Malato-Beliz, 1979).

Since both *Cistus* and *Thymus* occur on soils that are unsuitable for agriculture, the main threat is the tourist development which is taking place throughout this area. As uncontrolled urban expansion progressively invades the landscape it is urgent that steps are taken to protect these and other associated endemics.

Another coastal habitat is represented by that of *Linaria lamarckii* Rouy. This very beautiful *Linaria* from the littoral sands is, to a certain extent, similar to *Linaria caesia* var. *decumbens* Lange from the litoral dunes of northwestern Spain and northwestern Portugal. It differs, however, by its longer oblong spathulate to obovate leaves and by its smaller corolla and spur (Coutinho, 1939; Valdés, 1970). There seems to be a certain phytosociological correspondence between both species

Fig. 12. Cistus palhinhae Ingram

Fig. 13. Thymus cephalotos L.

since they occupy a very similar role in the coastal plant associations: *Linaria caesia* var. *decumbens* in the north and the centre and *L. lamarckii* southwards from the Tagus river.

L. lamarckii is comparatively frequent along the Portuguese littoral between Setúbal and the mouth of Guadiana river. It is an important component of the local plant associations of the area, and probably should be viewed as one of the characteristic species of a meridional vicariant alliance of the *Linario-Vulpion*.

It is not a rare plant, but once again the coastal area where this endemic lives has suffered much pressure from urban development in the past decades. Though *L. lamarckii* generally lives in the dunes, set back a short distance from the sea, the lack of rigid planning regulations covering vacation dwellings and the uncontrolled recreational use of beaches, have deeply influenced the sand-loving vegetation and have already destroyed many localities of the species. It is therefore urgent that a number of areas are recognized and set aside to eventually form protected reserves where the plant can reproduce itself.

3.10 Borderea chouardii *(Gaussen) Heslot (Dioscoraceae) a Pyrenean paleoendemic of intertropical origin*

The genus *Borderea* has only two species, *B. pyrenaica* Miégeville and *B. chouardii,* both from the central Pyrenees. Its importance has been commented upon by Braun-Blanquet (1948) and notably by Gaussen (1965, 1966): *Borderea* represents a true ancient relict whose relatives are to be found in Kenya, Ethiopia or Chile. The range of *B. pyrenaica* extends from Monte Perdido to Turbon, where it lives on loose calcareous screes. It does not seem to present any special conservation problems. By comparison, *B. chouardii* is restricted to the eastern end of the area where the genus is found, near Sopeira, Huesca Province, lives in rock crevices, and is considered as endangered.

According to Montserrat (personal communication), well-developed, mature specimens are becoming scarcer each year; they grow on inaccessible rocks, cave entrances, cliffs and rock ledges, the total population not being more than a few dozen. However, the number of juvenile plants may be 300 to 500, most of them forming an important subpopulation near a newly constructed dam.

Cliffs inhabited by *B. chouardii* are north - or northeast-facing, thus direct sunlight that might overheat the crevices is avoided. These crevices additionally collect water enriched with endolithic blue-green algae, lichens and animal life. This water, flows slowly but steadily, producing neither flooding nor total desiccation. The plants are, therefore, dependend upon shade and a measure of coolness in summer. Montserrat suggests that such a regulated water supply has made it unnecessary for the species to develop any genetic resistance to rotting. This is the reason why the plant is so very difficult to cultivate successfully. Gaussen and Montserrat have both attempted to do so, but root rot soon set in and killed all the plants.

There are several related reasons which cause us to be pessimistic about the conservation of this species. Apart from the ever - present risk from collectors - in this case somewhat diminished by the inaccessibility of much of the habitat - the construction of the dam has adversely affected the microclimatic conditions and lowered the average surface temperature of the cliffs where *Borderea* lives. Before the dam was constructed, channelled wind currents from the narrow Ribagorzana Valley sufficiently agitated cold air layers to prevent their accumulation. Such turbulence is now much weaker and, furthermore, a channel that formerly drained cold air has now been filled. The magnitude and exact effects of these changes are still unknown and difficult to gauge, at least in the extent that they can influence the microclimate of the crevices where the species lives. Though it will be never known with certainty, it is very probable that the construction of the dam directly destroyed several plant colonies. Finally, the road to the valley of Aran has now been widened after gaining importance following the opening of the Tunnel of Viella. The initial construction of this road and its recent enlargement have both been times of peril for *B. chouardii*. On one occasion, one of the subpopulations was partially buried under large amounts of gravel. The fall of debris over the population with the highest number of young plants still continues.

Acknowledgements

The data on the status of *Borderea chouardii* have been kindly supplied by Dr. P. Montserrat, from the Centro Pirenaico de Biología Experimental (C.S.I.C.) in Jaca, Huesca.

Thanks also are due to Miss L. Bermúdez-de-Castro and to Mrs. M.J. Cagiga for their valuable help in the study of local distributions of Iberian endemics, to Mrs. E. García-Mouton for her collaboration in the automatic proccessing of data, to Mr. A. Cadete for his drawings of eight endemic species and to Mrs. M.E. Tortosa for her help with other illustrations.

References

Anonymous (1983). List of rare, threatened and endemic plants in Europe. TPU-IUCN-CMC. Strasbourg: Council of Europe.

Asensi, A., & Díez-Garretas, B. (1977). Nota fitosociológica: *Centaurea lainzii* Fdez. Casas en la Sierra Bermeja de Estepona (Málaga). Anales Inst. Bot. Cavanilles 34: 183-188.

Ball, P.W. (1964). *Coronopus* Haller. In: Tutin et al. (Eds.), Flora Europaea 1: 333. Cambridge: Cambridge University Press.

Bernis, F. (1949). El género *Armeria* Willd. en Portugal. Bol. Soc. Brot. 23: 225-263.

Boissier, P.E. (1839–45). Voyage botanique dans le midi de l'Espagne pendant l'année 1837, Vol. 2. Paris: Gide et Cie.

Bolos, A. de (1962). Algunas novedades floristicas. Collect. Bot. 6: 357–362.

Bramwell, D. & Richardson, I.B.K. (1973). Floristic connections between Macaronesia and the East Mediterranean region. Monog. Biol. Canarienses 4: 118–125.

Braun-Blanquet, J. (1948). Les souches préglaciaires de la flore pyrénéenne. Collect. Bot. 2: 1–23.

Braun-Blanquet, J., Silva A.R.P. & Rozeira, A. (1964). Résultats de trois excursions géobotaniques a travers le Portugal septentrional et moyen III: Landes a cistes et ericacées (*Cisto-Lavanduletea* et *Calluno-Ulicetea*). Agron. Lusit. 23: 229–313.

Burtt, B.L. (1951). *Cistus palhinhae,* Cistaceae. Curtis Bot. Magazine 168, tab. 157.

Coutinho, A.X.P. (1939). Flora de Portugal (Plantas Vasculares), 2ª ed. Lisboa: Bertrand Irmaos.

Chater, A.O. & Walters, S.M. (1964). *Silene* L. In: Tutin et al. (Eds.), Flora Europaea 1: 158–181. Cambridge: Cambridge University Press.

Esteve-Chueca, F. (1965). Algunas novedades para la flora murciana. Anales Inst. Bot. Cavanilles 23: 173–186.

Fernández-Casas, J. (1975). Exsicattae quaedam a me nuper distributa. 1, 9, n. 52. Mimeographed.

Fernández-Casas, J. & Fernández-Morales, (1979). *Centaurea lainzii*, un triploide natural. Mem. Soc. Bot. Genève 1: 115–122.

Franco, J.A. (1971). Nova Flora de Portugal (Continente e Acores), Vol. 1. Lisboa.

Gaussen, H. (1965). Revision des *Dioscorea (Borderea)* pyrénéens. Doc. Cartes Prod. Végétales, sér. Pyrénées 3: 1–16.

Gaussen, H. (1966). Flore des Pyrénées: Fam. des Dioscoreaceae. Doc. Cartes Prod. Végétales, sér. Pyrénées 4: 1–15.

Gómez-Campo, C. (1977). Clinical variation and evolution in the *Hutera-Rhynchosinapis* complex of the Sierra Morena (South-Central Spain). Bot. J. Linn. Soc. 75: 119–140.

Gómez-Campo, C. (1980). Studies on Cruciferae: VI. Geographical distribution and conservation status of *Boleum* Desv. *Guiraoa* Coss. and *Euzomodendron* Coss. Anales Inst. Bot. Cavanilles, 35: 165–176.

Gómez-Campo C. (1981a). The taxonomic and evolutionary relationships in the genus *Vella* L. Bot. J. Linn. Soc. 82: 165–179.

Gómez-Campo, C. (1981b). Conservación de recursos genéticos. In: Ramos, J.L. (Ed.) Tratado del Medio Natural 2: 97–124. Universidad Politécnica de Madrid 2.

Gómez-Campo, C. Bermúdez-de-Castro, L., Cagiga, M.J. & Sánchez-Yélamo, M.D. (1984). Endemism in the Iberian Peninsula and Balearic Islands. Webbia 34: 101–107.

Gónzalez-Albo, J. (1934). *Hutera* Porta. Cavanillesia 6: 176–177.

Heywood, V.H. (1964) *Rhynchosinapis* and *Hutera*. In: Tutin et al. (Eds.), Flora Europaea 1: 340–342. Cambridge: Cambridge University Press.

Lucas, G. & Synge, H. (1978). The I.U.C.N. Plant Red Data Book. I.U.C.N.-T.P.C. Kew: Royal Botanic Gardens.

Malato-Beliz, J. (1975). Il faut sauvegarder la flore Méditerranéenne. Naturopa 22: 3–6.

Malato-Beliz, J. (1979). O barrocal algarvio: Flore e vegetacâo de Amendoeira (Loulé). Elvas.

Melderis, A. (1972). *Centaurium,* Hill In: Tutin et al. (Eds.), Flora Europaea 3: 56–59. Cambridge: Cambridge University Press.

Montserrat, P. & Villar, L. (1972). El endemismo ibérico, aspectos ecológicos y fitotopográficos. Bol. Soc. Brot. 46: 503–527.

Pau, C. (1922). Las herborizaciones del Sr. Gros por la región almeriense. Buttl. Inst. Catalana Hist. Nat. 22: 30–33.

Peinado-Lorca, M.G. (1980). Estudio florístico y fitosociológico de la cuenca del río Guardiana (prov. C. Real). Tesis Doctoral, Universidad Complutense, Madrid.

Porta, D.P. (1891). Vegetabilia in itinere iberico austro-meridionale lecta. Atti Accad. Agiati 9: 6.

Prentice, H. (1976). A Study in endemism: *Silene diclinis*. Biol. Conserv. 10: 15–29.

Prieto, P. (1973). Algunas notas sobre la conservación de la Naturaleza en Sierra Nevada. Las Ciencias 38: 163–168.

Prieto, P. (1975). Flora de la tundra de Sierra Nevada. Universidad de Granada, Granada.

Rigual, A. & Esteve, F. (1952). Algunas anotaciones sobre los últimos ejemplares de *Callitris quadrivalvis* Vent. en la Sierra de Cartagena. Anales Inst. Bot. Cavanilles 11: 437–476.

Rivas-Goday, S. (1959). Algunas especies raras o relícticas que deben protegerse en la España Mediterránea. Compt. Rend. Reunión Techn. d'Athens, U.I.C.N. 5: 95-101.

Rivas-Martinez, S. (1971). Bases ecológicas para la conservación de la vegetación. Las Ciencias 36: 131-148.

Rivas-Martinez, S. (1972). Relaciones entre los suelos y la vegetación: Algunas consideraciones sobre su fundamento. Anal. Real. Acad. *Farmacia* 38: 69-94.

Sáinz-Ollero, H. & Hernández-Bermejo, E. (1979). Experimental reintroductions of endangered plant species in their natural habitats in Spain. Biol. Conserv. 16: 195-206.

Sampaio, G. (1947). Flora Portuguesa, 2ª ed. Porto: Imprensa Moderna Lda.

Silva, A.R.P. da (1956). Plantes novas e novas áreas para a flora de Portugal, III. Agron. Lusit. 12: 11–48.

Tavares, C.N. (1953). Aspecto da protecao às espécies vegetais em Portugal. Naturalia 4: 5-7.

Tavares, C.N. (1959). Protection of the flora and plant communities in Portugal. Proc. I.U.C.N. Tech. Meeting, Athens 5: 86–94.

Templado, J. (1974). El araar, *Tetraclinis articulata* Vahl en las sierras de Cartagena. Bol. Estac. Central de Ecol. 5: 43-56.

Tutin, T.G., Heywood .V.H., Burges, N.A., Valentine, D.H., Walters, S.M. & Webb, D.A. (Eds.) (1964–1980). Flora Europaea. Cambridge: Cambridge University Press.

Valdés, B. (1970). Revisión de las especies europeas de *Linaria* con semillas aladas. Anal. Univ. Hispalense, Sevilla 7: 5-288.

Vasconcellos, J.C. (1970). Protecção à flora das praias e dunas. Protecção de Naturaleza 11: 20–24.

C. Gómez-Campo
Departamento de Biología
Escuela T.S. Ing. Agrónomos
Universidad Politécnica
E-28040 Madrid, Spain

J. Malato-Beliz
Dep. de Biología Analitica
Estacao Nacional de
Melhoramento de Plantas
Elvas, Portugal

CHAPTER 5

The Italian Peninsular and alpine regions*

S. FILIPELLO** and S. GARDINI-PECCENINI

1. The Italian situation

Many factors have to be considered when choosing examples that are truly representative of the state of Italian floristic conservation. First and foremost, such a choice depends upon our detailed general knowledge of the flora, which is still deficient in many respects today. Field and herbarium research, phytogeographical and taxonomic investigations are not as developed as they ought to be. This deficiency extends to the botanical institutions as well, where appropriate structures are insufficient or absent. Paid positions in these fields are scarce or non-existent, and student interest is equally limited. Little space in the high-school curriculum is left for natural history education. Thus the general situation remains more or less the same as it was when one of us presented our paper on this topic at the Florence OPTIMA Congress of 1977 (Filipello, 1979), and the level of floristic knowledge is similar to that expressed in Filipello et al. (1977).

Nevertheless, progress is being made. Our Institute at the University is often pressed by public societies (Advisory Bodies for Ecology, for the Environment, for Agriculture and Forestry, etc.) to give advice and assistance on environmental problems. Research also proceeds; one small step has been the establishment of the 'Floristic Advertisement' in the 'Informatore Botanico Italiano' sponsored by the Floristic Working Group of the Italian Botanical Society. This is a minor but important aspect, analogous to the classical series of 'Numeri cromosomico per la Flora italiana', and in a certain sense complementing them. The 'Advertisement' is a demonstration of goodwill, as well as an evidence of vigour.

The ecosystems that are of major floristic and vegetational interest are largely known: the two volumes of 'Censimento' (Pedrotti et al., Eds., 1971–1979) report a total of 563, which are widely representative of the different habitats in Italy.

A proposal concerning the solution of scientific and technical problems involved in the preservation of the floristic heritage of Italy was presented in December 1979 at a meeting organized in Florence by the National Research Council (Filipello, Ed., 1981). At that time it was affirmed that there were numerous improvements to be made at both the planning stage and at the stage of realization. Also, ideas on what has to be done are now more firmly settled and there is more agreement on these.

Secondly, there is a great disparity in the nature of support and in the method of implementation in programmes concerned with the protection of nature. In fact, the political body that deals with conservation in Italy does not have any central control. Jurisdiction over National Parks, enforcement of the laws that regulate the gathering of

* Contribution supported by a grant of C.N.R.P.F. 'Promozione della qualità dell'ambiente', Linea di ricerce 'Specie vegetali de proteggere'

** Deceased.

Gómez-Campo, C. (ed.), Plant conservation in the Mediterranean area.
© 1985, Dr W. Junk Publishers, Dordrecht.
ISBN 90 6193 523 7.

medicinal plants (Royal Decree No. 772, 26 May, 1932), and the enforcement of international programmes or conventions, all rest with the Central Government. The legislation of regulations concerning the protection of species is, on the other hand, delegated to the regions. This transfer of power can be justified by the need to promote a greater and more direct local interest in conservation. However, this responsibility has often been neglected by the regional authorities. Because of this negligence, many species of the Mediterranean flora are unprotected in Italy, especially in the southern regions, including Sicilia and Sardinia. Ratification of international agreements that concern the protection of endangered species could provide a partial remedy to this problem. For example, *Abies nebrodensis* (Lojac) Mattel, *Celtis aetnensis* Strobel and *Ribes sardoum* U. Martelli are more or less protected by the Washington Convention on the Trade of Endangered Species (CITES).

Thus, endangered species are not always protected whereas non-endangered ones may be.

Table 1 compares the results of four different censuses of threatened Italian species. The results differ remarkably from case to case. The first list (column A) is the result of research carried out by the Floristic Working Group of the Italian Botanic-

Table 1. Threatened and endemic plants in Italy: a comparison of four different censuses.

		A	B	C	D
Abies nebrodensis (Lojac.) Mattei (Pinaceae)	Si	0	0	E	*
Anchusa crispa Viv. (Boraginaceae)	(Sassari)	*		E	*
Anchusa littorea Moris (Boraginaceae)[a]	Sa	*			
Aquilegia kitaibelii Schott. (Ranunculaceae)	? It, Ju			(R)	*
Armeria elongata Hoffm. (Plumbaginaceae)	(Udine)	*			
Astragalus aquilanus Anzalone (Leguminosae)	It	*		E	*
Astragalus maritimus Moris (Leguminosae)	Sa			E	*
Astragalus verrucosus Moris (Leguminosae)	Sa			E	*
Brassica macrocarpa Guss. (Cruciferae)	Si	*		E	*
Buxus balearica Lam. (Buxaceae)	(Cagliari)	*			
Calendula maritima Guss. (Compositae)	Si	*		E	
Campanula petraea L. (Campanulaceae)	Ga, It	*		E (V)	
Campanula sabatia De Not. (Campanulaceae)	It			E	*
Caralluma europaea (Guss.) N.E. Brown (Asclepiadaceae)	(Lampedusa)	*		E (V)	
Celtis aetnensis Strobl (Ulmaceae)[b]	Si	*	*		
Centaurea horrida Badaro (Compositae)	Sa	*		E	*
Cheilanthes persica (Bory) Mett. ex Kuhn (Sinopteridaceae)	(Ravenna)	?Ex		E (R)	
Cistus clusii Dun. (Cistaceae)	(Foggia)	*			
Cyperus papyrus L. (Cyperaceae)	(Siracusa)	*			
Cyperus polystachyos Rottb. (Cyperaceae)	(Ischia)	*			
Cytisus aeolicus Guss. ex Lindl (Leguminosae)	Si			E	*
Erucastrum palustre (Pirona) Vis. (Cruciferae)	It	*		V	
Eryngium alpinum L. (Umbelliferae)	(Cuneo)	*		E (V)	
Euphrasia marchesettii Wettst. (Scrophulariaceae)	It, ?Ju	*		E (R)	*
Galium litorale Guss. (Rubiaceae)	Si			E	*

[a] Often included in *A. crispa* Viv.; [b] Often included in *C. australis* L.; [c] Often included in *N. radiiflorus* Salisb. or *N. poeticus* L. aggr.

Geographical territories of endemics (or provinces or islands in parentheses, for the more widespread species):
Au: Austria; Co: Corse; Ga: France; It: Italy; Ju: Jugoslavia; Sa: Sardinia; Si: Sicily; ?: doubtful occurence.

Category (in parentheses for Europe, where different):
E: endangered; Ex: extinct; I: indeterminate; Ik: insufficiently known; R: rare; T: threatened; V: vulnerable; ?: presumed

al Society. It lists the threatened species of Italy, independent of their rarity or their endemic character.

The Washington convention only considers commercially important species. It is therefore hard to understand why the three species, indicated in column B, are considered to be at greater risk than other species. Probably no species of the Italian flora is truly eligible to be included in this list.

With few exceptions, the species listed by the Threatened Plants Committee (column C) are endemic. Their evaluation is absolute, unrelated to local interests or the situations on the ground. This explains the difference from the 1972 list of the Floristic Working Group.

Finally, in Annex I of the Berne Convention (column D), most species earlier rated as endangered by the Threatened Plants Committee are considered to be 'strictement protegées'. *Muscari gussonei* (Parl.) Tod. and *Silene velutina,* Pourr. ex. Lois, a Sardinian - Corsican species whose presence in Sardinia remains unconfirmed, are also included.

On the basis of the information contained in Table 1, some important observations can be made.

It will again be noticed that, in Italy and elsewhere, the flora which has received the most damage in recent times, and is now exposed to the

		A	B	C	D
Gypsophila papillosa Porta (Caryophyllaceae)	It	*		E	*
Haplophyllum patavinum (L.) G. Don (Rutaceae)	(Padova)	*			
Hesperis oblongipetala Borb. (Cruciferae)	(Roma)	*			
Hymenophyllum tunbrigense (L.) (Hymenophyllaceae)	(Massa)	*			
Hypericum annulatum Moris (Guttiferae)	(Nuoro)	*			
Ipomoea stolonifera (Cyr.) J.F. Gmel. (Convolvulaceae)	(Ischia)	*		E (V)	
Kochia saxicola Guss. (Chenopodiaceae)	It	*		E	*
Lamyropsis microcephala (Moris) Dittrich ex W. Greuter (Compositae)	Sa	*		E	*
Leontodon siculus (Guss.) Fisch et P.B. Sell (Compositae)	It, Si			E	*
Lilium pomponium L. (Liliaceae)	Ga, It	*		Ik	
Muscari gussonei (Parl) Tod. (Liliaceae)	It, Si	*		I	*
Muscari kerneri Marchesetti (Liliaceae)	(Trieste)	*			
Narcissus stellaris Haw. (Amaryllidaceae)[c]	(Trieste)	*	z		
Potentilla pensylvanica L. (Rosaceae)	(Aosta)	*			
Primula apennina Widmer (Primulaceae)	It			E	*
Pteris vittata L. (Pteridaceae)	(Napoli)	*			
Ribes sardoum U. Martelli (Grossulariaceae)	Sa	*	*	E	*
Salicornia veneta Pignatti et Lausi (Chenopodiaceae)	It	*		E	*
Satureja thymbra L. (Labiatae)	(Cagliari)	*			
Sedum aetnense Tineo (Crassulaceae)	Si	*		E (V)	
Senecio candidus (Presl) DC. (Compositae)	(Palermo)	*			
Serapias orientalis Nelson subsp. *apulica* Nelson (Orchidaceae)	It	*			
Silene velutina Pourr. ex Lois. (Caryophyllaceae)	Co, ? Sa			(T)	*
Woodwardia radicans (L.) Sm. (Blechnaceae)	(Napoli)	*			
Wulfenia carinthiaca Jacq. (Scrophulariaceae)	Au, It, Ju	*			

A. (1972) Specie della flora italiana meritevoli di protezione. *Informatore Bot. Ital.* 4: 12-13.

B. (1975) Rules of *Convention on international trade in endangered species of wild fauna and flora.* Washington.

C. (1976) List of rare, threatened and endemic plants for the countries of Europe. I.U.C.N. – T.P.C. for the Council of Europe.

D. (1979) Annexe I à la Convention relative à la conservation de la vie sauvage et du milieu naturel de l'Europe. Berne.

greatest risk, is that of the Mediterranean zone. The Berne Convention enumerates six Sardinian endemics, five Sicilian endemics and nine endemics from the rest of Italy as endangered.

At the same time, it is necessary to consider the criteria which the compilers of these lists have followed and to understand the reasons why particular species were or were not included. The exclusion from the Berne Convention list of *Sedum aetnense* and *Calendula suffruticosa* subsp. *maritima,* both Sicilian endemics, may be merely an oversight; the exclusion of *Campanula petraea,* endemic to the southern side of the Alps, may be due to its known status on the French side of the border, where it is less threatened than it is in Italy. The fact that *Celtis aetnensis* was not mentioned is a consequence of it being regarded, at the present time, as a variant of *C. tournefortii* Lam. The inclusion of *Campanula sabatia,* a Ligurian endemic with a vigorous chance of survival, whose distribution has not changed recently, might be due to the fact that potential risks to the species, especially that of fire, were estimated to be high.

In contrast, it is sometimes difficult to estimate the extent to which a species is at risk, due to a lack of information upon which to base a decision. In these lists we find many recently discovered species, some so new to science that they have not been included in *Flora Europaea*, for example *Astragalus aquilanus* Anzalone, described in 1970. How can it be stated, in the cases of such species that their number of individuals has fallen to a critical level, or that their habitat has been drastically reduced? Certainly, it may be a question of rarity and recent discoveries in floristically well-known territories speak for themselves. Sometimes the attribute 'endangered' may derive from the emotional commitment of the author rather than any rational assessment.

2. Protection laws and natural reserves

It might be a useful exercise to check which of the species identified as endangered are actually protected in their respective areas. In fact, the regional laws, where these exist, have not always been created as the result of the intervention of botanists. Rather, these laws were concerned with natural history education and thus often concerned showy and more sought-after species rather than rarities that are interesting only from a scientific point of view. Among the plants listed in Table 1, the following are protected: *Potentilla pensylvanica* in Val d'Aosta; *Armeria elongata, Erucastrum palustre* and *Wulfenia carinthiaca* in Friuli-Venezia Giulia; *Haplophyllum patavinum* in Veneto; *Hesperis oblongipetala* in Lazio; *Astragalus aquilanus* in Abruzzo. Two other species are protected in an indirect way: *Aquilegia kitaibelii,* even though it is found elsewhere on Italian territory, is protected in Friuli, as are all the *Aquilegia* species; so is *Narcissus stellaris,* if included in the collective species *N. poeticus*, which is protected in Friuli.

The case of these three species from Friuli is special, since these species have the advantage of double protection. Their main stations have been declared protected areas of 'Natural Surroundings'. The *locus classicus* of *Wulfenia carinthiaca,* at the Pramollo pass, is included in the Regional Park of the Alpi Carniche. The cold environments that are refuges for the microthermic species *Armeria elongata* and *Erucastrum palustre* of Roggia di Varmo and Risorgive della Stella, are established as reserves. *Euphrasia marchesettii* also grows in the same place, neglected by the regional law for the protection of the flora, but similarly sheltered from further threats.

In other parts of Italy the creation of reserves has compensated for the lack of specific floral protection laws. A very noteworthy example is the reserve formed by the State Agency for Forests (A.S.F.D.) in the Vallone of the Ferriere (Salerno), to protect the habitats of *Woodwardia radicans* and *Pteris vittata.*

Needless to say, this type of protection of habitats is far preferable to other legislation (including the prohibition or limitation of the collection of specimens) in the defence of seriously endangered species.

Despite such clear-cut examples of protection, there are plenty of unsettled ones. In Liguria, for example, the picking of *Campanula sabatia* and *Lilium pomponium* is presently forbidden by a temporary ordinance, in anticipation of the establishment of regional parks and reserves that will include their stations. However, clear delimitation of these areas has not yet been produced, and some localities at which these species occur risk being excluded. In the province of Savona, the Finalese Park should include at least the *locus classicus* of *C. sabatia* at Capo Noli and also the separate station on Gallinara Island. In the province of Imperia, a system of reserves in the municipalities of Triora and Pigna (Ligurian Alps) should include all known Italian habitats of *L. pomponium.* The situation has grown temporarily worse for *Eryngium alpinum,* which is found in only one area of Italy, the province of Cuneo. During the last few years it has no longer been protected because the Piedmontese regional law for the preservation of the natural heritage is in process of being revised. The lists of protected species will soon be rewritten and *E. alpinum* will be included.

Among the cases considered above, many could constitute representative examples, but it must be clear that these only represent the tip of the iceberg, in a situation that is not always well known. For this reason it seemed proper to choose some case-histories from amongst lesser known species or, perhaps, less important ones, which are equally noteworthy and representative of floristic conservation. In our choice of species we have tried to reach a numerical balance between northern and southern species. We have also tried to consider, as much as possible, species from different habitats.

We describe below case histories of four endemic species and three non-endemic ones. Of these, four are northern or alpine and the remainder are from peninsular Italy. Figure 1 shows their geographical and Figure 2 their altitudinal distribution.

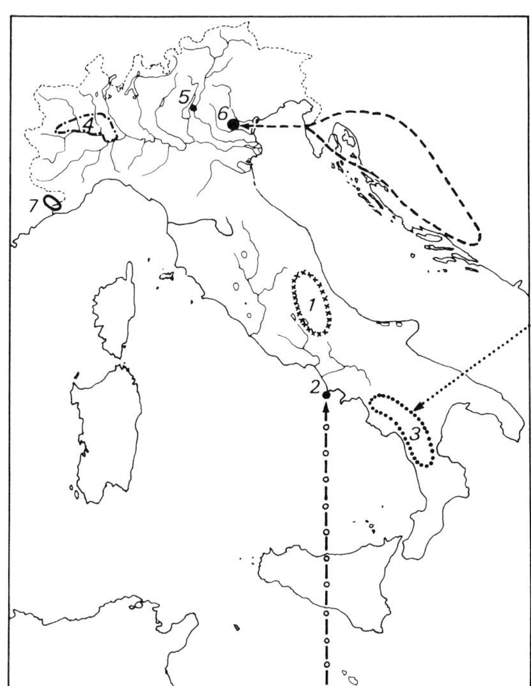

Fig. 1. Geographical distributions of the species mentioned in this article: 1. *Adonis distorta,* 2. *Cyperus polystachyos,* 3. *Pinus leucodermis,* 4. *Isoetes malinverniana,* 5. *Haplophyllum patavinum,* 6. *Gypsophila papillosa,* 7. *Saxifraga florulenta.*

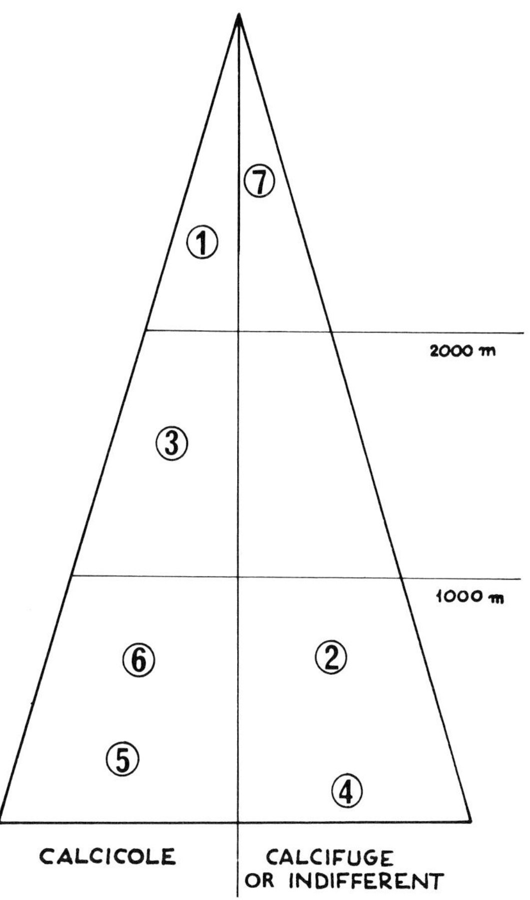

Fig. 2. Altitudinal distribution of the species mentioned in this article: 1. *Adonis distorta,* 2. *Cyperus polystachyos,* 3. *Pinus leucodermis,* 4. *Isoetes malinverniana,* 5. *Haplophyllum patavinum,* 6. *Gypsophila papillosa,* 7. *Saxifraga florulenta.*

Though it is recognized that non-endemics should have a lower conservation priority, the inclusion of three of these is justified by the fact that they are sufficiently well known to provide valuable and interesting information.

3. Some case histories of rare and threatened species in Italy

3.1 Adonis distorta *Ten. (Ranunculaceae), a rare endemic from the Apennines, central Italy (Fig. 3)*

Adonis distorta occurs in the high mountain Mediterranean vegetation belt, 'in the heaps of rocks crushed by the ice', at an altitude range of 2,000–2,400 m. It is a calcicolous pioneer species, which colonizes breccias and coarse gravel even where there is almost no soil, as long as there is a constant inflow of moisture. For this reason, the species is frequently present on northern slopes and near snowfields. Other species that are often present in the same habitat are *Adenostyles alpina* (L.) Bluff et Zingerh, *Anemone apennina* L., *Arabis alpina* L., *Aster alpinus* L., *Campanula rotundifolia* L., *Cynoglossum magellense* Ten., *Daphne alpina* L., *Isatis allioni* P.W Ball, *Linaria alpina* (L.) Miller, *Myosotis alpestris* F.W. Schmidt, *Pedicularis verticillata* L., *Ranunculus magellensis* Ten. and *Valeriana montana* L.

Adonis distorta occurs in the high mountain Mediterranean vegetation belt, 'in the heaps of rocks crushed by the ice', at an altitude range of 2,000–2,400 m. It is a calcicolous pioneer species, which colonizes breccias and coarse gravel even where there is almost no soil, as long as there is a constant inflow of moisture. For this reason, the species is frequently present on northern slopes and near snowfields. Other species that are often present in the same habitat are *Adenostyles alpina* (L.) Bluff et Zingerh, *Anemone apennina* L., *Arabis alpina* L., *Aster alpinus* L., *Campanula rotundifolia* L., *Cynoglossum magellense* Ten., *Daphne alpina* L., *Isatis allioni* P.W. Ball, *Linaria alpina* (L.) Miller, *Myosotis alpestris* F.W. Schmidt, *Pedicularis verticillata* L., *Ranunculus magellensis* Ten. and *Valeriana montana* L.

Adonis distorta is much sought by casual visitors for its attractive, yellow or white flowers, which are borne in summer. The plant is also of medicinal value, the whole plant being poisonous and used as a substitute for *Digitalis* as a heart-regulator, also as a diuretic and for inducing slight hypertension. Its status has declined due to road construction, the use of four-wheel drive vehicles and particularly by the development of ski resorts. The creation of ski-slopes, which are often cleared and levelled by heavy machinery, has seriously altered the habitat of this species, and that of other rare species as well.

The Mt. Sirente populations are included in a natural park formed for the protection of the fauna, but unfortunately the protection of the park's flora is almost non-existent. Throughout the area, *A. distorta* now enjoys double but illusory protection. It has been declared a medicinal species, as are all the Adonides, by the Royal Decree No. 772, 26 May 1932, and therefore its harvest and trade are regulated by law No. 99, 6 January 1931. Because of its great toxicity, private use is forbidden. In addition, law No. 47, 11 September 1979, of the Abruzzo Region, fully prohibits its harvest.

We shall be able to monitor the effects of this regional law from the summer of 1980 onwards. Though this law was essential, it is evidently insufficient to protect *A. distorta* while its habitat is still exposed to damage. The current situation is rather uncertain, and it will be necessary to monitor this species carefully because of its present vulnerability.

3.2 Cyperus polystachyos *Rottb. (Cyperaceae), in the fumarole environments of Ischia (Fig. 4)*

First collected on the island of Ischia by L.I Giraldi in 1772–75, and refound again by M. Tenore in 1802, this species has been known ever since. Around the same time, debate began on its origin, as to whether it was to be considered an ancient relict, or whether its presence was the consequence of a deliberate or accidental introduction. R.A. De Fillips (1980) in *Flora Europaea* accepts the latter hypothesis and considers *C. polystachyos* to be

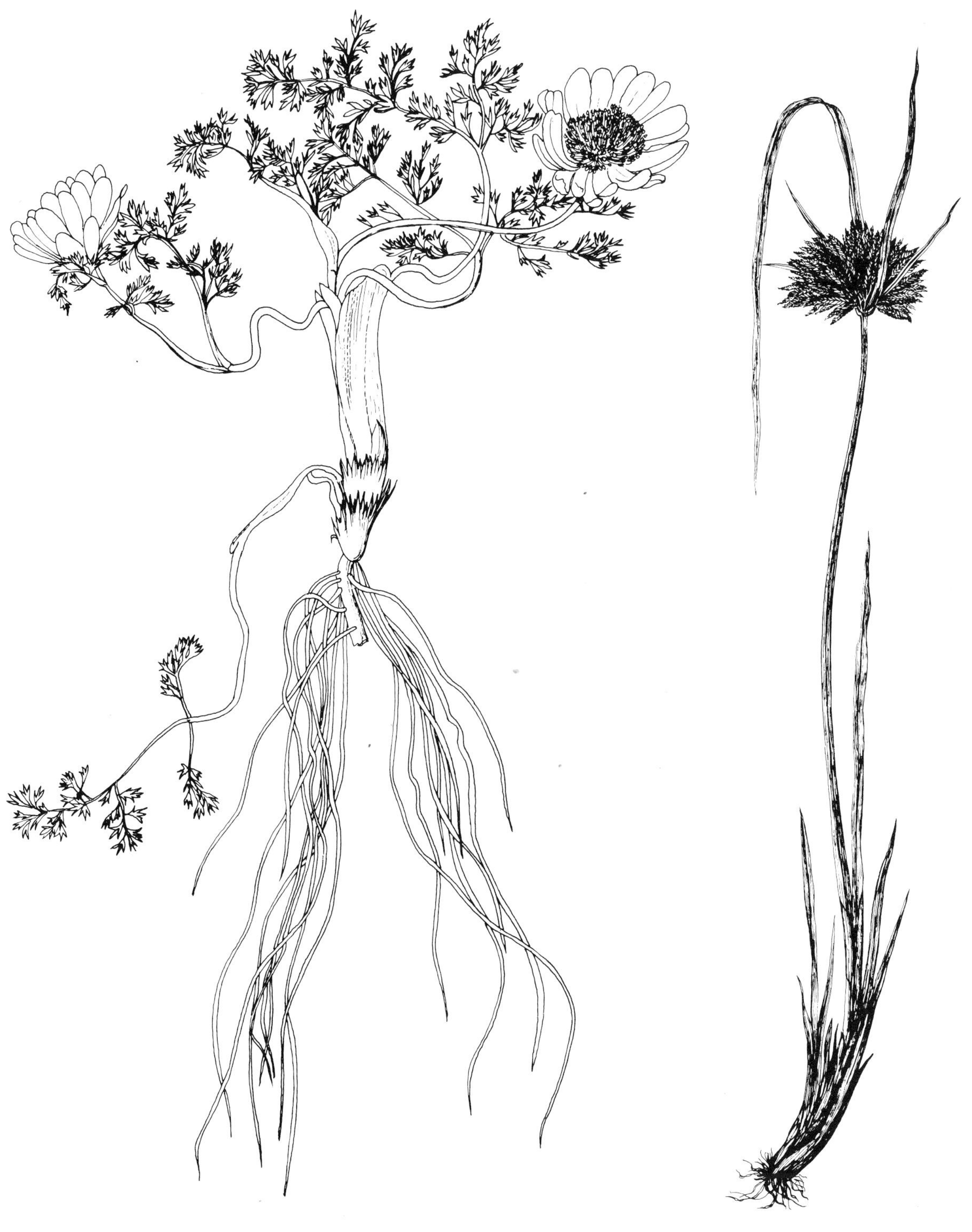

Fig. 3. *Adonis distorta* Ten.

Fig. 4. *Cyperus polystachyos* Rottb.

naturalized. This does not diminish its importance, since the species, which is universally common in a number of tropical countries, has a very peculiar ecology in Ischia. It is a strict hydrothermophile and does not stray far from fumaroles (Fig. 5) the only environment in our latitude that can assure conditions of continual heat and humidity. Fumaroles are phenomena of secondary volcanism which emit high temperature jets of vapour through a flue that humidify and heat surrounding soil and air. The thermal characteristics of the soil allow the survival of *C. polystachyos* only where sufficiently high temperatures are also found at the surface of the ground. The epigeal parts as well as the roots need protection from low temperatures; this protection is afforded only when the temperature of the soil remains 6–10°C higher than that of the first 2–3 cm above the ground, so that even in the coldest winter the survival of the buds and of many of the leaves is assured. According to Merola (1957), ideal conditions for *C. polystachyos* in Ischia are established when the temperature at ground level does not fall below 15°C (the roots can penetrate the ground deeply to avoid temperatures below 15°C) and when the relative humidity is about 80%. As the vapours emerge at a temperature of nearly 100°C, the plants occur at a distance from the opening and the roots become more superficial as ground temperature rises towards the vent. The very hot nature of this environment prevents most arboreal and shrub species dominant on the island from developing their own roots in depth and this lack of competition allows *C. polystachyos* to form large stands. Under conditions of high humidity the following species join *C. polystachyos: Briza minor* L., *Cyperus esculentus* L., (*C. aureus* Ten.), *C. rotundus* L., *Centaurium tenuiflorum* (Hoffmans et Link) Fritsch, *Digitaria sanguinalis* (L.) Scop. (*D. gracilis* Guss.), *Heliotropum suaveolens* subsp. *bocconei* (Guss.) Brummit, *Lotus subbiflorus* Lag. subsp. *subbiflorus* and *Pteris vittata* L. In the surrounding forested area are found *Arbutus unedo* L., *Asparagus acutifolius* L., *Cistus salviaefolius* L., *Erica arborea* L., *Myrtus communis* L., *Quercus ilex* L., *Ruscus aculeatus* L., *Smilax aspera* L., *Viburnum tinus* L., and other species.

C. polystachyos produces large quantities of seed, which is dispersed by the wind. Also, mature individuals spread vegetatively during the summer by producing several lateral buds, which rapidly develop into additional plants. These in turn, flower, set seed and produce new buds at their bases. In

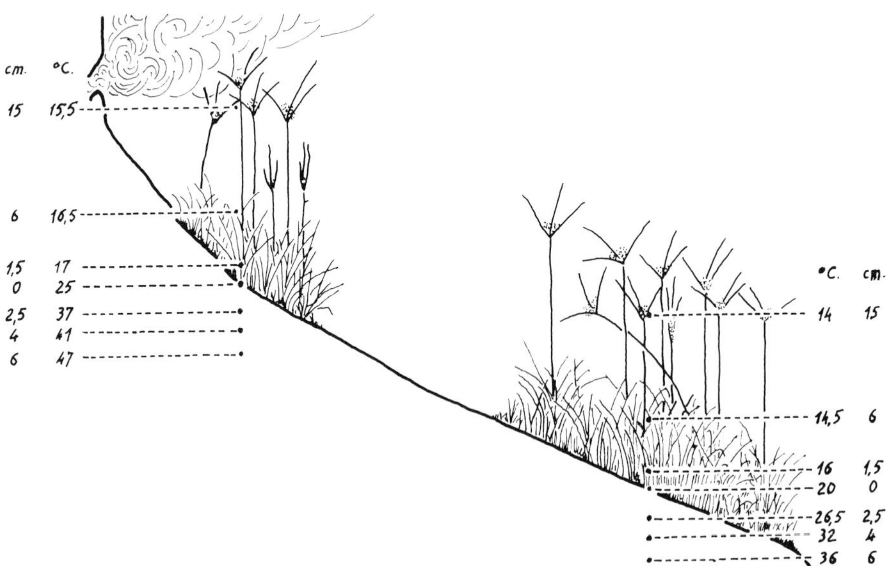

Fig. 5. Temperatures are pointed out both in and above the ground, in correspondence with two colonies of *Cyperus polystachyos* on a slope only a short distance from the fumarole opening (upper left). This demonstrates both the thermal character of the ground and the heating of the air immediately above and around the plants (from Merola, 1957) Normal air temp. is 11–12°C.

view of its biology and its reproductive vigour, *C. polystachyos* does not appear to be threatened on Ischia. Indeed, it readily establishes itself in all places where conditions are appropriate, and it is also easily cultivated.

However, *C. polystachyos* is confined to very special surroundings, such as the fumaroles, in generally small sites, covering only a few thousand square meters. In this situation ever-increasing human influence justifies concern for its survival.

In 1972 the Floristic Working Group of the Italian Botanical Society included *C. polystachyos* in the group of species of the Italian flora that are most vulnerable when the risk of extinction is considered. Protection has been sought (in Pedrotti et al. Eds., 1971-1979) for at least two populations, one close to Faiano and the other close to Fango, among the ten which were known in 1957. Since then, however, no action seems to have been taken, while human influence has continued to increase.

3.3 Pinus leucodermis *Aiton (Pinaceae), in Italy (Figs. 6 and 7)*

The distribution of *Pinus leucodermis* is centred on the Balkans (Albania, Yugoslavia and Greece), with a disjunct area in the Tyrrhenian part of southwestern Italy, consisting of three main groups of localities.

The most northerly Italian locality recently discovered by Hofman (1958), is on Mt. Groce, near Avellino, in Campania. The localities on the Pollino Massif, in Basilicata and Mt. Palanuda, east of Orsomarso, and Montea in Calabria, have been known for a long time. A common characteristic of these sites is their location in zones of strong erosion, where calcareous dolomite or calcareous marl rocks outcrop and the soil is poorly developed.

P. leucodermis is a pioneer, xerophytic, basophilous species. It occurs on southern slopes at 850 - 2270 m. While occurring at the same altitude as oak, the requirements of *P. leucodermis* are directly opposite with regards to soil, climate and exposure. *P. leucodermis* tends to form pure stands with low coverage values, with usually 40 - 50 mature trees per hectare and rarely more than 90 - 100. It is associated with other tree species only at the altitudinal limits with oak at the upper limit and black pine, ash and rarely holm oaks at the lower limit.

The *P. leucodermis* woods cannot be typified phytosociologically. The species may be present in various assemblages where it plays, at most, a differential role. The herb layer of these woods is characterized by *Sesleria tenuifolia* Schrad, or *Sesleria nitida* Ten. In addition *Achillea erba-rotta* subs. *calcarea* (Porta) I.K.B. Richardson, endemic to the southern Appennines, is also frequent.

According to Pennacchini & Bonin (1975), *P. leucodermis* forests belong to *Elymo-Seslerietea* Br.-Bl. 1948, due to the presence of *Alyssum diffusum* Ten., *Anthyllis vulneraria* L. subsp. *pulchella* (Vis.) Bornm., *Biscutella laevigata* L., *Bunium bulbocastanum* L., *Festuca circummediterranea* Patzke, *Linum perenne* subsp. *alpinum* (Jacq.) Ockendon, and *Paronychia kapela* (Hacq.) Kerner and to the *Seslerietalia tenuifoliae* Horvat, 1930 due to the presence of *Carex macrolepis* DC., *Centaurea parlatoris* subsp. *tenorei* (Guss. ex Lacaita) Dostal, *Edraianthus graminifolius* (L.) A.DC., *Linum capitatum* Kit. ex Schuttesland, *Sesleria nitida* Ten. The same authors recognize an influence of *Seslerion tenuifoliae* Horvat, 1930 as demonstrated by the presence of *Avenula versicolor* subsp. *pretutiana* (Parl ex Arcangeli) Holub., *Acinos alpinus* subsp. *meridionalis* (Numan) P.W. Ball, *Sesleria tenuifolia* Schrad, and with *Scabioso-Astragalion* as demonstrated by the presence of *Asperula gussonii* Boiss., *Pimpinella anisoides* Briganti, and *Senecio doronicum* L. (var. *pseudoarachnoideum* Fiori). In such woods some mediterranean-montane garrigue species are also found: *Asphodeline lutea* (L.) Reichenb., *Cytisus spinescens* (C. Presl) Rothm., *Euphorbia myrsinites* L., *Lavandula angustifolia Miller, Sideritis syriaca* L. and *Stipa pennata* L.

Pinus leucodermis is a long-lived and very slow-growing species which reaches considerable heights, up to a maximum of 30 m. Some specimens live to 60 - 100 years, generally in sites that proviced favourable conditions. Older and more majestic trees, such as some at the Serra di Ciavola

Figs. 6. and 7. Stands of *Pinus leucodermis* Ait. in Southern Italy. ⟶

of 600 years age, are visibly damaged by the weather and the conditions of partial desiccation. (This phenomenon of great-age attainment is not limited to this species but is rather common in isolated trees growing at their altitudinal limits.)

Natural regeneration is barely sufficient to ensure the survival of the species, which, fortunately is well suited for cultivation. During the years 1962 - 1975, over 60,000 three-year-old trees have been transplanted from the Campotenese forestry nurseries at Mirano Calabro. The success rate is estimated to be around 80%. In spite of its slow growth, the adaptability and colonizing ability of *P. leucodermis* make it a suitable species for use in soil stabilization and hydrogeological schemes. Until now, new plantations have been largely limited to the areas where the species formerly existed, but results have been encouraging enough to suggest wider utilization of the species.

P. leucodermis was in decline until a few years ago, but is now increasing and is no longer threatened. However, this improvement has been largely due to artificial propagation. Natural survival could be ensured through protection of certain areas. Several indications of this are provided by 'Biotope censuses' such as those for Ponte di Mola, Serra della Spina and Monte Alpi in the Potenza Province, and for the Orsomarso and Verbicaro Mountains, Cosenza Privince (in Pedrotti et al, Eds., 1971, 1979). A final solution could be

provided by the establishment of the Mt. Pollino National Park at present being planned under the auspices of the World Wildlife Fund and the National Research Council.

3.4 Isoetes malinverniana *Cesati et De Notaris (Isoetaceae), in the Po Valley*

This preglacial relict species is a rare example of an aquatic plant, endemic to the northwestern sector of the Po plains, a territory without clear-cut natural boundaries.

Described in 1858, from specimens gathered 'in aequeductibus' (sic) by Greggio and Oldenico and from near Vercelli by A. Malinverdi, German authors of that time considered it an exotic species, imported with rice. Mattirolo (1912), cast doubt on this assumption by discovering new sites. These widened the known distribution to the west as far as Pianezze, up to the border of the Rivoli's morainic amphitheatre, and to the lower Canavese in the province of Turin; and to the east, to Isarno and Vignale, in Novara Province between Agogna and Terdoppio Rivers. Half a century then elapsed before the discovery of new sites by F. Corbetta. The first discovery in 1965 was once again in Vercelli Province, but quite far from previous finds. Soon after, in 1967 and 1968, ten more localities near Lomellina, in Pavia Province, were found (Corbetta, 1968). Thanks to these latter sites the known distribution has been substantially enlarged to the east and southeast. Recently, Rosenkrantz & Tosco (1979) added another new site between Front Canavese and Busano, which extends the area to the north of the known lower Canavese range. In summary, the habitats of *Isoetes malinverniana* are the moraines at the foot of the alpine amphitheatre, in the alluvial terraces of the Lomellina.

Mattirolo (1912) and Corbetta (1968) disagree somewhat on the habitat requirements of this species. According to Mattirolo (loc. cit.), *I. malinverniana* lives exclusively in spring-waters where the temperature is rather low. Corbetta (loc. cit.), claims that it is more frequent in a number of canals derived from the Po river, such as the Dora Baltea, the Sesia or the Agogna. Often these waters are not clear and they sometimes receive drainage. The two agree, though, that *I. malinverniana* prefers rather cold waters.

Mattirolo and Corbetta also disagree on the effect that canal-cleaning has on the species. According to Mattirolo, annual dredging has damaging effects on the growth of *I. malinverniana,* particularly when slime is removed. Corbetta, instead, maintains that dredging does not have much effect since the deepest plants are untouched by it anyhow.

Mattirolo defines *I. malinverniana* as a 'delicate, lone antisocial species', an opinion with which Rosenkrantz & Tosco (1979) agree. Corbetta claims that it generally, and rather consistently, associates with several species.

Mattirolo states that *I. malinverniana* is doomed to disappear in a not-too-distant future because of its low 'adaptability to variations in conditions of its localities'. Corbetta reports it is 'true that in certain areas it is very scarce, but it is also true that, elsewhere, its presence is so massive and its growth so vigorous as to often displace other neighbouring species'. In winter, *I. malinverniana* occurs abundantly, with cover values approaching 100%. However, values above 60% are not infrequent in other seasons.

Among the species that are associated most frequently with *I. malinverniana,* are *Agrostis stolonifera* L. (floating, sterile variant), *Alisma plantago-aquatica* L., *Callitriche stagnalis* Scop., *Eleocharis carniolica* Koch, *Glyceria fluitans* (L.) R. Br., *Elodea canadensis* Mich., *Juncus conglomeratus* L. *Nuphar lutea* (L.) Sibth. et Sm., *Ottelia alismoides* (L.) Pers., *Potamogeton crispus* L., *P. natans* L., *P. pectinatus* L., *P. perfoliatus* L., *Ranunculus aquatilis* L., *Sagittaria sagittifolia* L., *Vallisneria spiralis* L., *Veronica anagallis-aquatica* L., and the moss *Fontinalis antipyretica* Hedw. However, the associations of *I. malinverniana* with any one of the species above are too irregular to identify with any phytosociologial association except the order *Potametalia* W. Koch 1926 and the class *Potametea* Tx. et Preising 1942.

To summarize our present knowledge, *I. malinverniana* does not appear to be in serious danger and only a radical change in present conditions will threaten its survival. Even if one takes into account the deterioration of the waterways since the beginning of the century due to changing agricultural practices and techniques of cleaning and draining ditches, the newly discovered sites give additional reason to be optimistic. Luckily, the fact that *I. malinverniana* does not flower makes it aesthetically insignificant and unattractive to those such as Sunday hikers who might disturb showier flowers.

Attempts at cultivating *I. malinverniana* have been very few, although Mattirolo reports a specimen that survived for several years in the Turin Botanic Garden.

The Threatened Plant Committee's evaluation of *I. malinverniana* as 'threatened' appears, therefore, to be somewhat exaggerated. Nevertheless it seems very important that at least one site of this species is given the status of a reserve. This possibility has not been considered so far by Piedmont or Lombardy.

3.5 Haplophyllum patavinum *(L.) Don fil (Rutaceae): Italian localities (Fig. 8)*

This plant was first described as a species of *Ruta* by Linneus, who probably studied the specimen collected by Micheli at Sassolungo near Arquà (now Arquà Petrarca) on the Euganean hills of Padua Province. G. Don transferred it to *Haplophyllum* because of its ternate leaves. While Sassolungo has remained the *locus classicus*, it later became clear that this was only a disjunct population. It is now known that *H. patavinum* is common to the mountainous areas of Dalmatia and Herzegovina and from there extends, with decreasing abundance, into Italy, Montenegro, Albania, Greece and southwestern Rumania. Within the Euganean hills, it has been found in other calcareous, southern areas such as Monte Cero, near Arquà and Mondonego near Valsanzibio.

H. patavinum is a calcicolous species throughout its range. It is found from sea-level to over 1,000 m in dry, stony cultivated areas, abandoned fields, scrub, roadsides, on stony, sunny slopes, ravines and landslides, and on riverbeds. Its preference for unstable soils evidently is a characteristic which gives it an advantage over other species when colonizing abandoned fields. However, *H. patavinum* is also found in more developed vegetation, such as this olive groves by Monte Cera, and occasionally in true forests. Beck von Mannagetta (1901) in fact considered *H. patavinum* as a characteristic component of Karstic woods in Dalmatia.

H. patavinum is associated with a *Xerobrometum*-like pioneer vegetation comprising *Acinos arvensis* (Lam.) Dandy, *Anthyllis vulneraria* L., *Artemisia campestris* L., *Asperula cynanchica* L., *Brachypodium pinnatum* (L.) Beauv., *Bromus erectus* Hudson, *Dianthus caryophyllus* L.,

Fig. 8. Haplophyllum patavinum (L.) G. Don

Euphorbia cyparissias L., *Galium verum* L., *Globularia vulgaris* L., *Helianthemum nummularium* subsp. *obscurum* (Celak) J. Holub, *Hieracium pilosella* L., *Hippocrepis comosa* L., *Koeleria macrantha* (Ledeb.) Schult., *Linum tenuifolium* L., *Picris hieracioides* L., *Poa bulbosa* L., *Potentilla erecta* (L.) Räuschel, *Prunella laciniata* (L.) L., *Salvia pratensis* L., *Sanguisorba minor* Scop., *Scabiosa triandra* L., *Spartium junceum* L., *Stachys recta* L., *Teucrium chamaedrys* L., and *T. montanum* L.

It is also frequently associated with agricultural weeds, particularly segetal communities on calcareous soils, with other thermophilous species, with communities associated with ruins of human habitation, and with other micellaneous species of open communities.

In the Venetian region *H. patavinum* occurs in habitats which are becoming progressively more modified by agriculture and construction. Besides the deleterious effects of human disturbance, *H. patavinum* seems to be doomed to local extinction in a not-too-distant future from natural causes: *viz.* 'Species with a limited and/or highly discontinuous distribution, in which the geographical isolation forces crossings between only a few individuals, living in uniform ecological conditions, often show a tendency to further reduce their areas. A species in such conditions appears to have an impaired power of expansion depending on a number of factors, which must be ascertained in each single case' (Cappelletti, 1929).

The single most important cause of decline of the Venetian populations of *H. patavinum* appears to be their almost complete inability to produce fertile seeds. In the Euganean area, multiplication is almost exclusively rhizomatous; seed sterility is mainly due to the effects of *Penicillium* mycorrhiza. Whether by depriving the host plant of specific nutrients or by somehow, the result of this interaction is in effect, the induction of total sterility. This sterility prevents *H. patavinum* from colonizing new sites; it can only form dense, but precarious, stands in places where the plant already occurs. This ready multiplication perpetuates the damage since the annual growth of the rhizome does not extend the new shoots enough to avoid the infected soil. Its preference for loose soils might assure its survival, since earth movements and light agricultural practices upset the normal microbiological processes of the soil, including the mycorrhizal ones. For these reasons it is not difficult to cultivate *H. patavinum* in sunny, sheltered, well-drained soils. However, seed fertility continues to be very low.

However, a potentially favourable loosening of the soil can become destructive, as in the current agricultural practice of deep plowing. Cappelletti (1929) sounded the first alarm, reiterated in 1957, when he realized that the Mondonego population had been destroyed by agricultural practices.

Only two sites now survive in the Euganean hills, both evidently debilitated and also threatened by construction. For this reason the Regional law No. 53 of 15 November 1974, prohibiting collection of *H. patavinum,* is an insufficient measure. To prevent yet another loss to the Italian flora, the two remaining stations - in particular the *locus classicus* - must be protected. While landscape protection and the institution of suitable, as yet unspecified, reserves has been urged for the whole Euganean hills (Pedrotti et al., 1971), no specific measures have been considered so far for the two surviving *H. patavinum* sites in Italy.

3.6 Gypsophila papillosa *Porta (Caryophyllaceae) a severely restricted endemic in the lake Garda region*

This species is endemic to a small area on the eastern shore of Lake Garda a few ten square metres in area, consisting of two small localites above the town of Garda.

G. papillosa was discovered in 1902 by Rigo, who did not assign it to species rank, but chose to aggregate it with *G. fastigiata* L., an East-Central European species as var. *benacensis.* Shortly after, Béguinot (1905) became convinced that he was dealing instead with *G. hispanica* Willk. This author paid special attention to the phytogeographical problem posed by its disjunct distribution, in particular to the significance of this for such a small and isolated population. Finally, Porta, again in 1905, described Rigo's *Gypsophyla* as a new species, initially as *G. glandulosa,* and then with the current specific epithet *G. papillosa,* since the previous epithet was already in use. He acknowledged the affinities of the new species with both *G. fastigiata* and *G. hispanica.*

G. papillosa lives in a xerothermic 'island' with a mean annual temperature of around 14°C. The soil is formed by rough, calcareous detritus that in some places slopes gently, and in other places, steeply. These habitat conditions are reflected in the flora of the sites, which is composed of xerophilic, thermophilic, steppe or ruin-associated species such as *Artemisia absinthium* L., *A. alba* Turra, *Conyza canadensis* (L.) Cronq., *Cotinus coggyria* Scop., *Daucus carota* L., *Dichanthium ischaemum* (L.) Roberty, *Euphorbia nicaeensis* All., *Fraxinus ornus* L., *Globularia cordifolia* L., *Helianthemum canum* (L.) Baumg., *Melampyrum arvense* L., *Ononis natrix* L. and *Thymus pulegioides* L.

According to Bianchini (1974) the area in which the species is found has not diminished since its discovery. However, the whole belt of hills surrounding Lake Garda is being irreparably damaged by exceptionally intense and nearly indiscriminate building activity in connection with tourism.

Since the very small range is, by its very nature, a source of danger, *G. papillosa* was included in the list of highly threatened species compiled by the Floristic Working Group of the Italian Botanical Society, the Threatened Plants Committee and the Berne Convention.

However, according to the census of the Italian Botanical Society, these warnings were insufficient to include this ecosystem among those worthy of interest, and to prohibit the collection of specimens (largely by foreign botanists). This could easily have been done through the floral laws of the Veneto Region. As a result, the future of this species is very uncertain.

3.7 Saxifraga florulenta *Moretti (Saxifragaceae), crevice-dwelling species of the Maritime Alps (Fig. 9)*

This Tertiary relict is of interest for a number of reasons, including its systematic position within the genus, its floral biology and its rarity. The species was first described in 1823 by G. Moretti, who acknowledged that it had already been reported several years previously. The type samples had been given to Moretti by G. Biroli, who, in turn, had received them from C.A.L. Bellardi. These changes of hand perhaps account for the vague localization of the description: 'Maritime Alps, near Nice'. The region then belonged to the Kingdom of Sardinia, which justified the inclusion

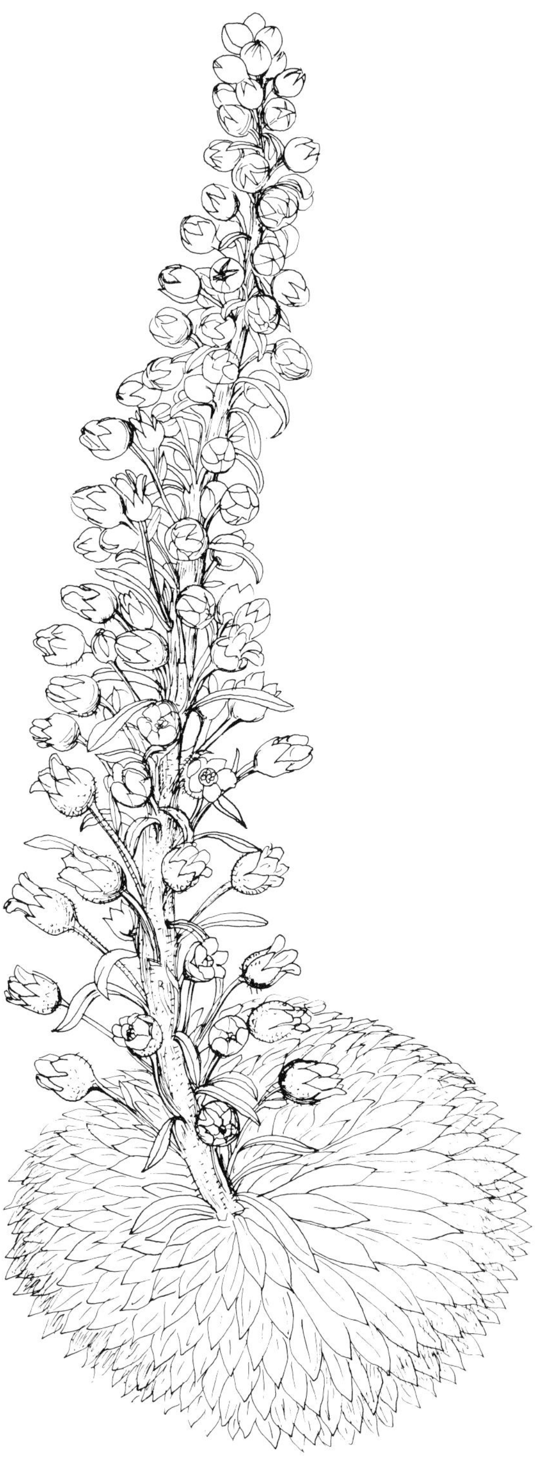

Fig. 9. Saxifraga florulenta Moretti

of the new species in the *Saggio di una monografia delle saxifrage italiane*. The area has a maximum width of about 50 km between its extremities, the Pontebernardo Valley to the northwest and the Fontanalba Valley to the southeast. The Italian-French border now crosses this area. Reports of *S. florulenta* are rather numerous in several localities - the Ténibres group, the upper Tinée Valley, the Pietraporzio Valley, the Tête de Malinvern, the upper Gesso Valley, the Fremamorta, Mercantour and Argentera mountain tops, Col de Fenêtre, Mts. Gelas, Clapier and Valmasca.

The species is nevertheless rare because of its characteristic, strict habitat requirements. *S. florulenta* lives in the fissures of siliceous rocks (granite and gneiss) with some shelter from exposure (Guinochet, 1938; Quézel, 1951), at heights between 2215 and 3290 m (Burnat, 1901). Here the steepness of the sites prevents the snow from covering the plants despite the high altitude.

S. florulenta is characteristic of the association '*Galium tendae* Reichenb. fil. and *S. florulenta*' Guinochet, 1938 (*Androsacetalia multiflorae* Br.-Bl. 1926). Due to its habitat and its tendency to form small colonies or almost pure stands, *S. florulenta* is associated with other species, such as (with no phytosociological distinction) *Silene cordifolia* All., *Artemisia eriantha* Ten., *Androsace vandellii* (Turra) Chiov., *Eritrichium nanum* (L.) Schrader, *Saxifraga pedemontana* All., *S. retusa* Gouan, *Rhodiola rosea* L., *Draba dubia* Suter, *Primula marginata* Curtis, *Saxifraga paniculata* Miller, *Globularia repens* Lam., *Silene rupestris* L., *Asplenium viride* Hudson, *A. septentrionale* (L.) Hoffm.; *Cystopteris fragilis* (L.) Bernh., *Arabis alpina* L., *Phyteuma globulariifolium* Sternb. et Hoppe, *Juncus trifidus* L., *Saxifraga moschata* Wulfen., *Silene acaulis* subsp. *exscapa* (All.) J. Braun, *Potentilla valderia* L., *Festuca ovina* L., *F. varia* Haenke, *Sempervivum arachnoideum* L., *Cerastium alpinum* L., *Carex atrata* L., *Campanula rotundifolia* L., *Juniperus communis* subsp. *nana* Syme, *Cardamine resedifolia* L., *Achillea erba-rotta* All., *Pulsatilla alpina* (L.) Delarbre, *Saxifraga aspera* L., *Carex sempervirens* All., *Poa nemoralis* L., *Luzula lutea* (Vill.) DC., L. *spicata* (L.) DC., *Sempervivum montanum* L. and *Poa alpina* L.

S. florulenta is monocarpic and incapable of vegetative reproduction. It does not flower until after 10–12 years of growth and then produces a prominent, most beautiful panicle. This saxifrage is very delicate, is difficult to cultivate and is easily damaged during transplanting. It is extremely hard to bring into flower outside its natural habitat. In spite of these difficulties or perhaps because of the challenges that it presents, *S. florulenta* has been much collected since its discovery by botanists, collectors and particularly by specialist gardeners and nurserymen. As a result, whole populations of the species have been destroyed. In 1967 Barbero and Bono warned that 'ce saxifrage est devenue très rare dans les localités oú il était signalé au siècle dernier'. More recently the Meraviglie Valley area, in French territory, has been under threat of serious damage from a proposed uranium mine. However, this territory belongs to the Parc National des Alpes Maritimes and this should prevent mining activity.

At present this species is vulnerable. However there are firm plans to turn the Italian side of the valley into a National Park and this should improve the situation. For the time being, due to a *vacatio legis* in the Piedmont region, collecting is allowed. However, *S. florulenta* is given absolute protection by a new law which is in the process of ratification. Thus, apart from an element of uncertainty with regard to the uranium mine, the species appear to have a safe future both in Italy and in France.

It is nevertheless necessary to oppose, by all possible means, the collection of living specimens. It should be made clear, particularly to horticulturalists, that *S. florulenta* is one of those rare species for which cultivation is of no use as a means of ensuring their survival.

In conclusion, it may be interesting to examine the representative values of the chosen examples of species.

Cyperus polystachyos is comparable in a number of respects, such as ecological affinity and status, with *Cyanidium caldarium* Geitl. This little blue-green alga also lives in habitats that are characterized by secondary volcanism, and is similarly endangered by the effects of recreational developments. *C. polystachyos* also has similarities with *C. papyrus* L., with which it shares a similar distribution, in Sicily and elsewhere.

Tourism, the most important Italian industry, often causes serious alterations to natural habitats. Besides *C. polystachyos* there are other species in the Gulf of Naples, that have had analogous fates. For example, *Ipomoea stolonifera* (Cyr.) J.F. Gmelin of which the last locality in Ischia has been utilized to build a seaside resort with a swimming pool (M. Ricciardi, pers. comm.). A similar example is provided by *Kochia saxicola* Guss., whose only site in Capri has been covered since 1922 beneath 3–4 m of calcareous gravel from the building of the road to Anacapri (Gundagno, 1931). In the absence of other reliable and recent data for Capri and Ischia, one must unfortunately assume that *K. saxicola,* a rare endemic of the lower Tyrrhenian Sea, is now reduced to a few specimens on Strombolicchio (Ferro & Furnari, 1968). Evidently, it would not be wise to rely on its survival. Although Strombolicchio is not a likely candidate for new quarries or reservoirs, there is little room for optimism.

Saxifraga florulenta conditions have parallels with those of many species of rocky habitats, some endemic, some not, such as *Telekia speciossisima* (L.) Less., *Primula palinuri* Petagna, *Physoplexis comosa* (L.) Schur., *Androsace alpina* (L.) Lam., and *Eritrichium nanum* (L.) Schrader. Other parallels can also be seen between *Isoetes malinverniana* and several aquatic and swamp species such as *Osmunda regalis* L., *Iris pseudacorus* L. and many Orchidaceae; or between *Adonis distorta* and *Papaver rhaeticum* Ler. and *Linaria tonzigii* Lona. The case of *Pinus leucodermis* might be considered to be a stalemate situation, but one that can be improved. Indeed, that which has been achieved so far for *P. leucodermis* should also be attempted for *Abies nebrodensis*. The *Gypsophila papillosa* example is a model for any very restricted endemic species in habitats which afford little or no inherent protection, a situation which is unfortunately more serious in this case than for other species of rocky habitats. Moreover, *G. papillosa* is also illustrative of the innumerable cases of species over collected for herbarium material. The case of *Haplophyllum patavinum* displays two principal features that of a species (not necessarily phenerogamous) whose life-cycle is conditioned by interspecific interactions or allelophaty, and that of a species damaged by modern agricultural practices such as chemical weeding (*Tulipa sylvestris* L.) the application of fertilizers (*Centaurea cyanus* L.) and drainage schemes (*Gladiolus palustris* Gaudin).

Perhaps one should also consider other special situations such as cytodemes, ecotypic variants, monospecific genera. The Italian (including Sardinia and Sicilia) flora consists of over 6,000 species, about 1,000 subspecies and about 1,300 varieties of which around 10% are endemic. It is evident that many of them correspond to as many distinct habitats. Clearly this brief treatment can neither illustrate nor give an exhaustive list of such taxa.

Acknowledgements

The authors wish to express their gratitude to Mr. and Mrs. Scudo for their help in preparing the English translation of this account.

References

Barbero, M. & Bono G. (1967). Groupement des rochers et éboulis siliceux du Mercantour-Argentera et de la Chaine ligure. Webbia 22: 437–467.

Beck Von Mannagetta, G. (1901). Die Vegetationsverhältnisse in den illyrischen Ländern. Leipzig: Engelmann.

Béguinot, A. (1905). Intorno a due *Gypsophila* della Flora Italiana. Bull. Soc. Bot. Ital., pp. 6–12.

Bianchini, F. (1974). *Gypsophila papillosa* Porta endemismo puntiforme. Boll. Mus. Civ. St. Nat. Verona 1: 531–534.

Burnat, E. (1901). Flore des Alpes Maritimes. Lyon: Georg. et Co.

Cappeletti, C. (1929). Sterilità di origine micotica nella *Ruta patavina* L. Annali Bot. 18: 145–166.

Corbetta, F. (1965). Osservazioni relative ad una nuova stazione di *Isoetes malinvernianum.* Natura et Montagna 5: 57–61.

Corbetta, F. (1967). Nuovi dati sulla distributione di *Isoetes malinvernianum* in Lomellina. Giorn. Bot. Ital. 101: 290–291.

Corbetta, F. (1968). Nuovi dati sulla distribuzione di *Isoetes malinvernianum* in Lomellina. Giorn. Bot. Ital. 102: 107–112.

De Filipps, R.A. (1980). *Cyperus* L. In: Tutin et al. (Eds.), Flora Europaea 5: 287. Cambridge: Cambridge University Press.

Del Grosso, F. & Pogliani, M. (1971). Studio cariologico di *Adonis distortus* Ten. Lavori Soc. Ital. Biogeogr. 2: 69–79.

Ferro, G. & Furnari, F. (1968). Flora e vegetazione di Stromboli (Isole Eolie). Arch. Bot. Biogeogr. Ital. 44: 59–85.

Filipello, S. et al. (Eds.), (1977). Carta delle conoscenze floristiche d'Italia. Informatore Bot. Ital. 9: 281-284.

Filipello, S. (1979). Projects, problèmes et aboutissements de la conservation de la flore et de la végétation en Italie. Webbia 34: 63–69.

Filipello, S. (Ed.), (1981). Atti del Seminario sulli Problemi scientifici et tecnici della conservazione del patrimonio vegetale. Collana Programma finalizzato Promozione qualità ambiente, AQ/ 1/96–110. OPTIMA Leaflet No. 114. Pavia.

Guadagno, M. (1931). Flora Capraearum Nova. Arch. Bot. 7: 7-38 and 145–176.

Guinochet, M. (1938). Etudes sur la végétation de l'étage alpin dans le bassin supérieur de la Tinée (Alpes Maritimes). Bosc. et Riou, Lyon.

Hofmann, A. (1958). Sull'ecologia di una nuova stazione avellinese di Pino loricato. Italia for. mont. 13: 67–76.

Mattirolo, O. (1912). Sull'endemismo dell'*Isoëtes malinvernianum* di Cesati et De Notaris. Annali Bot. 10: 129–146.

Merola, A. (1957). Ecologia del *Cyperus polystachyos* Rottb. nelle sue stazioni eterotopiche dell'Isola d'Ischia, Delpinoa 10: 21–92.

Pedrotti et al. (Eds.), (1971-1979). Censimento dei biotopi di rilevante interesse vegetazionale meritevoli di conservazione in Italia, Vols. 1. and 2. Soc. Bot. Ital., Camerino.

Pennanichini, V. & Bonin, G. (1975). *Pinus leucodermis* A. t. et Pinus nigra Arn. en Calabre Septentrionale. Ecologia Mediterranea 1: 35–61.

Porta, P. (1905). Appendix Florulae nostrae Tridentine, finitimisque in regionibus. Atti. Accad. Agiati 11: 208–216.

Quézel, P. (1951). L'association à *Galium baldense* var. tendae et *Saxifraga florulenta* Guinochet dans le massif de l'Argantera-Mercantour. Le monde des plantes, 274–275; 3–4.

Rosenkrantz, D. & Tosco, U. (1979). Le stazioni di *Isoetes malinverniana* Cesati et De Notaris del Basso Canavesse (Piemonte). Allionia 23: 155–160.

Tenore, M. (1830). Flora Napolitana. Napoli: Stamperia francese.

Zodda, G. (1957). La flora teramana: supplemento I. Webbia 13: 229–270.

Zodda, G. (1958). La flora teramana: supplemento II. Webbia 14: 213-242.

Istituto di Botanica
Università di Pavia
Pavia, Italy

CHAPTER 6

The Greek mountains

A. STRID and K. PAPANICOLAOU

1. Introduction

On the basis of *Flora Europaea,* Webb (1978) gave the total number of species of vascular plants in Greece as 4,100–4,250 (excluding apomictic taxa). Crete, which was considered separately in *Flora Europaea,* was estimated to have 1,700–1,850 species. There are no exact figures as to how many of these are absent from the rest of Greece, but a reasonable guess is about 250. The East Aegean Islands, which were not included in *Flora Europaea,* are likely to add another 100–200 species, and finally there are at least 150 species that are known to occur in Greece, although not listed in *Flora Europaea.* This brings the total number of species in Greece to just below the figures for Italy and Spain, to approximately the same level as those for Jugoslavia and France, and ahead of those for Bulgaria, Rumania, Austria and Albania.

The dissected topography of Greece, (Fig. 1), its complicated geological history and multitude of rock substrates (limestone, schist, granite, serpentine, etc.) have created a great variety of habitat conditions which form the basis for a rich and diversified flora. In absolute figures, the highest number of species are found in lowland habitats (maquis, phrygana) of the mainland, a fact which may be surprising to someone accustomed to seeing Greece in the summer when the plains and hills are baked by the sun and appear dry and desolate. In a study of the flora of Mt. Olympus, about 1,700 species were recorded on the mountain slopes (Fig. 2) and the alluvial plains between the Aegean Sea and the eastern foothills. Fully 1,100 of these, or nearly two thirds, are restricted to altitudes below the 1,200 m contour and only some 600 are found between 1,200 m and the summit, which is at 2,917 m. In fact the discrepancy is likely to be even larger, as the lowland habitats are generally less well explored than those at montane and alpine levels. In particular, it is the large number of spring-flowering annuals and geophytes that contribute to the rich flora of the plains and foothills. Generally speaking, these are species with a wide distribution, often occurring throughout the Mediterranean area, whereas species of montane levels often belong to the central European forest element, and the flora above the tree-line includes many species with a more restricted distribution (regional or local endemics).

Although they may be relatively poor in species in absolute terms, it is often islands and mountains that attract botanists, and with good reason, for here is the greatest chance of finding the special plants that occur nowhere else. They are great natural laboratories for the forces of evolution. Geographical and ecological isolation of small populations has procured numerous neo-endemics, and has segregated species that have evolved unique combinations of traits, either in the process of adaptation to special environmental conditions or as a result of random genetic or population drift. At the same time, islands and mountains have afforded shelter to ancient species that have other-

Gómez-Campo, C. (ed.), Plant conservation in the Mediterranean area.
© 1985, Dr W. Junk Publishers, Dordrecht. *ISBN 90 6193 523 7.*

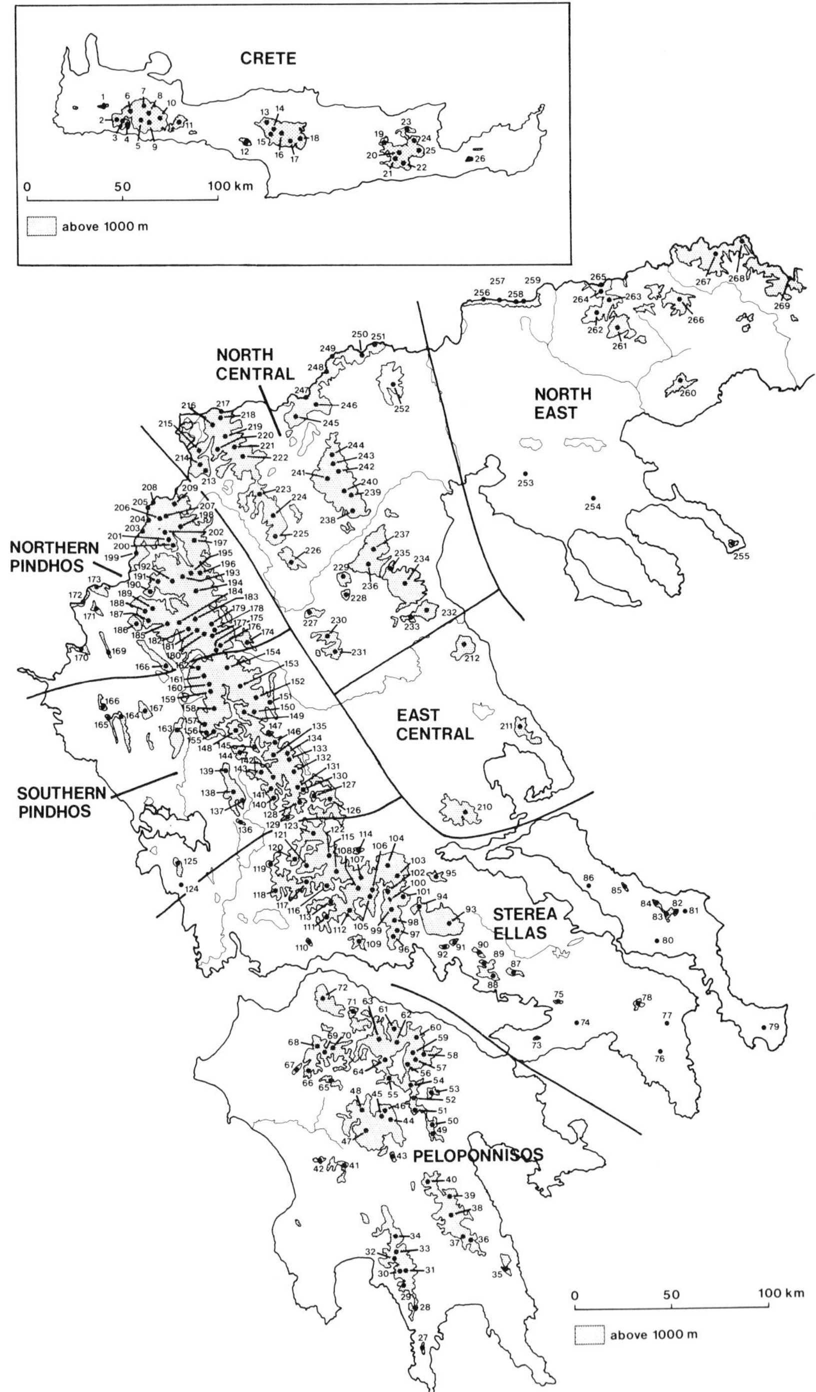

Fig. 1. Map of mainland Greece and Crete prepared for the Mountain Flora of Greece project. The numbers refer to a list of mountain names (only a few are mentioned in the text).

arc and the Cyclades is thus about 237.

With the completion of Volume 1 of the *Mountain Flora of Greece* (Strid, 1984), it is now possible to make reasonably accurate calculations concerning distribution patterns and endemism of Greek mountain plants. The *Flora* uses two categories of taxa:

a) 'Full members', defined as species and subspecies chiefly found above 1,800 m or in open, treeless habitats above 1,500 m.

b) 'Associate members' (which are more summarily treated), defined as species and subspecies ascending above 1,800 (1,500) m, but mainly found at lower altitudes.

Volume 1 of the *Flora* includes 926 numbered taxa (species and subspecies), 768 of these being 'full members' and 158 'associate members'. The sequence of families, genera and species follows that of *Flora Europaea;* our Volume 1 covers vascular cryptograms to Plumbaginaceae, i.e. Vol. 1 (1964), Vol. 2 (1968) and the beginning of Vol. 3 (1972) of *Flora Europaea.* A comparison shows that of our 926 taxa no less than 304, or c. 33%, were either not recorded from Greece in *Flora Europaea* or were recorded under a different name and/or with a different rank. If 'full members' only are considered, the figures are 282 out of 768, or about 37%. Finally, 30 taxa listed for Greece in *Flora Europaea* have been excluded, either because the records have been demonstrated to be erroneous or because they have been reduced to synonymy. Since we have seldom used narrower concepts of species and subspecies than in *Flora Europaea,* these figures serve to demonstrate the considerable increase in floristic and taxonomic activity that has taken place in Greece since - and largely inspired by the publication of the early volumes of *Flora Europaea.*

The Greek mountain flora is composed of a number of phytogeographical elements. Many widespread arctic-alpine or boreal species reach their southernmost localities on granitic or schistose mountains in northern Greece (e.g. *Saxifraga stellaris* L., *Viola palustris* L. and *Juncus trifidus* L.), and more rarely on limestone mountains (e.g. *Dryas octopetala* L.). Others are widespread oro-Mediterranean species which, in some cases, extend from the Pyrenees to the Caucasus (thus often variable and taxonomically problematic). Of particular interest in this context are taxa (especies and subspecies) of restricted distribution, data on which have been summarized in Table 1. These data should be qualified as follows:

a) They are based only on Volume 1 of the *Mountain Flora of Greece.* As this represents approximately 50% of the total flora (one more volume is planned), the absolute numbers should be roughly doubled, whereas the percentage figures (in parentheses) are likely to be essentially correct;

b) The data are based on 'full members' (768 taxa) and on those 'associate members' for which the distribution in Greece has been determined with reasonable accuracy (48 taxa), or 816 taxa in all. The true absolute numbers should thus be slightly higher than those given in the Table, but again the percentage figures will scarcely be affected.

For the purpose of the *Mountain Flora,* Greece has been divided into eight more or less natural geographical regions as shown in Figure 1 (note, for instance, that our Sterea Ellas does not quite correspond to the administrative unit with the same name). Problems arise in subdividing the more or less continuous chain of mountains running SSE to NNW on the mainland. The border between Sterea Ellas and S. Pindhos follows a valley system just north of Timfristos, and the border between south and north Pindhos is close to the Katara Pass, and more or less coincides with the border between a limestone area in the south and a serpentine area in the north. The small figures on the map refer to a list of mountain names (not included here).

The Table should be read as follows: The figures 14 and (21; 9) in the upper left-hand corner indicate that on Crete 14 taxa belong to the element 'Balkan Peninsula + Anatolia'. This constitutes 21% of the 'narrowly distributed' taxa found on Crete (68 in all), or 9% of the total mountain flora of Crete (149 taxa). The figures 53 and (13; 6) in the lower lefthand corner similarly indicate that 53 out of 418 (816) taxa, or 13 (6)%, belong to this element when the total mountain flora of Greece is considered. Some details should be pointed out:

Table 1. Endemism in the Greek mountain flora. Further explanation in the text.

Area in Greece	Phytogeogr. elem. Balkan Penins. + Anatolia [a]	Balkan Penins. + Italy [b]	Balkan endemics [c]	Greek endemics	Single-area endemics	Single-mountain endemics [d]	Total
Crete	14 (21; 9)	6 (9; 4)	4 (6; 3)	7 (10; 5)	26 (38; 17)	11 (16;7)	68 -149
Peloponnisos	23 (17; 7)	17 (12; 5)	35 (26; 11)	41 (30; 13)	10 (7; 3)	11 (8; 4)	137-311
Sterea Ellas	24 (14; 6)	26 (15; 6)	61 (35; 15)	46 (27; 11)	9 (5; 2)	7 (4; 2)	173-418
S Pindhos	17 (14; 5)	28 (23; 8)	60 (50; 17)	16 (13; 5)	0 (0; 0)	0 (0; 0)	121-348
N Pindhos	19 (12; 4)	30 (19; 7)	84 (53; 19)	17 (11; 4)	5 (3; 1)	4 (3; 1)	159-439
East Central	10 (20; 5)	8 (16; 4)	25 (51; 13)	6 (12; 3)	0 (0; 0)	0 (0; 0)	49-190
North Central	23 (13; 5)	25 (14; 5)	98 (54; 20)	13 (7; 3)	1 (1; 0)	21 (12; 4)	181-495
North East	26 (22; 7)	12 (10; 3)	65 (55; 17)	7 (6; 2)	1 (1; 0)	8 (7; 2)	119-388
Total	53 (13; 6)	37 (9; 5)	155 (37; 19)	59 (14; 7)	52 (12; 6)	62 (15; 8)	418-816

[a] Including taxa extending slightly beyond the borders of E or S Antatolia.
[b] As defined here, Italy includes Sicily, Sardinia and Corsica.
[c] Including taxa extending locally to SW Romania or the Carpathians.
[d] Large massifs such as Levka Ori and Taygetos have been defined as a 'single mountain'. Taxa endemic to border mountains such as Kajmakcalan have been classified as 'single-mt. endemics' even if they occur on both sides of the border.

a) The 'Balkan Peninsula + Anatolia' element is approximately equally strong in Crete and in North East - 21 (9)% and 22 (7)%, respectively - indicating that the southern and northern connections between the mountain floras of Greece and Anatolia are of roughly equal importance.

b) The 'Balkan Peninsula + Italy' element is considerably stronger in the Pindhos than in Peloponnisos and Crete, indicating that connections with Italy are mainly via a northern route (the more southerly taxa are often likely to represent an ancient oro-Mediterranean relict. element).

c) Balkan endemics constitute the largest element, 155 taxa or 37 (19)%. In North East, North Central, north Pindhos and south Pindhos 50–55 (17–20)% of the taxa belong to this element. It decreases in importance in Sterea Ellas and Peloponnisos, and is very poorly represented in Crete.

d) Greek endemics are particularly abundant in Peloponnisos and Sterea Ellas. Often taxa are restricted to these two regions, and the Gulf of Korinthos appears to be a very weak phytogeographical barrier. The serpentine endemics of the northern Pindhos, which also have fairly small distribution areas, usually extend into Albania (and sometimes further north) and have thus been classified as Balkan endemics. The same applies to limestone species restricted to northeastern Greece and southwestern Bulgaria.

e) Crete is far ahead of the other regions with respect to the percentage of single-area endemics (if lowland taxa were included the difference would certainly be even larger). Although Peloponnisos and Sterea Ellas are rich in narrow endemics, they are frequently found on both sides of the Gulf of Korinthos (see above), and the numbers of single-area endemics are therefore not particularly high.

f) Single-mountain endemics are chiefly found on some of the large and/or isolated limestone massifs such as Levka Ori (Crete, nos. 2–11), Taygetos (Peloponnisos, nos. 28–34), Olimbos (North-Central, no. 234) and Athos (Northeastern, no. 255), less frequently on serpentine (Smolikas, N Pindhos, nos. 191–193; Vourinos, North-Central, no. 226), granite (Varnous, North-Central, nos.

216–219) or micaceous schist (Kajmakčalan, North-Central, no. 247).

4. Habitat of the endemics

The largest concentration of endemics in Greece is found in rock crevices and screes on limestone or serpentine. Triassic and Jurassic limestones are the most widespread substrate in the Greek mountains. Limestone mountains are generally dry; rain-water or snow melt-water disappears quickly into cracks and fissures, and sheer rocks and scree fields are the dominant habitats. Moderately sloping ground with a mixture of fine soil, gravel and coarser material is subject to solifluction: in the spring when the uppermost layers are watersoaked and the underlaying strata still frozen there is a slow downward sliding of the surface soil material, resulting in terracing. Such slopes support grassland communities with tussocks of wiry grasses growing on the edges of the terraces, the horizontal 'steps' being almost devoid of vegetation. A continuous grass cover is found only in patches in shallow depressions or on nearly horizontal ground where a layer of fine-particled soil has accumulated. The various grassland communities are generally poor in endemics, particularly in local endemics.

Serpentine, a dark mineral containing hydrous magnesium silicate and often marked by shades of green or purple, occurs in large quantities in the mountains of northwestern Greece and scattered at lower altitudes in Sterea Ellas and on the island of Euboea. The serpentine areas of northwestern Greece extend through Albania and are more or less contiguous with the large serpentine areas of Serbia and Bosnia. Extensive scree fields and a flora relatively poor in species, but rich in endemics, is characteristic of the serpentine mountains.

Gneiss, granite and micaceous schist make up some mountains, chiefly in northern Greece. The vegetation often forms a dramatic contrast to that of limestone or serpentine mountains. The substrate has a much greater water-holding capacity; brooks and seepage meadows are frequent at alpine levels and there is generally a continuous grass cover. There are relatively few endemics. The flora has a central European character and several arctic-alpine species reach their southernmost limit of distribution on granitic or schistose mountains in northern Greece.

Levka Ori or the White Mountains on the map in southwestern Crete certainly have the largest concentration of endemics anywhere in Greece. It is a typical limestone mountain appearing exceedingly dry and desolate in the summer but is nevertheless something of a botanist's paradise. The famous Samaria gorge (which is now a national park) and the mountains above it abound in rare and interesting species, many of them extremely local, for example, *Silene variegata* (Desf.) Boiss. & Heldr., *Dianthus juniperinus* Sm., *Sanguisorba cretica* Hayek, *Onobrychis sphaciotica* W. Greuter, *Satureja spinosa* L., *Anchusa cespitosa* Lam., *Scabiosa albocincta* W. Greuter, *Asperula idaea* Halácsy, *Symphyandra cretica* DC. and *Crepis auriculifolia* Sieber ex Springel. It is a characteristic feature of Crete that many of the endemics occur at low altitudes and that a montane zone is virtually lacking.

The large limestone massifs of the Greek mainland, among them Taiyetos, Killini or Zyros, Chelmos or Aroania, Parnassos Giona, Timfi, Olympus and Athos, also have their fair share of local endemics, mostly in rock crevices and screes at alpine levels. Also a relatively small but isolated limestone mountain such as Dirfys on Euboea has a surprising number of local endemics (10–15 species, several of then at rather low altitudes and thus not included in the survey in Table 1); the vegetation in the summit area (1,500–1,743 m) bears a striking resemblance to that of Mount Athos (2,033 m), although the species are largely different. Olympus, which at 2,917 m is the highest mountain in Greece and consists almost entirely of limestone, has about 21 well-defined local endemic species and Athos, which is much smaller but more isolated, has about 16. In both cases several more have been described, but a critical review has either reduced them to subspecific status or revealed that they occur on other mountains as well. On both Olympus and Athos, about half of the local endemic species are confined to rock-crevices and screes at alpine levels; the others are found in a variety of habitats, but usually in dry, rocky places.

A number of interesting species occur in the serpentine areas of northwestern Greece. They are rarely confined to a single mountain, however, but usually extend into Albania and often as far as the serpentine areas of Serbia and Bosnia. Among such species are: *Silene schwarzenbergeri* Halácsy, *Bornmuellera baldaccii* (Degen) Heywood, *B. tymphaea* Hausskn, *Peltaria emarginata* (Boiss.) Hausskn, *Alyssum heldreichii* Hausskn, *Fumana bonapartei* Maire & Petit, *Cistus albanicus* Warb. ex Heywood, *Thymus teucrioides* Boiss & Spruner, *Campanula hawkinsiana* Huasskn & Heldr. and *Fritillaria epirotica* Turrill ex Rix. Among the relatively few local endemics are *Cerastium smolikanum* Hartvig, on Smolikas (191–193), and *Onosma elegantissima* Rech. fil & Goul, on Vourinos (226). This last mountain, which is of relatively low altitude (1,866 m) has one of the richest serpentine floras in Greece, which includes several local endemics.

Mountains of granite, micaceous schist and other acid rocks in northern Greece are rich in species but poor in local endemics. Such mountains are Varnous or Peristeri (216–219), Voras or Kajmakcalan (247), Pieria (237) and Kerkini or Belles (256–259). These mountains are rich in central Balkan endemics and have many phytogeographical connections with the Carpathians. There are also many central European species and some arctic-alpine species at their southernmost limit of distribution, e.g. *Cardamine impatiens* L., *Drosera anglica* Hudson, *Chrysosplenium alternifolium* L., *Viola palustris* L., *Ligusticum mutellina* (L.) Crantz, *Vaccinium vitis-idaea* L., *Veronica bellidiodes* L., *Melampyrum sylvaticum* L., *Centaurea nervosa* Willd., *Polygonatum verticillatum* (L.) All., *Festuca gigantea* (L.) Vill., *Juncus trifidus* L., *Eriophorum latifolium* Hoppe., *E. vaginatum* L., *Rhynchospora alba* (L.) Vahl. and *Carex sempervirens* Vill. Some local endemics have recently been described from Kajmakcalan *(Dianthus kajmaktzalanicus* Micevski), *Silene horvatii,* Micevski, *Ranunculus cacuminis* Strid & Papanicolaou), but considering the richt and varied flora of this mountain (which is made up chiefly of micaceous schist), the number of local endemics is low.

5. Some case histories of rare and threatened species in the Greek mountains

5.1 Helichrysum sibthorpii *Rouy (Compositae). (Fig. 3)*

Syn.: *Gnaphalium virgineum* Sibth. & Sm., *Helichrysum virgineum* (Sibth. & Sm.) Griseb.

The genus *Helichrysum* has about 20 species in the Mediterranean area. Many related species occur in the winter-rainfall areas of South Africa and Australia. Some are popular in rock gardens and everlasting flowers in dried flower arrangements. *H. sibthorpii* is a small species with woolly leaves and a few white capitula per stem. It grows in rock crevices in the summit area of Mt. Athos (no. 255 on map) where it was discovered by Sibthorp. It is illustrated in the classic work *Flora Graeca* of Sibthorp and Smith (Vol. 9, tab. 860, 1839).

5.1.1 Affinities

Helichrysum sibthorpii belongs to the section *Virginea* DC.) Fiori which comprises at least five species in the eastern and central Mediterranean area; all are local endemics and apparently relicts. In addition to *H. sibthorpii,* these are *H. amorginum* Boiss & Orph. from the eastern Cyclades (Amorgos and adjacent small islands), *H. doerfleri* Rech. fil. from eastern Crete and *H. frigidum* (Labill.) Willd. from the mountains of Corse and Sardegna. A fifth species occurs in Lebanon and still another, *H. taenari* Rothm., has been described from southern Peloponnisos. The latter has only been collected once and the material has been lost; it may not be specifically distinct from *H. amorginum.*

5.1.2 Distribution and ecology

H. sibthorpii is a very rare plant, restricted to the alpine region of Mt. Athos between c. 1,750 and 2,000 m. It grows in small crevices of nearly vertical limestone rocks, together with species such as *Viola delphinantha* Boiss., *Arabis bryoides* Boiss., *Aethionema orbiculatum* (Boiss.) Hayek and *Linum elegans* Spruner ex Boiss. It flowers from the middle of June to the middle of August. The

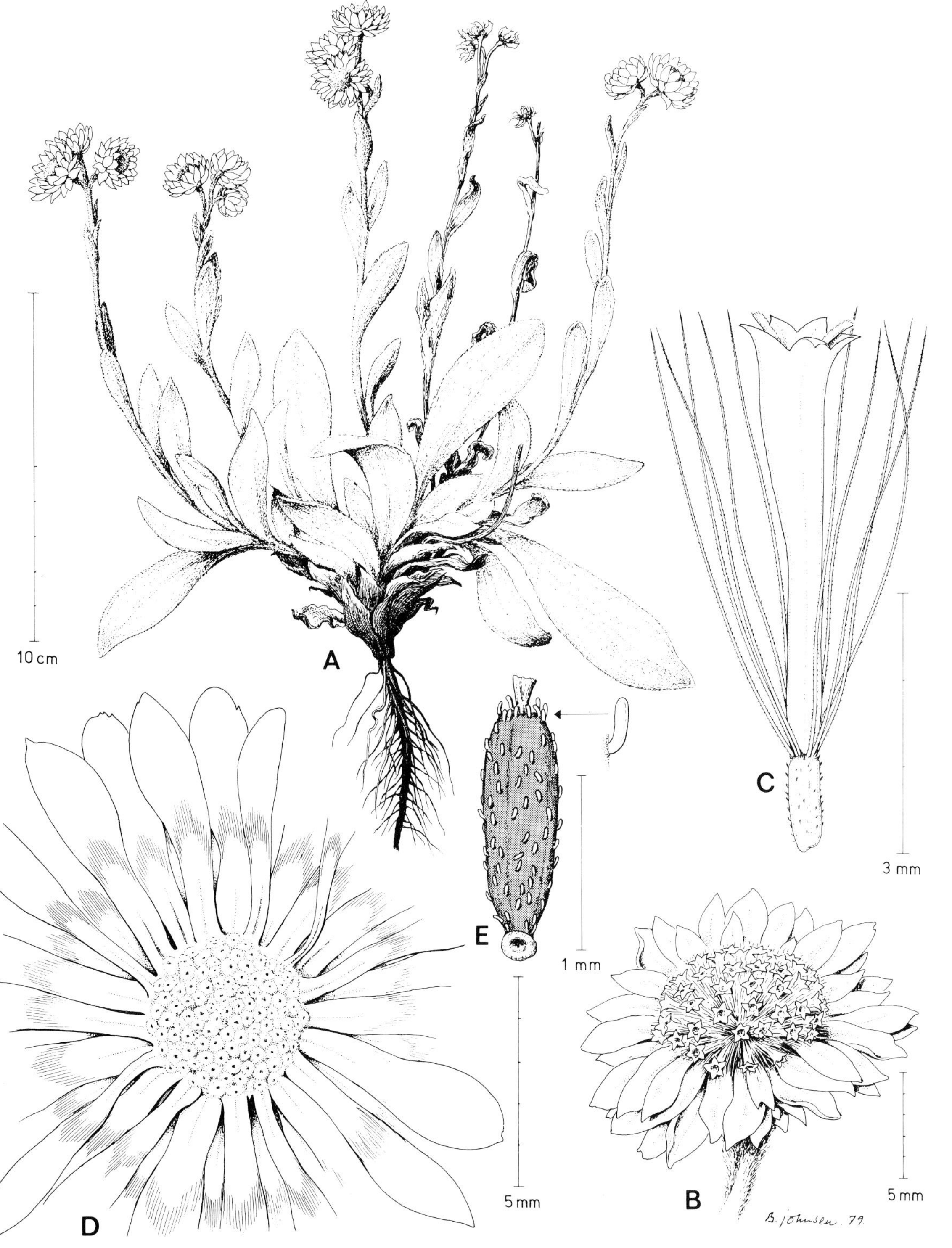

Fig. 3. Helichrysum sibthorpii Rouy. A: Habit. - B: A flowering capitulum, completely expanded. - C: Floret. - D: Fruiting capitulum (achenes removed) to show the structure of the receptacle. - E: Achene with scattered hyaline glands.

other species of the section *Virginea* are found in similar habitats.

5.1.3 Conservation status

H. sibthorpii is one of the rarest endemics of Mount Athos. It is scattered in the summit area and never found in large quantities. Because of the special status of the Athos peninsula, as a self-governing monastic community, the number of outside visitors is limited, and under present conditions there is no immediate threat to the mountain's endemics. The collection of *H. sibthorpii* should be avoided; there is already ample material in several herbaria and it is cultivated in some botanical gardens. The local name of the plant is *amarandos* ('not shrivelling'); the monks occasionally pick it for decoration in churches.

This species is relatively easy to cultivate in an alpine greenhouse and flowers readily. Outdoors in Copenhagen, it suffers from the wet Danish climate, especially in the winter, and will be killed if covered by wet snow. Many other Mediterranean chasmophytes, especially those with tomentose leaves, show the same reaction to an Atlantic climate and, if they are to be grown in a rock garden, have to be covered by glass or otherwise protected from excessive moisture.

5.2 Silene orphanidis *Boiss. (Caryophyllaceae) (Fig. 4)*

Silene (catchflies and campions) is one of the largest genera of flowering plants in the Mediterranean area, where there are many local endemics. *S. orphanidis* is another Mt. Athos endemic that is restricted to a few sites in the summit area, where it forms loose cushions in rock crevices. It has relatively large flowers on slender, 1- to 3-flowered stems. The petals are white above and pinkish beneath.

5.2.1 Affinities

There are several related species in the Balkan Peninsula, although *S. orphanidis* is nevertheless fairly distinct. It has a characteristic habit, being rather densely caespitose with woody stem bases of a peculiar gnarled appearance. Other characteristic features are the small uppermost leaves inserted just below the calyx, the relatively long calyx and large petals, long glabrous carpophore and exserted capsule. Its closest relatives are *S. pindicola* Hausskn. An endemic of northwestern Greece found chiefly on serpentine, and *S. waldsteinii* Griseb, which is fairly widespread in the central parts of the Balkan Peninsula (Albania, Greece, Jugoslavia and Bulgaria) and grows chiefly on schistose rocks. All belong to the *S. multicaulis* group in a wider sense.

5.2.2 Distribution and ecology

S. orphanidis is one of the rarest endemics of Mt. Athos. It is found in the alpine region between c. 1,750 and 2,000 m and is restricted to a few sites on the southwest and northeast sides. It grows in deep, slightly shaded rock crevices or at the base of larger rocks. Accompanying species are several Athos endemics, e.g. *Helichrysum sibthorpii, Anthemis sibthorpii* Griseb, *Asperula suberosa* Sibth. & Sm. *Campanula albanica* ssp. *sancta* (Hayek) Podl., *Viola athois* W. Becker, *Aethionema orbiculatum, Acinos alpinus* ssp. *nomismophylla* (Rech. fil.) Lebl., as well as other interesting species such as *Viola delphinantha, Potentilla speciosa* Willd., *Prunus prostrata* Labill, *Arabis bryoides, Anthyllis montana* L. and *Campanula orphanides* Boiss. The related *S. multicaulis* Guss. ssp. *genistifolia* (Hal.) Melzh. (= *S. genistifolia* Hal.) may occur at alpine levels but is more often found at lower altitudes. It was believed to be an Athos endemic, but plants from Thessaly, Euboea and the Northern Sporades are now considered to belong to the same taxon (Melzheimer, 1977). It grows on grassy slopes and rock ledges rather than rock crevices, and can be distinguished by the loosely caespitose habit, non-woody stem bases, shorter calyx, smaller petals and shorter, pubescent carpophore. *Silene orphanidis* flowers in late, summer from mid-July to the end of August; *S. multicaulis* ssp. *genistifolia* somewhat earlier.

5.2.3 Conservational status

There is no immediate threat to *S. orphanidis* other than the fact that it is very rare, restricted to a single mountain-top and consequently vulner-

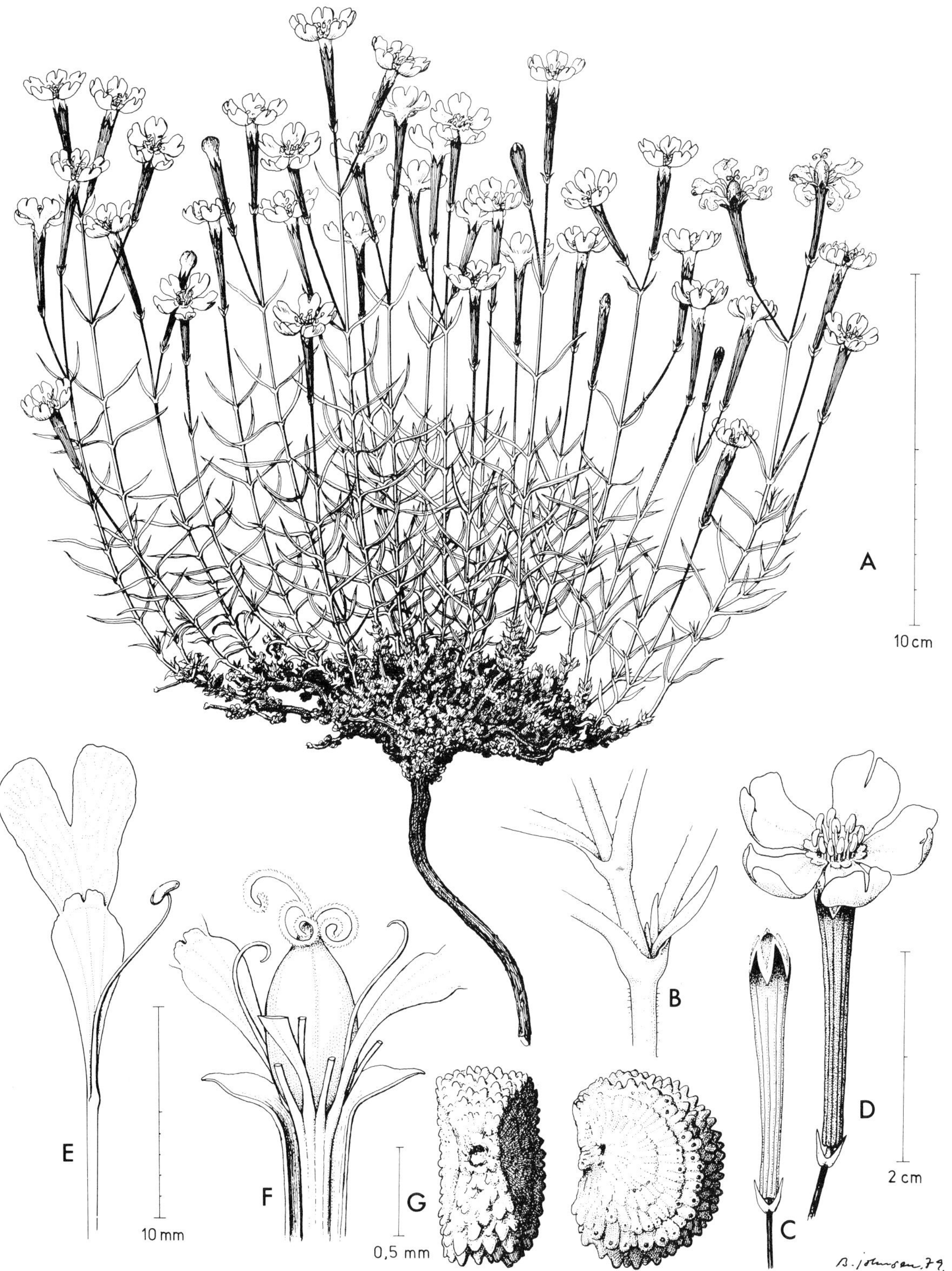

Fig. 4. Silene orphanidis Boiss. A: Habit: note basal parts. - B: Branching of stem. - C and D: Bud and flower; note the pair or small leaves just below the calyx. - E: Petal and stamen. - F: Young capsule; calyx opened and some petals and stamens removed - G: Seed seen in ventral and lateral view.

able to extinction by collecting. It is easily grown from seed and has flowered well for summers in a rockery in the Copenhagen Botanical Garden. In cultivation it has a tendency to be biennial and it is thus necessary to collect seeds for its propagation.

5.3 Anthemis sibthorpii *Griseb. (Compositae) (Fig. 5)*

Syn.: *Santolina montana* Sibth. & Sm., *Anthemis orientalis* (L.) Degen ssp. *sibthorpii* (Griseb.) Hayek

Anthemis (chamomiles) is a taxonomically difficult genus with many local variants in the Mediterranean area. *A. sibthorpii* is characterized by completely glabrous leaves and capitula without ligules. It is found in the summit area of Mt. Athos, where it prefers shady places at the base of rocks. It is a short-lived perennial and relatively easy to keep in cultivation.

5.3.1 Affinities

A. sibthorpii belongs to the widespread and very polymorphic *A. carpatica-cretica* complex, which extends from the Pyrenees to the Caucasus. The taxa of this complex have been variously treated at specific or infraspecific level by different authors. *Flora of Turkey* (Davis, 1975), for instance, lists no less than 11 subspecies under *A. cretica* L. It has been suggested that *A. sibthorpii* should be reduced to a subspecies of *A. carpatica* Willd., but it has a number of very distinctive features which seem to warrant its recognition as a separate species. Among these characters are: whole plant glabrous, capitula always lacking ligules, receptacle conical and involucral bracts with light brown margins.

The most closely related taxon appears to be *Anthemis carpatica* ssp. *petraea* (Ten.) Fernandes, which occurs in Italy (central Apennini) and has recently also been collected on Mt. Timfi in Northwestern Greece. This is also glabrous, but has well-developed ligules, attenuate outer involucral bracts, longer receptacular scales and somewhat larger (2. 5–3 mm) achenes.

A superficially similar species, often lacking ligules, is *A. tenuiloba* (DC.) Fernandes which is fairly widespread in the Balkan Peninsula. It is appressed pubescent with greyish leaves, narrow hyaline margins of the involucral bracts, acutely conical receptacle and entire obtuse receptacular scales.

5.3.2 Distribution and ecology

A. sibthorpii is found in shady, slightly damp places, usually at the base of north-facing rocks from c. 1,700 m to the highest summit. Immediately north of the summit at 2,020–2,030 m it grows together with *Saxifraga juniperifolia* ssp. *sancta* (Griseb) D.A. Webb, *Campanula albanica* ssp. *sancta, Arenaria biflora* L., *Galium demissum* Boiss. and *Arabis alpina* L.

5.3.3 Conservational status

Anthemis sibthorpii is rare and seems to be restricted to a few sites in the alpine zone. Like the other Athos endemics it is scarcely threatened unless some drastic development takes place or collecting increases significantly. It is easily propagated from seeds and grows well in our rock garden in Copenhagen; being a plant of rather damp habitats it has no difficulties in adjusting to an Atlantic climate. In cultivation it is usually biennial but self-propagates vigorously.

5.4 Ligusticum olympicum *F.A. Novak (Umbelliferae) (Fig. 6)*

This is a rare inconspicuous umbelliferous plant of damp calcareous screes in the summit area of Mt. Olympus (no. 234 on map). The stems are creeping or ascending and only up to 15 cm long. Each umbel is 7–12 mm wide with 2–4 primary rays and tiny, white or pinkish flowers. The fruit is ellipsoid and has conspicuous winged ridges.

5.4.1 Affinities

Ligusticum olympicum is taxonomically isolated and phytogeographically interesting. Its closest relatives appear to be a small group of species occurring in Afghanistan and Soviet Central Asia, viz *L. gayoides* (Regel & Schmalh.) Korov., *L. irramosum* Rech. fil. & Riedl and *L. steineri* Podlech. Having these far eastern affinities *L. olympicum* is likely to be an ancient and formerly more

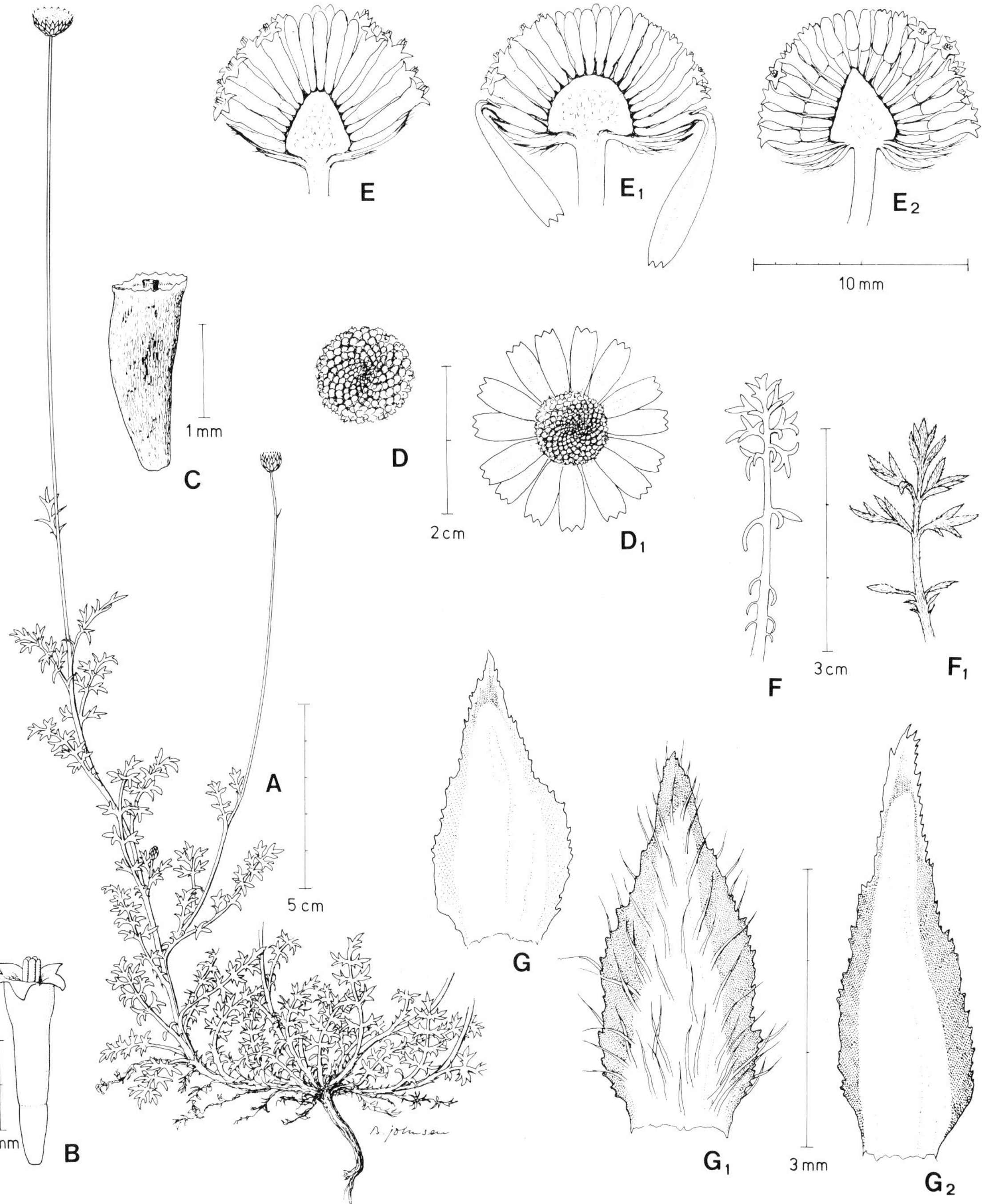

Fig. 5. Anthemis sibthorpii Griseb. A: Habit. - B: Floret. - C: Achene. - D: Capitulum seen from above. - E: Longitudinal section of capitulum to show shape of receptacle. - F: Basal leaf. - G: Involucral bract. - D_1 - G_1: Corresponding parts of *Anthemis carpatica* Willd. (Strid & Papanicolaou 12619 from Mt. Kajmakcalan) for comparison. - E_2: Longitudinal section of a capitulum of *Anthemis tenuiloba* (DC.) Fernandes (Hartvig et al. 6717 from N Pindhos). - G_2: Involucral bract of *Anthemis carpatica* Willd. ssp. *petraea* (Ten.) Fernandes (Strid & al. 15559 from Mt. Timfi).

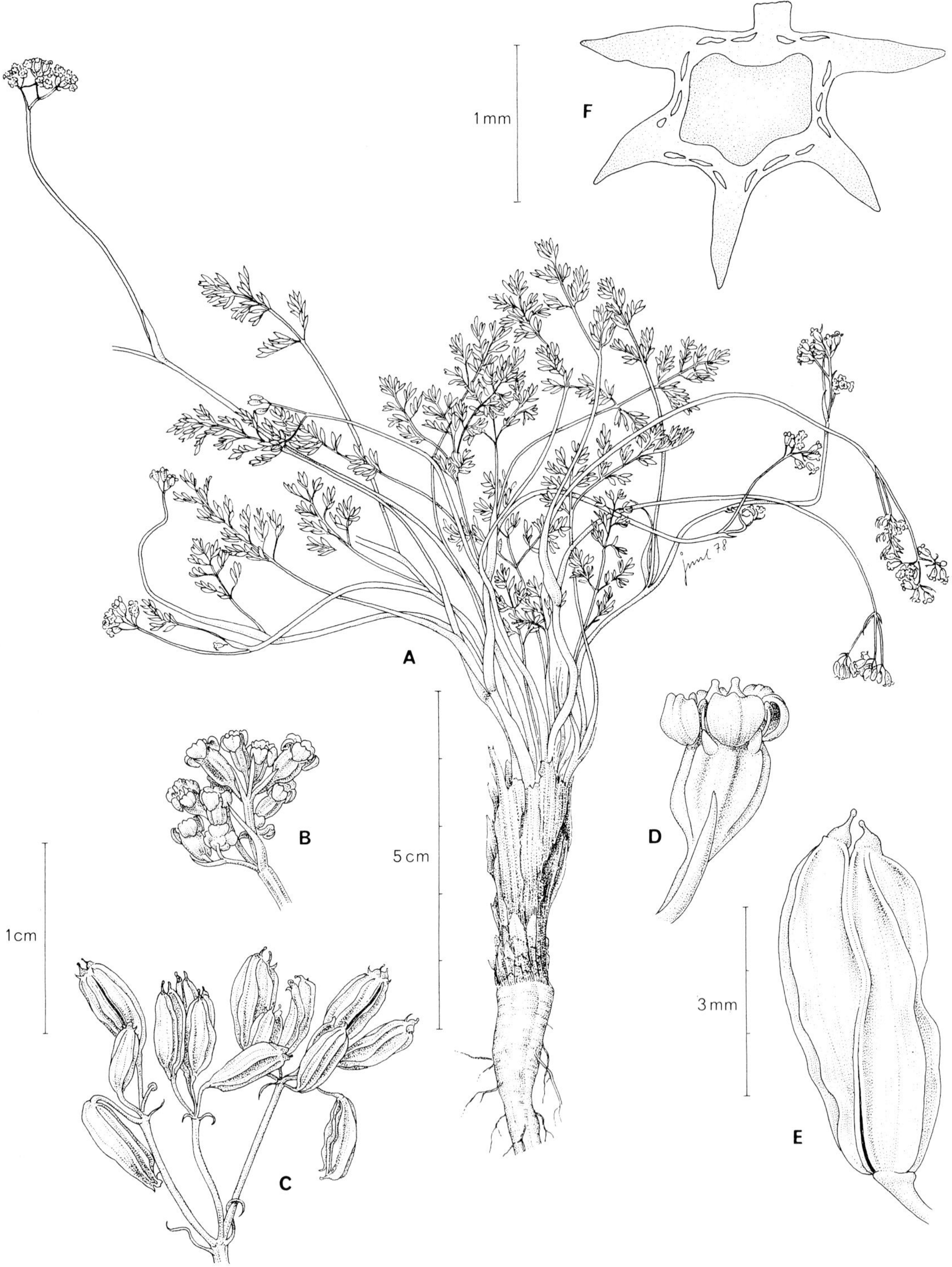

Fig. 6. Ligusticum olympicum F.A. Novák. A: Habit. - B: Flowering umbel. - C: Fruiting umbel. - D: Individual flower. - E: Fruit. - F: Transverse section through a mericarp; note the resin canals.

widely distributed relict endemic. From a phytogeographical point of view it may be compared to some recently discovered cases of Irano-Turanian species with isolated occurrences in the Greek mountains, such as *Trigonella strangulata* Boiss. and *T. velutina* Boiss. (Greuter, 1975), *Thesium brachyphyllum* Boiss. (Aldén, 1976), *Thlaspi kotschyanum* Boiss. & Hohen (Gustavsson, 1978) and *Veronica bornmuelleri* Hausskn. (Hartvig, 1979).

L. olympicum was first collected by H. von Handel-Mazzetti in 1927, but was misidentified as *Carum heldreichii* (cf. Hayek, 1928 p. 290). Somewhat later the collection came to the attention of F.A. Novák who realised that it represented a new species of *Ligusticum*. His description (in *Acta Botanica Bohemica* 8 : 5, 1929) has been overlooked by subsequent authors, and the species was not mentioned in *Flora Europaea*.

5.4.2 Distribution and ecology

L. olympicum is endemic to Mt. Olympus and confined to somewhat damp calcareous scree and gravel between c. 2,550 and 2,750 m. It is found in association with species such as *Arenaria conferta* Boiss., *Ranunculus brevifolius* Ten., *Corydalis bulbosa* ssp. *blanda* (Schott) Chater., *Cardamine carnosa* Waldst & Kit., *Sesleria korabensis* (Küm. & Jav. Deyl., and *Poa pirinica* (Staj. & Acht.). The taproot is reminiscent of that of species such as *Beta nana* Boiss. & Heldr. and *Trinia guicciardii* (Boiss. & Heldr.) Drude, which occur in the same area but prefer places where the scree and gravel is mixed with larger quantities of fine soil.

L. olympicum flowers in June and July; ripe fruits may be found from the middle or end of July. This easily overlooked species has been collected only a few times.

5.4.3 Conservational status

Being an inconspicuous plant *Ligusticum olympicum* is unlikely to be picked or even observed by the casual mountaineer. Though vulnerable because of its restricted distribution, it is under no immediately threat.

5.5 Cerastium theophrasti *Merxm. & Strid (Caryophyllaceae) (Fig. 7)*

This recently described species of mouse-ear chickweed belongs to the arctic-alpine *C. alpinum* group and differs from other species in this complex, e.g.

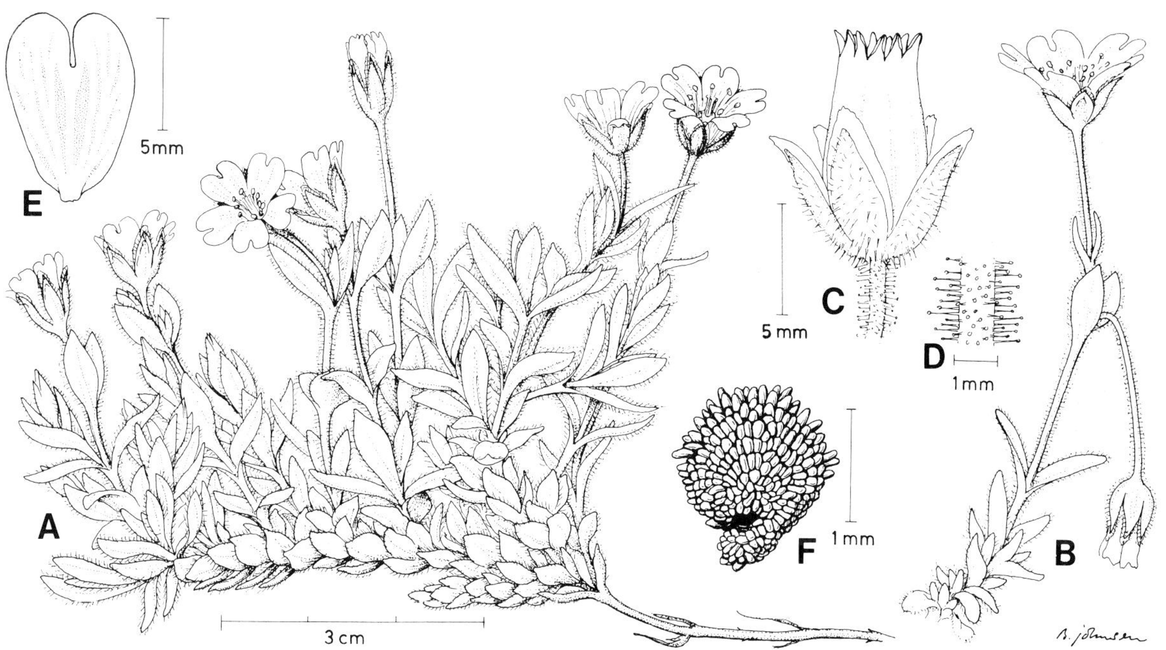

Fig. 7. Cerastium theophrasti Merxm. & Strid. A: Habit. - B: Inflorescence; not deflexed pedicel after flowering. - C: Capsule. - D: Part of pedicel to show the patent glandular hairs. - E: Petal. - F: Seed.

in the glandular indumentum and larger seeds as well as in the diploid chromosome number (all other species are polypoid). It occurs on screes and in rock crevices in the alpine zone of Mt. Olympus (2,600–2,900 m).

5.5.1 Affinities

C. theophrasti is a recently described species (Merxmüller & Strid, 1977) of the arctic-alpine *C. alpinum* group, a circum-polar polyploid complex which is also represented in most mountains of central Europe. *C. theophrasti* has a number of distinctive characters, *viz.* the large seeds with long acute tubercles, the complete absence of long soft eglandular hairs, the relatively short petals, the often single-flowered inflorescences and the deflexed fruiting peduncles. Another important feature is the fact that *C. theophrasti* is diploid (2n = 36), whereas all populations of *C. alpinum* and the related *C. arcticum,* as far as is known, are polyploid (2n = 54, 72 or 108).

The recently described *C. runemarkii* Möschl & Rech. fil., known from montane levels on the south Aegean islands of Naxos and Euboea, is also a diploid belonging to the *C. alpinum* group in a wide sense; it is distinguished from *C. theophrasti,* by for instance its smaller seeds and completely different indumentum. *C. hekuravense* Jáv., described from Mt. Hekurave in Albania, may also be related.

Phytogeographically *C. theophrasti* and *C. runemarkii* are of considerable interest as they are local species representing the previously unknown diploid chromosome number in a widespread polyploid complex and are thus likely to be relict taxa.

5.5.2 Distribution and ecology

C. theophrasti grows on calcareous screes, and occasionally in rock crevices, in the summit area of Mt. Olympus (2,600–2,900 m). It is associated with other Olympian endemics such as *Alyssum handelii* Hayek., *Rhynchosinapis nivalis* (Boiss. & Heldr.) Heywood., *Potentilla deorum* Boiss. & Heldr. *Achillea ambrosiaca* Boiss. & Heldr.) Boiss., and *Festuca olympica* Vetter. Other accompanying species are: *Minuartia verna* ssp. *idaea* (Halacsy) Hayeck., *Paronychia rechingeri* Chaudhri, *Cardamine carnosa, Saxifraga spruneri* Boiss., *S. sempervivum* C. Koch., *Euphorbia capitulata* Reichenb., *Thymus cherlerioides* Vis., *Linaria alpina* (L.) Miller., *Veronica thessalica* Bentham, *Galium degenii* Bald. ex Degen and *Sesleria korabensis.* It flowers from the end of June to the beginning of August.

5.5.3 Conservational status

C. theoprasti is widespread summit area of Mt. Olympus, although it is not particularly abundant. Nevertheless it is scarcely threatened at the present time. It is fairly easy to grow and is usually biennial in cultivation. Plants cultivated outdoors in Copenhagen flower from the end of May to Mid-June. The flowers are protandrous and have a faint, but distinct, sweet scent. Seed-setting occurs rapidly after flowering; in cultivated material ripe seeds are produced in about three weeks.

5.6 Aethionema carlsbergii *Strid & Papanicolaou (Cruciferae) (Fig. 8)*

Syn.: *Ae. orbiculatum* (Boiss.) Hayek auct., *Eunomia orbiculata* Griseb. auct., non *Crenularia orbiculata* Boiss.

This species is known only from the summit area of Mt. Taiyetos (nos. 28–34 on map) in Peloponnesos. It was first found here in 1829 but was misinterpreted and believed to be conspecific with *Ae. orbiculatum* from Mt. Athos. It was not rediscovered until 1979 and then proved to be a distinct species. It is a small, short-lived perennial of damp screes, with inconspicuous pinkish flowers and orbicular siliculas containing a single seed.

5.6.1 Affinities

Ae. carlsbergii is a recently described species (Strid & Papanicolaou, 1980) known only from the summit area of Mt. Taiyetos. It was apparently collected in 1829, however, by the French botanist Despréaux, whose material later came to the attention of Boissier. The collection seems to have consisted of some rather poor flowering specimens, and Boissier (*Flora Orientalis* 1: 337, 1867) believed the Taiyetos plants to be conspecific with *Crenularia orbiculata (Eunomia orbiculata,*

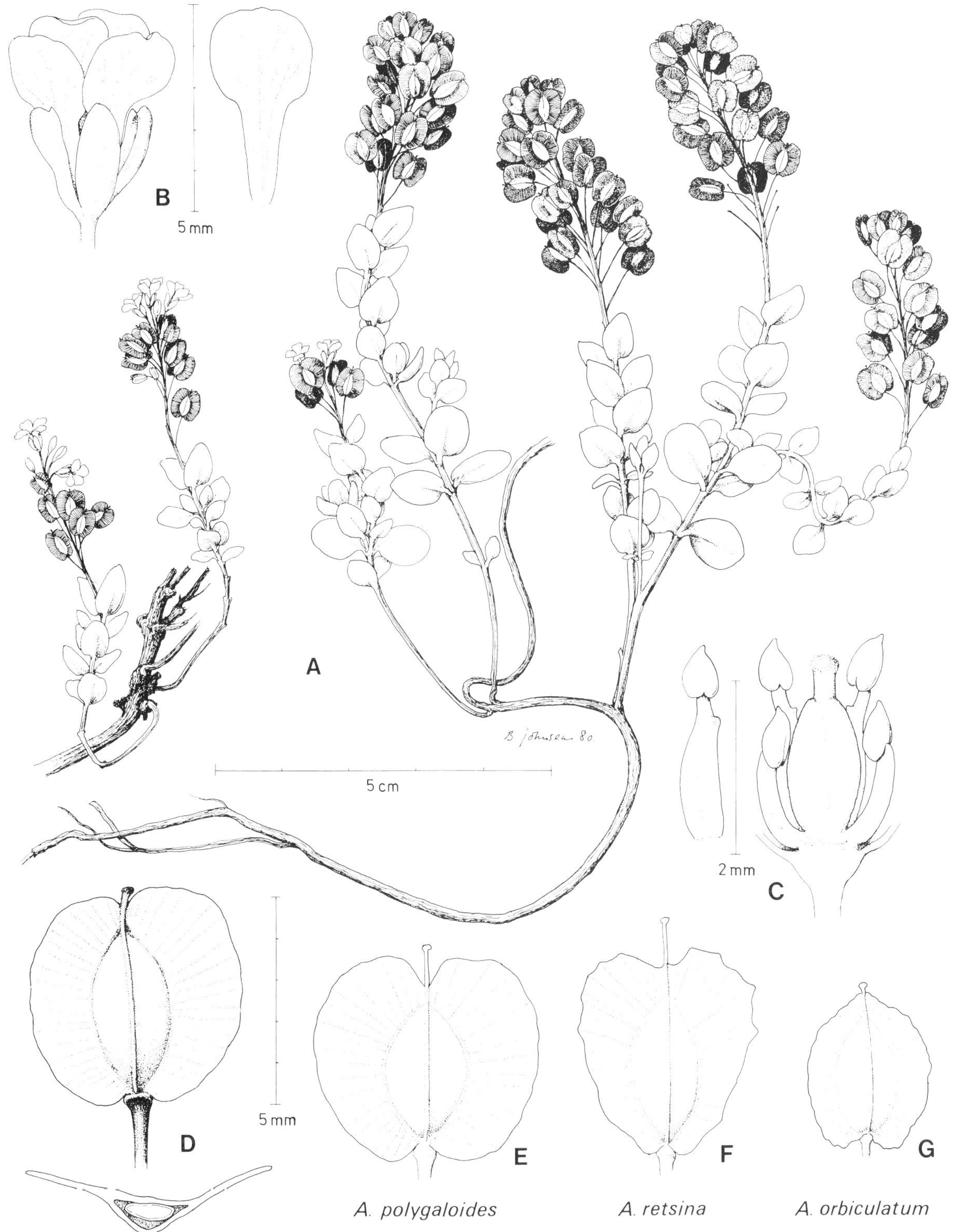

Fig. 8. Aethionema carlsbergii Strid & Papanicolaou. A: Habit. - B: Flower and petal. - C: Flower with sepals, petals and 2 of the inner stamens removed. Left: One inner (longer) stamen. - D: Silicula (cross section below). - E: Silicula of *A. polygaloides,* type specimen (G-DC). - F: Silicula of *A. retsina* (redrawn from Phitos & Snogerup 1973 p. 145). - G: Silicula of *A. orbiculatum* from Mt Athos (Strid & Papanicolaou 15955).

Aethionema orbiculatum) from Mount Athos, a mistake which was repeated by subsequent authors (Halácsy, 1901; Hayek, 1925; Rechinger, 1950; Chater, 1964). When the plant was rediscovered by us in 1979 (after 150 years!) it became clear that it was a distinct species totally distinct from *Aethionema orbiculatum.*

Aethionema carlsbergii seems to be most closely related to *A. polygaloides DC.,* a somewhat obscure species described from the east Aegean island of Chios and also reported from the west Aegean islands of Poros and Euboea. It does not seem to have been gathered in Greece recently, but a collection from Antalya province in south-western Anatolia is cited in *Flora of Turkey* (Hedge, 1965). *A. polygaloides* differs from *A. carlsbergii* in the alternate, narrower leaves and longer style on the silicula. More distantly related are the recently described *A. retsina* (Phitos & Snogerup, 1973) from Skiros and adjacent small islands, and *A. eunomioides* (Boiss.) Bornm. Boiss. from Turkey, as well as the Athos endemic *A. orbiculatum* (Boiss.) Hayek.

5.6.2 Distribution and habitat

Ae. carlsbergii is known only from the summit area of Mt. Taiyetos where we collected it on July 1979, just west of the main ridge and south of the highest peak, at an altitude of c. 2,300 m (flowering and fruiting material). It was observed in a small area of about 300 × 100 m, growing scattered on somewhat damp, but rather barren, calcareous scree, together with species such as *Ranunculus subhomophyllus* (Halàcsy) Vierh., *R. brevifolius* Ten and *Geocaryum peloponnesiacum.* It was not seen in other places and is likely to be strictly local since it has been overlooked by all previous visitors for a century and a half. The seeds were unfortunately not ripe enough to germinate. It is likely to be a short-lived perennial or biennial, possibly flowering in the first year.

5.6.3 Conservational status

Being a rare and very local species, *Ae. carlsbergii* is certainly vulnerable, though inconspicuous and not likely to be picked by the casual mountaineer. It should be sought in suitable habitats in other parts of the extensive Taiyetos range.

5.7 Symphyandra wanneri *(Rochel) Heuffell (Campanulaceae) (Fig. 9)*

This species was believed to be endemic to Bulgaria, Romania and eastern Jugoslavia, but has recently been found in the Greek part of the Rhodopi mountains. It is considerably smaller than *S. cretica* and has bluish-violet flowers. It grows in shaded rock crevices and is a monocarpic perennial, forming a leaf rosette in the first 2–4 years, then producing a flowering stem and dying after flowering.

5.7.1 Affinities

S. wanneri is a distinct species. Its closest relative may be *S. hofmannii* Pant. which differs, e.g. in having deflexed appendages between the calyx lobes and yellowish-white corolla.

5.7.2 Distribution and ecology

This species is an endemic of the central and northern parts, of the Balkan Peninsula previously known from Romania, E Jugoslavia and Bulgaria and recently discovered in NE Greece (NNW of Xanthi, summit area of Mt. Surpika, WNW of the village of Dimarion, 1,500 m (*Strid & Georgiadou 13601*). The species appears to be rather uniform. It was found in shaded crevices of more or less vertical rocks (*Fagus* forest extends to the top of the mountain on the northern side). Accompanying species (but often in somewhat more exposed situations) were *Silene lerchenfeldiana* Baumg. and *Potentilla haynaldiana* Janka, which have a similar distribution, as well as *Asplenium septentrionale* (L.) Hoffm., *Genista carinalis* Griseb, *Cotoneaster integerrimus* Medicus, *Saxifraga paniculata* Miller, *Sedum album* L., tec. Phytogeographically *Symphyandra wanneri* belongs to a rather large group of central and northern Balkan endemics often extending to the Carpathians and reaching their southernmost limit of distribution by the Greek/Bulgarian and Greek/Jugoslavian borders. Other such taxa are *Dianthus petraeus* ssp. *noeanus* (Boiss.) Tutin, *Silene waldsteinii,* Griseb., *S. lerchenfeldiana, Cardamine raphanifolia* ssp. *acris,* (Griseb) O. E. Schulz., *Saxifraga heucherifolia* Griseb & Schenk., *S. pedemontana* ssp. *cymosa*

Fig. 9. Symphyandra wanneri (Rochel) Heuffel. - A: Habit. - B: Non-flowering leaf rosette. - C: Flower. - D: Staminal apparatus. Note anthers united in a tube around the style (generic character of *Symphyandra). -* E: Fruit seen from above. - F: Fruit seen from below to show the basal pores.

Engler., *Geum coccineum* Sibth & Sm., *Trifolium velenovskyi* Vandas, *Potentilla haynaldiana, Pedicularis orthantha* Griseb, *P. limnogena* A. Kerner, *Pinguicula balcanica* Casper, *Achillea atrata* ssp. *multifida* (DC) Heim. *A. chrysocoma* Friv. *Cirsium appendiculatum* Griseb., and *Senecio abrotanifolius* ssp. *carpathicus* (Herbich) Nyman.

5.7.3 Conservational status

Symphyandra wanneri seems to be fairly rare and scattered throughout its area of distribution, but is probably not threatened other than at a local level. It is easy to cultivate and makes an attractive rock garden plant.

5.8 Jankaea heldreichii *(Boiss.) Boiss. (Gesneriaceae) (Fig. 10A)*

This species, the most famous endemic of Mount Olympus, was discovered by Heldreich in 1851. *Jankaea heldreichii* is found in crevices of damp, shaded limestone rocks, chiefly in the beech forest zone and especially in the vicinity of a stream. It forms rosettes of softly- and silvery-pubescent leaves with a short scape appearing from the centre of each rosette and bearing a few blue, broadly bell-shaped flowers.

5.8.1 Affinities

Jankaea heldreichii is a very distinct species and is now placed in a monotypic genus. It is one of the most remarkable paleo-endemics of the Balkan Peninsula. It belongs to Gesneriaceae, a family principally of the wet tropics, with only five species in Europe (four in the Balkan Peninsula, one in the Pyrenees). The other Balkan species are *Haberlea rhodopensis* Friv. of Northeast Greece and S. and C. Bulgaria, *Ramonda nathaliae* Pancit & Petrovic which occurs in N. Greece and S. Jugoslavia, and *R. serbica* Pancic which is somewhat more widespread.

Jankaea heldreichii was first discovered by Heldreich on an excursion to Mt. Olympus in 1851. In a letter to Boissier in Geneva, dated 28 August 1851, he gives an account of this highly interesting trip and, among other things, mentions that he has found 'une Gesneriacée: (*Haberlea Rhodopensis* Friv.?, par malheur seulement en fruits)'. When material reached Boissier the species was first described as *Haberlea heldreichii,* but later, when flowering plants had been collected, it became clear that it had to be placed in a genus of its own and in fact is closer to *Ramonda* than to *Haberlea.* The diagnostic characters of *Jankaea* are the 4-merous corolla with relatively long tube and subequal lobes, as well as the included stamens with free anthers.

5.8.2 Distribution and ecology

Jankaea heldreichii is endemic to Mt. Olympus, where it is chiefly found in valleys and ravines on the northern and eastern sides at c. 600–1,800 m (occasionally down to 400 m or up to 2,400 m). At the lowest altitudes it flowers in mid-May, and at c. 2,000 m at the end of July. Though it has a wide altitudinal range, it is nevertheless very particular in its choice of habitat. It is found in slightly damp, shaded crevices of more or less vertical limestone rocks, usually in the vicinity of a permanent or semi-permanent stream. It is particularly abundant in the Papa Rema and Xerolakki Rema, two deep ravines on the northern side of the mountain, where it may cover large rock faces in the beech forest zone. It is less common in the Enipevs Valley on the eastern side where scattered individuals are found from just below the Monastery of Ag. Dionysios (c. 600 m) to up as far as the vicinity of the Hellenic Alpine Club (E.O.S.) Refuge at 2,100 m.

The other Balkan species of Gesneriaceae are found in the same type of habitat. The one that occurs in greatest abundance is probably *Haberlea rhodopensis,* which covers extensive areas of shaded rock faces in deep ravines in the Rhodopi mountains, both on the Greek and the Bulgarian sides of the frontier.

→

Fig. 10A: Jankaea heldreichii (Boiss.) Mt. Olympus. E of the Papa Rema ravine, 810 m. 10.6.1976. – B: *Fritillaria epirotica* Turril ex Rix. Mt Smolikas, E. Ridge, 2100 m.

5.8.3 Conservational status

This remarkable Olympus endemic is likely to have survived in its present habitat for millions of years and is one of the finest examples of a Tertiary relict species in Europe. In spite of its small seeds, which could be expected to be easily dispersed, it has not been able to spread to the neighbouring mountains of Ossa, Pieria of Vermion, where suitable habitats are not uncommon. It is quite rare in the Enipevs Valley which is the most frequently visited part of Mount Olympus, but occurs in adequate quantity in inaccessible places on the northern side of the mountain where visitors rarely come. Unless drastic measures are taken in terms of development of forestry or tourism *Jankaea* will persist safely in these refuges. Cultivation is difficult. The seeds germinate erratically and the seedlings grow very slowly and must be constantly kept under suitable moisture conditions. Air humidity must be high, but the plants should be kept on a sloping calcareous substrate where they are never allowed to be wet for long periods. The other European species of Gesneriaceae, especially *Haberlea rhodopensis* are easier to cultivate, and generally grow well outdoors in an Atlantic climate under suitable edaphic conditions.

5.9 Fritillaria epirotica *Turrill ex Rix. (Liliaceae) (Fig. 10B)*

The genus *Fritillaria* of the lily family has many species in Greece and the Near East and is popular in cultivation. *F. epirotica* was described in 1975 and is possibly the smallest species, attaining a height of only 5–10 cm and bearing a single, nodding, brownish-purple flower. It grows on serpentine screes at montane and alpine levels in the Mt. Smolikas area of northwest Greece.

5.9.1 Affinities

F. epirotica is related to *F. graeca* Boiss. & Sprun. and *F. messanensis* Rafin. but differs from both in its low stature (which is retained in cultivation), relatively broad, alternate lower leaves, uniformly brownish-purple perianth which is not expanded at the mouth large nectaries and short and broad capsule.

This species was described recently (Rix, 1975) from a single collection made by E.K. Balls in alpine screes on Mt. Smolikas at 2,600 m in 1937. It was believed to have arisen as an alpine variant of the polymorphic *F. messanensis*. This seems unlikely, however, in view of the great morphological differences. *F. epirotica* from Smolikas and *F. messanensis* from Olympus, grown side by side in our rock garden in Copenhagen, differ conspicuously, for instance in stature (4–10 cm vs. 25–45 cm), width of lower leaves (8–15 mm vs. 3–7 mm), perianth (uniformly brownish-purple, not expanded at the mouth) and not least in the capsule (about as wide as long vs. twice as long as wide). *F. epirotica* is probably more closely related to *F. graeca,* but seems to be a distinct species.

5.9.2 Distribution and ecology

F. epirotica is restricted to serpentine screes in northwest Greece at altitudes of between c. 1,400 and 2,600 m (most other *Fritillaria* species are lowland species). It has recently been collected on Mt. Smolikas (nos. 191–193 on map) and Mt. Bouchetsi (200), and on Mt. Milea (178). It is found in deep mobile screes where very few other species can grow, among them *Viola magellensis* Porta & Ligo ex Strobi, *Aethionema saxatile,* (L.) R. Br., *Iberis epirota* Halácsy (*I. pruitii sensu-lato*) and a variant of *Minuartia verna* (L.) Hiern.

5.9.3 Conservational status

This species is rare and scattered and may be actually threatened if excessively collected by bulb growers. Its best protection is the fact that it is a small, inconspicuous plant that grows in remote areas. The flowers, which are borne near ground-level, are remarkably difficult to spot against the dark serpentine rock.

Acknowledgements

Our studies of the Greek mountain flora have been chiefly financed by grants from the Danish Natural Science Research Council and the Carlsberg Foundation. The drawings were made by Bengt Johnsen, except Figure 6, which was made by Poul

Juul. The map was prepared by Lars-Åke Gustavsson for the *Mountain Flora of Greece* project. Information on *Fritillaria epirotica* was supplied by Per Hartving.

References

Alden, B. (1976). Floristic notes from the high mountains of Pindhos, Greece. Bot. Notiser 129: 297-321.

Chater, A. (1964). *Aethionema* R. Br. In: Tutin et al. (Eds.) Flora Europaea 1: 322. Cambridge: Cambridge University Press.

Davis, P.H. (Ed.) (1975). Flora of Turkey and the East Aegean Islands, Vol. 5. Edinburgh: Edinburgh University Press.

Greuter, W. (1971). Betrachtungen zur Pflanzengeographie der Südägäis. Opera Bot. 30: 49-64.

Greuter, W. (1975). Quisquiliae floristicae graecae, 1-3. Candollea 30: 323-330.

Gustavsson, L.A. (1978). Floristic reports from the high mountains of Sterea Ellas, Greece. Bot. Notiser 131: 201-213.

Halacsy, E.V. (1901). Conspectus Florae Graecae, Vol. 1. Leipzig: Guilelmi Engelman.

Hartvig, P. (1979). *Veronica bornmuelleri* (Scrophulariaceae), new to Europe. Bot. Notiser 132: 367-370.

Hayek, A.V. (1925). Prodomus Florae Peninsulae Balcanicae, Band 1, Lief. 3. Feddes Repert. Beih. 30 : 1.

Hayek, A.V. (1928). Ein Beitrag zur Kenntnis der Vegetation und der Flora des thessalischen Olymp. Beih. Bot. Zentralbl. 45: 220-328.

Hedge, I. (1965). *Aethionema* R. Br. In: Davis P.H. (Ed.), Flora of Turkey and the East Aegean Islands 1: 314-322. Edinburgh: Edinburgh University Press.

Heldreich, Th.V. (1851). Letter to E. Boissier, Geneva - In the library of Conservatoire Botanique de Genève, unpubl.

Melzheimer, V. (1977). Biosystematische Revision einiger *Silene Arten* (Caryophyllaceae) der Balkan-Halbinsel (Griechenland). Bot. Jahrb. Syst. 98: 1-92.

Merxmüller, H. & Strid, A. (1977). A new species in the *Cerastium alpinum* group from Mt. Olympus, Greece. Bot. Notiser 130: 469-472.

Phitos, D. & Snogerup, S. (1973). A new species of *Aethionema* from Skiros, Greece. Bot. Notiser 126: 142-145.

Rechinger, K.H. (1950). Grundzüge der Pflanzenverbreitung in der Ägäis. Vegetatio 2: 55-119, 239-308, 365-386.

Rechinger, K.H. (1951). Phytogeographia Aegaea. Denskschr. Österr. Akad. Wiss., Math.-Nat. Kl. 105, 2 - 2: 1-208.

Rix, E.M. (1975). Notes on *Fritillaria* (Liliaceae) in the Eastern Mediterranean region. III. Kew Bull. 30: 153-162.

Strid, A. (Ed.) (1984). Mountain Flora of Greece, Vol. 1. Cambridge: Cambridge University Press. pp c. 700 (in press).

Strid, A. & Papanicolau, K. (1980). New species of *Aethionema* and *Peucedaneum* from the Greek mountains. Bot. Notiser 133: 521-526.

Webb, D.A. (1978). Flora Europaea: A retrospect. Taxon 27: 3-14.

Institute of Systematic Botany
University of Copenhagen
140, Gothersgade
DK-1123 Copenhagen K
Denmark

CHAPTER 7

The Anatolian Peninsula

H. DEMIRIZ AND T. BAYTOP

1. Endemism in the Turkish flora

Turkey contains an extremely diverse range of vegetation-types. This is a consequence of the country's geographical position, its situation at the intersection of three major phytogeographical regions, and the various climatic and edaphic conditions that affect these regions. The number of vascular plants in the flora is very high in proportion to Turkey's total area of 780,576 km^2. Realistic counts for these species will be revealed in the supplement to *Flora of Turkey* (Davis, 1965–84, 1979), although a tentative assessment suggests a total of over 8,000 species.

Certain other factors, related to the geological past of the region, have led to the development of a high level of endemics. A very high percentage of Turkey's flora is, in fact, endemic. Of the native species treated in the first three volumes of *Flora of Turkey*, about 30% are endemic (Davis, 1971). 36% of the Turkish members of the Compositae, to which the fifth volume of the *Flora* is devoted (Davis, 1975), and 39% of the species treated in the sixth volume (Davis, 1979) are also endemic to Turkey.

On the basis of the published list of Turkish endemics (Lucas, 1980), the greatest number of endemics is to be found in the following families: Compositae (335 species), Leguminosae (332 species) and Scrophulariaceae (232 species), closely followed by Caryophyllaceae (including Illecebraceae) (188 species), Cruciferae (164 species), Umbelliferae (108 species) and Boraginaceae (107 species).

The Compositae owes its high proportion to the genus *Centaurea*, of which 104 species (61% of the genus) are endemic. The highest diversity in the Leguminosae occurs in *Astragalus*, which is represented by 209 endemics (56%), while *Verbascum* (Scrophulariaceae) has 166 endemic species (73%). In Caryophyllaceae, the proportion of endemics in *Paronychia* is 71% and in *Gypsophila* 59%; in Cruciferae, in *Isatis*, it is 77% and in *Alyssum* 54%; in Boraginaceae in *Alkanna* it is 77% and in *Paracaryum* 67%.

The highest instance of endemism is found in *Ebenus* (Leguminosae), of which all 14 species are Turkish endemics.

There are twelve monotypic, endemic genera in Turkey. They were collated and published, together with a map of their distribution, by Davis (Davis, 1971). They are: *Phryna ortegioides* (Fish. et Mey.) Pax et Hoffm. and *Thurya capitata* Boiss. et Bal. (Caryophyllaceae), *Cyathobasis fruticulosa* (Bunge) Aellen and *Kalidiopsis wagenitzii* Aellen (Chenopodiaceae), *Physocardamum davisii* Hedge and *Tschihatchewia isatidea* Boiss. (Cruciferae) (Fig. 1), *Anatropostylia koeieana* (Rech. fil.) Kupicha and *Sartoria hedysaroides* Boiss. et Heldr. (Leguminosae), *Crenosciadium siifolium* Boiss. et Heldr. (Fig. 2), *Microsciadium minutum* (D'urb.) Briq. and *Olymposciadium caespitosum* (Sm.) Wolff (Umbelliferae), and *Necranthus orobanchoides* Gilli (Orobanchaceae).

Gómez-Campo, C. (ed.), Plant conservation in the Mediterranean area.
© 1985, Dr W. Junk Publishers, Dordrecht. *ISBN 90 6193 523 7.*

Fig. 1 Tchihatchewia isatidea Boiss. (from N. Yakar-Tan).

Turkish endemics occur over a wide area of the country, although many are very local in particular regions. These are largely concentrated in the Irano-Turanian floristic zone (central and eastern Anatolia), and the mountainous boundaries between this and other regions. There are few in the north, west and south of Anatolia. Thus, only a comparatively small fraction of Turkish endemics could be properly considered to be Mediterranean elements.

2. Conservation problems and solutions

2.1 Main threats to rare and endemic species

Although the level of population and of industrial development in Turkey is well behind that of many European countries, laws and decrees concerning nature conservation were established at an early period in its history. The first decree concerning nature conservation was made in 1894 (as compa-

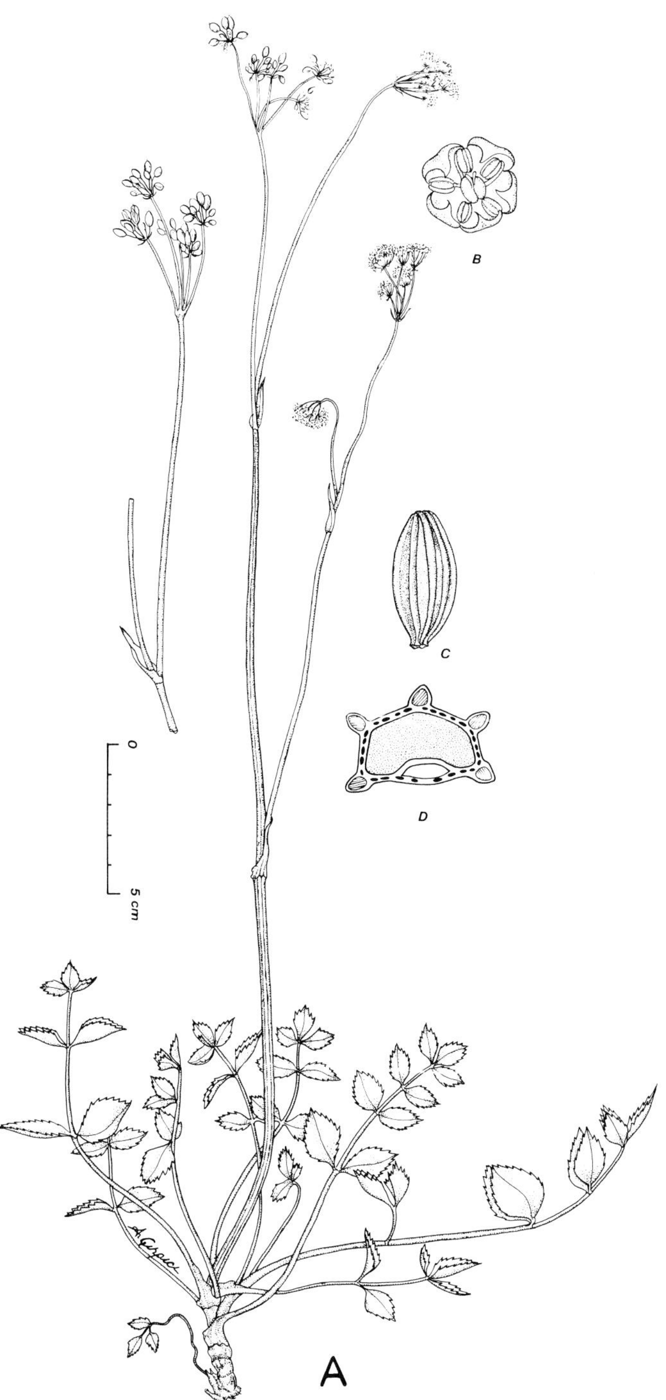

Fig. 2. Crenosciadium siifolium Boiss. et Heldr. (from A. Çirpici). A. Habit, B. Flower, C. Fruit, D. Section of fruit.

red with 1872 in U.S.A.), prohibiting the pollution of water channelled through aqueducts in the Belgrad forest near Istanbul. Activities, such as hunting or felling trees, that damaged the natural forest environment, were forbidden.

However, the rapid increase in Turkey's population, which began in the 1930s (15 million inhabitants in 1930, rising to 45 million by 1980) has resulted in a parallel increase in the destruction of

Turkey's flora and vegetation (Baytop, 1959; Birand, 1959; Demiriz, 1981; Anonymous, 1981). The present conservation situation, and the main threats to rare and endemic species, can be summarized as follows.

The rapid increase in population has meant that arable land in the hands of the small-holder is insufficient to support or even to feed the rural population. The small-holder's requirement for arable land has been met by local deforestation and shrub clearance. The new areas brought into cultivation by these methods produced high yields during the first few years but mismanagement, and the progressive processes of wind and flood have later led to the kind of erosion common in Anatolia. The resulting decrease in yield led in turn to the destruction of further areas of forest and shrub, producing a widespread pattern of devastation to the flora and vegetation. Although the present Forestry Law (No. 6831) has forbidden the cultivation of woodlands and has established severe penalties for the destruction of forests, it has proved an insufficient deterrent to deforestation.

The rise in population has also brought a parallel increase in demand for animal products, and a consequent rise in the price of these. This has encouraged the development of animal husbandry in those areas, throughout Anatolia, suited to this form of farming, with the result that Turkey now has 20 million head of goats and 45 million head of sheep. This places Turkey's flocks amongst the ten largest national livestock quotas in the world, a figure that is far above what can be supported by the grazing lands of Anatolia. Pastures have thus been severely damaged, and alpine meadows especially have lost many of their former characteristic features.

One of the main reasons for the destruction of Turkey's flora and vegetation can also be found in the systematic clearing of shrubby areas for agricultural purposes. This has upset the moisture balance, and tends to produce extreme drought conditions in summer, and floods in spring.

A large number of lakes have dried up in some localities as a result of changes orientated towards further acquisition or improvement of arable land by irrigation. This has led to the complete disappearance of populations of aquatic plants (and also of aquatic animals). Two of the many lakes destroyed in this way are Lake Amik, in Hatay province, and Lake Söğüt, in Antalya province.

The introduction of mechanized farming into Turkey from 1950 onwards has also influenced the damage to Turkey's flora. The wide use of tractors (in 1930 there were 1,400 tractors, and by 1980 the figure had risen to 370,000) has opened up many previously unutilized areas. This has totally destroyed both the vegetation of these areas, and to a large extent the whole ecosystem, leaving only very limited possibilities for the survival of natural vegetation. Mechanized deep-ploughing has also destroyed the roots of many species, causing their gradual disappearance from the fields. Species abundant on arable land before the introduction of mechanized farming, such as *Gypsophila paniculata* L., *G. venusta* Fenzl, *Glycyrrhiza echinata* L. and *G. glabra* L. are now very scarce. In contrast to these, some bulbous species, such as *Tulipa bithynica* Griseb., *T. sintenisii* Baker, *T. undulatifolia* Boiss. and *Crocus olivieri* Gay, have been able to survive unaffected by modern farming methods.

Many side-effects have resulted from the use of chemicals for weed control. Some species, especially those infesting wheat, such as *Papaver rhoeas* L. and *Hyoscyamus reticulatus* L., have been noticeably reduced by the use of 'clean' and uncontaminated seed.

The widespread export of bulbs has led to a considerable decrease in the number of many bulbous species. Turkey has been an exporter of bulbs to Europe for hundreds of years and their sale is steadily rising. The export figures for the last four years are shown in Table 1.

The harvesting and export of certain plants for medicinal purposes or the extraction of oil has further damaged the flora and vegetation and is seriously threatening certain species. An alarming decrease in the numbers of *Gentiana lutea* L. is an example. The widespread harvesting of, as yet, abundant species, such as *Laurus nobilis* L., *Origanum heracleoticum* L. and *O. onites* L. for the export of their leaves and flowers suggests that soon they will eventually be threatened by extinction.

Table 1. Turkey's bulb exports; quantity exported, in kg (*Annual Foreign Trade Statistics, 1975-78*).

Countries	1975	1976	1977	1978
Denmark	941	–	975	1,632
United Kingdom	21,907	10,804	6,501	616
F.R. Germany	77,351	50,634	28,315	7,796
Jordan	–	16,500	30,000	102,243
Kuwait	–	16,510	–	–
Lebanon	–	211,884	50,000	–
Netherlands	125,235	142,463	149,798	184,506
Norway	225	–	–	–
Switzerland	8,615	4,196	4,001	–
Syria	–	29,970	10,000	–
U.S.A.	4,865	1,180	4,304	644
Total	239,138	484,141	288,897	297,437

Finally, the construction of industrial installations, dams and new roads, and the practice of drainage to acquire new settlement areas, are constant sources of damage to the flora and vegetation. The growth of industry and technology is especially significant in a developing country such as Turkey.

2.2 Steps to protect rare and endemic species

The nine above-mentioned factors influencing the destruction of the flora and vegetation in Turkey are probably similar to those to be observed in other developing countries. Differences may exist, however, in the implementation of preventive measures to remove, or at least reduce, the effect of these destructive factors. Such measures, as taken and implemented in Turkey, are listed below. Although in some cases there has been only a short period of implementation, some measures already show signs of producing results. We list only those which have proved appropriate to the prevailing conditions in our country.

After the recognition of the failure to prevent illegal deforestation and destruction of scrubland by a penal deterrent, a policy of employing rural people in their local forestry reserves was adopted. Villagers have been allocated exclusively the jobs in forestry - reforestation, lumbering, transport of lumber and road-building. This has given them access to a regular income and has been a positive move towards the prevention of the destruction of the forests. The establishment of animal-feed factories has also been encouraged, to provide additional support to animal husbandry, and a reduction of severe over-grazing is expected.

The destruction of the flora by flooding and inundations are partially prevented by some efforts aimed at the restriction of flood sensitive areas. To prevent the extinction of bulbous and tuberous plants, their export has been placed under strict control and is now restricted. A resolution has been drafted to allow the export of cultivated bulbs only in the future. Lastly a law with basic rules for nature conservation has been elaborated.

2.3 Nature reserves

For Turkey, a country with approximately 8,000 vascular plant taxa, of which 30% are endemic, the publication of lists of the rare or endemic species of its flora is not likely to contribute to their preservation. On the contrary, it is felt that the publication of such lists, complete with coloured illustrations (as in Switzerland), would tend to draw attention to these species and to increase the likelyhood of their annihilation by collectors in an even shorter time.

We feel, therefore, that the most acceptable method of preservation would be the creation of a suitable environment, through the selection of well-protected localities to conserve the species in their native habitats. This method is being adopted in Turkey at the moment. After 1976 the problem was taken in hand officially by the incorporation of such areas into the jurisdiction of the General Directorate of National Parks and Wildlife, under the aegis of the Ministry of Forestry. At the present time, Turkey has 16 well-protected, fully staffed National parks (Fig. 3), and 42 additional localities have been included under a programme for conversion into National Parks.

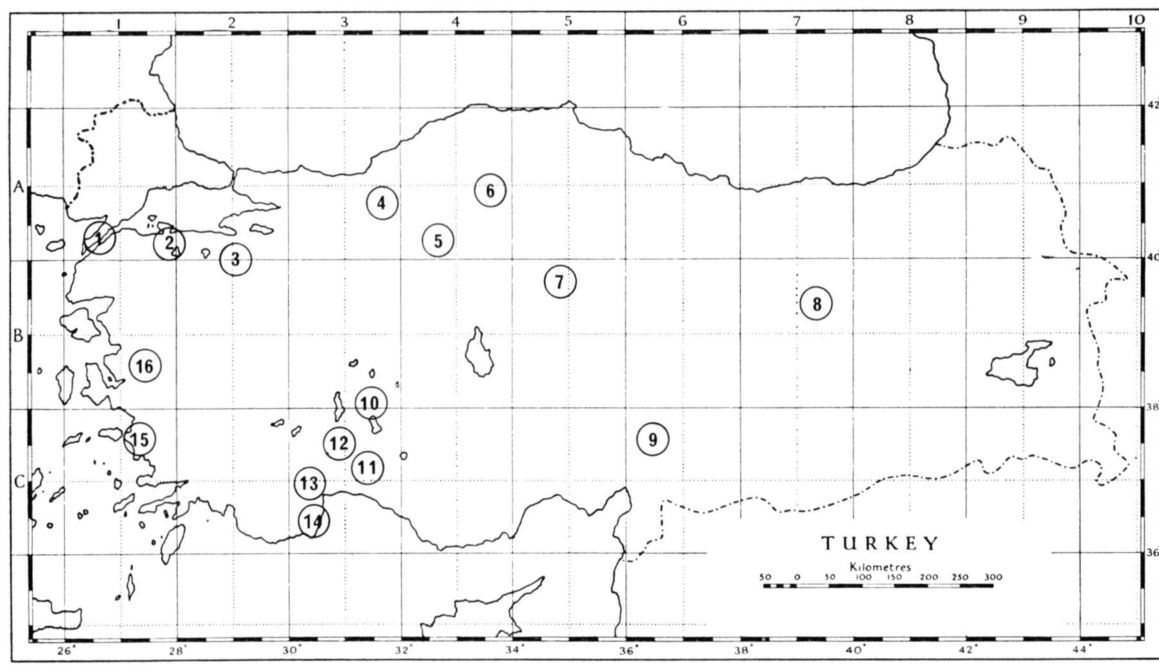

Fig. 3. National Parks in Turkey (Map prepared by H. Demiriz and A. Çirpici). 1. Gelibolu Peninsula, 2. S. Kuşcenneti (Kuş gölü), 3. Uludağ, 4. Yedigöller, 5. Soğuksu, 6. Ilgaz Daği, 7. Yozgat çamliği, 8. Munzur Valley, 9. Karatepe – Aslantaş, 10. Kizildağ, 11. Köprülü Canyon, 12. Kovada gölü, 13. Termessos, 14. Olimpos – Baydağlari, 15. Dilek Peninsula, 16. Sipil Daği.

3. Some case histories of rare and threatened species in Turkey

3.1 Pteridophytes around Istanbul

It cannot be said that overall the ferns in the Turkish flora are under serious threat. However, the noteworthy exploitation of some species - as, for example, of *Polystichum setiferum* (Forsk.) Woynar for wreath-making by Istanbul florists, and of *Pteridium aquilinum* (L.) Kuhn for the decoration of melon stalls - may be significant. The latter, which is widespread and abundant, is of course under no threat, but the wide use of the former will undoubtedly lead to the progressive extinction of the species.

There are, however, many cases of species which can be said to be under threat through conversion of their rocky habitats to quarries and to other types of land-uses. Some species that may be so threatened with extinction, or extreme reduction of their habitat, are *Cheilanthes corsica* Reichst. et Vida, *Anogramma leptophylla* (L.) Link, *Asplenium obovatum* Viv., *A. septentrionale* (L.) Hoffm. and *Polypodium australe* Fée, which survive, sometimes in hidden refuges, around the Kayidağ, Yakacik, Dragos and Büyük Ada areas of Istanbul (Demiriz et al., 1969, 1977).

Cheilanthes corsica Reichst. et Vida is known only in the region of Viranbağ (Paliambelo), Büyükada. The locality of this chasmophytic species is limited to a narrow strip 30–40 m above sea-level. The destruction of its rocky environment would mean the extinction of this rare species of the Istanbul region.

Anogramma leptophylla (L.) Hoffm. is extremely sparsely distributed in remote parts of the Istanbul area at Kayişdağ, Dragos and Büyükada. It was collected for the first and last time on Büyükada by G.V. Aznavour in 1899. It is now literally struggling for survival at its locality at Kayişdağ. The population on the hills of Dragos is threatened by a likelihood of quarrying for building materials, so it now appears highly probable that this species will shortly disappear altogether from the Istanbul area.

Asplenium obovatum Viv. is distributed along the same strip, at Aydos Daği, Yakacik and Büyükada, in as yet inaccessible areas, undisturbed by human activity. Like other insular *Asplenium* species in similar localities, there are certain taxonomic and cytological problems yet to be solved in connection with this species. Its distribution around Istanbul is still under investigation, but it can be said that changes in land-use and the introduction of stone-quarrying, could represent a threat to its existence. The same can be said of *Polypodium australe* Fée, which is found at Yakacik and Dragos.

The station for *Asplenium septentrionale* (L.) Hoffm. nearest to sea-level is on Aydos Daği (400–500 m). It will not be endangered as long as there is no increase in the destruction of rocky and bushy localities, and if its populations are not decimated by botanists.

In summary, the example of the ferns of the Istanbul area shows that the activities of florists and the development of stone-quarrying for building materials may lead to the eventual disappearance of some rare or endemic species.

3.2 Threats from urban development

The expansion of housing developments, the rise in the number of coastal holiday resorts, and the consequent disruption of the landscape, have led to steady uncontrolled pressure on the flora not only around large towns but also in dunes, shrubby areas and meadows. While this is a situation common to many developing countries, for a country with such a diverse indigenous flora this is more likely to produce irreparable loss of characteristic elements of the natural environment.

A case in point is Istanbul, with its peculiar phytogeographical situation between Asia and Europe. This city is a significant part of both the historical and cultural heritage of Turkey, and its economic and business life, and houses a tenth of the country's total population. Its rapid development constantly threatens the rare and endemic plants which are part of the natural flora of its surroundings (Baytop & Demiriz, 1980).

In places such as Kisirkaya, Kilyos, Riva and Şile on the Black Sea coast north of Istanbul, the natural vegetation has been deeply affected by the growth of bathing areas and resorts. Some of the plants most threatened by human activity are typical dune plants such as *Isatis arenaria* Azn., *Verbascum haussknechtii* Heldr. ex Hal., *Centaurea kilaea* Boiss. and *Jurinea albicaulis* ssp. *kilaea* (Azn.), Kozuharov all endemic to this area.

The conversion of meadows and *maquis* into arable land around the city, the construction of new roads, and the use of scrub for firewood will eventually exceed the tolerance level of various rare species in the Istanbul area. As an example, we may cite *Onosma propontica* Azn. which has survived west of European Istanbul in scrubby road-side localities; any attempt to widen the roads or further conversion of scrub into arable land would be disastrous for this species.

Centaurea hermanni Herm., a local endemic of the Istanbul area, *C. amplifolia* Boiss. et Heldr. and *C. inermis* Velen., thrive in some limited areas that would be adversely affected by the destruction of maquis communities.

Some of the localities of *Colchicum micranthum* Boiss., an endemic of the Istanbul region and of Turkey-in-Europe, have been completely destroyed through the conversion of these areas to urban development. The autumn-flowering *C. turcicum* Janka, once widely distributed in the eastern part of the Balkan peninsula, was common in the meadows around Edirne and Tekirdağ (Turkey-in-Europe). Mechanized farming in these areas in recent years has completely wiped out these communities, and the species is now only infrequently found in open clearings in forest areas.

3.3 The use of Astragalus *species as fuel*

Astragalus species are used for fuel in villages and towns of the mountainous regions of eastern Anatolia. *Astragalus* species endemic to these regions are seriously threatened by the continuous uprooting of plants, as the roots are also used for fuel. The disappearance of these species, in turn, is removing the protective cover of shrubs that save other plants from over-grazing. This situation can be observed most clearly in the mountains of Kayseri and Van provinces.

3.4 *Plants harvested for food*

In south and east Anatolia, the corms of *Crocus cancellatus* Herbert and *C. kotschyanus* C. Koch are widely harvested for food, and are sold in the local markets. While this widespread use does not yet constitute a major threat to these species, it may well do so in the future.

The threat to certain geophytes through uprooting of their tubers can be illustrated by the case of some orchids (Orchidaceae) in western and northern Anatolia, which are subject to widespread harvesting for use in the making of 'salep' (a milk-based drink). The survival of these plants, despite centuries of uprooting, can possibly be explained by their high yield. Examples of orchid species at present being collected in large numbers are *Orchis anatolica* Boiss., *O. italica* Poiret, *O. tridentata* Scop., *Ophrys bombyliflora* Link, *O. scolopax* Car. subsp. *cornuta* (Steven), Camus, *O. ferrum-equinum* Desf., *O. fuciflora* (F.W. Schmidt) Moench, *O. fusca* Link, *O. sphegodes* Miller subsp. *mammosa* (Desf.) Soo ex. Nelson, *Himantoglossum hircinum* (L.) Sprengel, *Aceras anthropophorum* (L.) Aiton fil. (Baytop & Sezik, 1968).

3.5 *Medicinal plants*

The extraordinarily varied flora of Turkey includes many plants with medicinal value. The general abuse of the environment is affecting these species in the same way as it does others in Turkey's flora. Some examples of affected species follow.

Some species of *Gypsophila* (especially *G. paniculata* L. and *G. venusta* Fenzl.) are gathered for the export of their roots. Widespread harvesting of these species and deep ploughing by mechanical methods have greatly reduced their numbers in central Anatolia, so that collectors are now concentrating on the plants which are found in eastern Anatolia.

Similarly, *Gentiana lutea* L. has been widely harvested for the medicinal value of its roots in the mountain localities of western Anatolia (Uludağ, Bozdağ, etc.), where it was abundant before 1940. Consequently, this species can now be found only in very isolated, precipitous localities and then very infrequently.

Digitalis grandiflora Miller, widely distributed throughout eastern and central Europe, is represented in Turkey-in-Europe by one isolated colony, restricted to the locality of Kirklareli. This species was collected here in large quantities for chemical and pharmacological research. Today, ten years after that intensive harvesting, the population has been unable to recover its former strength and only a few individual specimens are to be encountered.

Some biennial species of *Papaver* are in a similar situation. The collection of practically whole populations of these species for chemical research is leading to their extinction.

3.6 *Ornamental plants*

One of the most important threats to the Turkish flora is the wide-spread collection of many tuberous, rhizomatous and bulbous plants for ornamental purposes, particularly for export. This is done despite a number of legal restrictions and controls.

For the tubers of some of the Turkish endemic *Cyclamen* species (*C. cilicium* Boiss. et Heldr., *C. mirabile* Hildebr., *C. parviflorum* Pobed. and *C. pseudoibericum* Hildebr.), the export figures are not yet sufficiently alarming to suggest a serious threat to the species, but efforts are being made to keep these exports under constant surveillance.

The bulbs of *Galanthus elwesii* Hooker fil., which is widely distributed throughout western Anatolia and the East Aegean islands, are being harvested in significant numbers by bulb exporters. As a result, this once common species has been greatly diminished, especially in the Taurus region and, in some cases, entire populations have disappeared. The species as a whole is not faced with complete extinction; however, whole populations are being decimated by this intense collection. The same threat is facing the Black Sea endemics in the eastern region, *G. latifolius* Rupr. and *G. rhizehensis* F.C. Stern.

Sternbergia candida Mathew & T. Baytop is an endemic species recently described from Turkey, discovered and published in 1979 (Mathew & Baytop, 1979). It is easily distinguished from other species of *Sternbergia* by its white flowers. To protect this species from bulb harvesters, its area of distribution over a narrow strip of western Anatolis (near Fethiye) was deliberately left unspecified in the original description. Despite this precaution, in the autumn of the same year of publication, a German bulb collector offered 1,000 bulbs of this species for sale to a Dutch bulb merchant. This is an interesting case, which illustrates a close relationship between the publication of a new bulbous species and its incorporation into the bulb market.

Arum italicum subsp. *albispathum* (Steven) Prime is a species of the eastern Black Sea region, whose bulbs are subjected to constant collection for export. However, the ready reproduction of this species from tubers, and the harvesting of only tubers that have reached a suitable size, has counteracted the effects and dangers of export on a large scale.

References

Anonymous (1980). Environmental Profile of Turkey. Ankara: Environmental Problems Foundation of Turkey.

Baytop, T. (1959). Les plantes rares de l'Anatolie et les précautions prises en vue de leur protection. Proc. IUCN 7th Tech. Meeting, Brussels 5: 197–198.

Baytop, A. & Demiriz, H. (1980). Rare plants and endemics in Turkey-in-Europe. Ist. Üniv. Fen. Fak. Mec. Seri B, 45: 109-111.

Baytop, T. & Sezik, E. (1968). Recherches sur les saleps de Turquie. J. Fac. Pharm. Istanbul 4: 61-68.

Birand, H. (1959). La végétation anatolienne et nécessité de sa protection. Proc. IUCN 7th Tech. Meeting, Brussels 5: 192–196.

Davis, P.H. (1965-1984). Flora of Turkey and the East Aegean Islands, Vols. 1–8. Edinburgh: Edinburgh University Press.

Davis, P.H. (1971). Distribution patterns in Anatolia with particular reference to endemism. In: Davis, P.H., et al. (Eds.), Plant Life of South-West Asia. Edinburgh: Edinburgh University Press.

Davis, P.H. (1975). Turkey: Present state of floristic knowledge. In: 'La flore du bassin méditerranéen: Essai de systematique synthetique. Colloques Intern. du C.N.R.S. 235: 93-113.

Davis, P.H. (1979). Towards a supplement for the Flora of Turkey. Webbia 34: 135-141.

Demiriz, H. (1981). Nature conservation in Turkey. Threatened Plants Committee Newsletter 7: 21-22.

Demiriz, H., Tutel, B. & Aydin, A. (1969). Studia ad Floram et Vegetationem Turciae pertinentia: IV. New materials to the Pteridophytes of Turkey: Filicales. Ist. Üniv. Fen. Fak. Mec. Seri B 34: 137-181.

Demiriz, H., Tutel, B. & Aydin, A. (1977). Studia ad Floram Turcicam: VII. New materials to the ferns of Turkey: 2. Ist. Üniv. Fen. Fak. Mec. Seri B 42: 71-79.

Lucas, G. Ll. (1980). First preliminary draft of the list of rare, threatened and endemic plants for the countries of North Africa and the Middle East. Kew: IUCN-TPC.

Mathew, B. & Baytop, T. (1979). A new white *Sternbergia*. The Garden 104: 302-303.

Department of Botany
University of Istanbul
Istanbul

Department of Pharmacology
University of Istanbul
Istanbul

CHAPTER 8

The arid eastern and south-eastern Mediterranean region

L. BOULOS

1. Introduction

It is rather difficult for a botanist in 1985 to trace the history of conservation problems in an area such as the arid eastern and south-eastern Mediterranean region. The area embraces one of the richest and most variable floristic regions of the basin, one which has witnessed some of the oldest civilizations on earth and where diverse human activities, especially those connected with the natural vegetation and the domestication of plants and animals - grazing, agriculture and settlement - have taken place for a very long time. Moreover, the aridity or extreme aridity of the greatest part of the East Mediterranean countries has exerted a considerable stress on the richer coastal, as well as montane, vegetation, beyond which lies a vast wilderness of almost sterile land. In Egypt arid regions represent 96% of the area of the country; in Libya 97%; a large proportion of Syria, Jordan and Israel is also arid.

In general, the different habitat types within our area may be approximately differentiated into: deserts, oases, littoral and coastal lands, wetlands and hilly and mountainous regions.

2. Deserts and oases

The deserts represent the major part of the Eastern and South-eastern Mediterranean, with a climatic range from arid, usually close to the coast, to extremely arid inland regions, where low rainfall is the limiting factor to plant growht, in addition to the extremely high temperatures.

Apart from natural disasters such as the occurrence of several successive seasons with lower than average rainfall, excessive human activities and over-exploitation of these deserts play the greatest role in the establishment of unfavourable conditions for certain plant species, particularly those which already survive in delicate or fragile equilibrium with their habitats. Over-exploitation of the desert's natural vegetation through over-grazing or over-collection of shrubs and trees for fuel or for other purposes (cf. Batanouny, 1983) may gradually make common species less common, rare species endangered and endangered species extinct.

Reclamation of some desert areas for agriculture or the establishment of industrial or tourist centres, together with their accompanying urban communities, may lead to the disappearance of rare species or local endemics. *Salsola tetragona* Del. (= *S. pachoi* Volkens & Aschers.), a rare dwarf shrub, restricted to the limestone area of Abu Rawash near the Giza Pyramids in the Western Desert of Egypt, is severely threatened (or even near to extinction) by the heavy air pollution from an industrial plant, recently established in the area to process quarried limestone into cement for building purposes.

In the newly reclaimed desert region west of the Nile Delta, new agricultural development is based on the traditional method of irrigation by digging new canals and using water of the Nile. Many

Gómez-Campo, C. (ed.), Plant conservation in the Mediterranean area.
© 1985, Dr W. Junk Publishers, Dordrecht. ISBN 90 6193 523 7.

weedy species were introduced with crops, and some indigenous desert plants have become less common or rare; among these are threatened or endangeredones.

Medicinal desert plants which are collected from their natural habitats on a commercial scale may be subject to a similar threat, especially when the part collected is the seed, fruit or in some instances the whole plant. A common desert annual, *Anastatica hierochuntica* L. (Cruciferae), is collected for its use in folk medicine. Mature plants, with seeds, are uprooted, which deprives the plant's natural habitat of seeds that should be left in the soil to await the next winter rain, in order to germinate and produce new plants. The species is becoming less and less common, especially when the desert is subjected to drought for prolonged periods, sometimes for several seasons. Another example is *Ducrosia flabellifolia* (Umbelliferae), a perennial herb, collected in extensive amounts by the dwellers of the deserts in Jordan for its value in folk medicine. The species, although locally common in El Jafr-Bayir desert (Boulos, 1977), is however, less common on other desert habitats of Jordan and the western desert of Iraq near the frontier with Jordan. Over-collection of this plant may be a threat to the survival of these populations of this interesting species, which are already separated from each other by rather long distances (Boulos & Al-Eisawi, 1977b).

Some *Acacia* species (Leguminosae), e.g. *A. raddiana* Savi and *A. tortilis* (Forssk.) Hayne, especially in some isolated parts of the deserts of Egypt and Libya, are cut down and used as firewood, to make charcoal or for carpentry. The branches are occasionally pruned, especially during the dry summer, to offer the foliage as green fodder to goats and sheep. Camels are able to browse and satisfy their needs even if the trees are comparatively high. The few *Acacia* seedlings which succeed in germinating are also subjected to grazing, which makes the regeneration of these desert trees difficult. Moreover, ripe *Acacia* pods falling under the trees constitute a food favoured by desert gazelles, which usually find welcome shelter in the shadow of the trees during the heat of the summer.

Oases are depressions which receive very little or almost no rain but are 'green patches' within the surrounding sterile desert. Their size varies from a fraction of a square kilometre to hundreds or thousands of square kilometres. Oases owe their greenness to their perennial underground water supply, which appears on the soil surface in the form of springs or is pumped from wells. Some oases are inhabited and their human population lives in villages that depend mainly on agriculture and agricultural commercial products. Others are now depopulated because they have run out of fresh underground water and consequently have been abandoned. Depending on available natural resources, some oases support a seasonal population who collect dates, firewood, etc., or graze their herds.

I shall give a few striking examples from different oases in our region, particularly of trees which are either extinct, endangered or threatened. An example of a tree which has been known from the oases in Palestine and is now extinct is *Acacia laeta* R. Br. The last record for the tree in Ghor es Safieh was made by Hart in 1891 (Zohary, 1972; Dafni & Agami, 1976). The tree is principally Sudanian in distribution (Zohary, 1972) and this locality seems to have been the northernmost limit of its distribution.

The Qasr 'Amra are near Azraq Oasis in Jordan includes, among other interesting trees and shrubs, *Pistacia atlantica* Desf. (Anacardiaceae). Many female trees grow there, some as much as 1,000 years old, but no male trees occur. As a result of the lack of pollen, only seedless fruits are formed and, consequently, the plant is unable to regenerate. It is therefore proposed to introduce male trees to this most interesting locality to assure fertilization and to provide a natural means by which the trees can regenerate.

The fan palm *Medemia argun* (Mart.) Württemberg ex Wendl. (Palmae) was known to occur in Ancient Egypt and the fruits have been found in tombs, among offering gifts, almost as frequently as those of the date palm (*Phoenix dactylifera* L.) and dom palm (*Hyphaene thebaica* (L.) Mart.) (Täckholm & Drar, 1950), suggesting that the plant was abundant in the oases, and most probably along the Nile banks, of Upper Egypt.

According to Täckholm & Drar (1950), the most recent period when fruits of *Medemia argun* palm were found in Egypt (Monastry of Epiphanus at Thebes) was the sixth to seventh century A.D.; no living palms were recorded from the country from that date. However, Täckholm & Drar (loc. cit.), refer to a doubtful record, an unbranched fan palm with small fruits called 'Doleib' growing at Nakhila near Kurkur Oasis. They add: 'to be looked out for! The name Doleib is usually applied in the Sudan to *Borassus* which, has much larger fruits'. Therefore, the presence of living *Medemia* palms remained doubtful and the tree was thought to be extinct in Egypt. In 1963, the late Professor Vivi Täckholm and the present author visited Dungul, a formerly inhabited oasis in the Nubian desert of Egypt, about 220 km southwest of Aswan. As expected, the palms growing there were dom and date palms. However, it was a remarkable sensation to see a fan palm with an unbranched stem about 12 m high (Fig. 1), together with a few small young plants, and an immense amount of brown-violet fruits on the ground and others hanging down from the crown in large clusters. The rediscovery of a living *M. argun* fan palm in Egypt was therefore realized (Boulos, 1968). In 1964, Dr Bahay Issawy of the Egyptian Geological Survey discovered one more *M. argun* palm at Nakhila Oasis, about 200 km west of Aswan, the same locality from which the doubtful record was quoted above. Dr. Issawy also reported five *M. argun* palms cut down at Nakhila. Apparently, Dungul and Nakhila oases had been visited by nomadic desert dwellers, who cut down these five palms, collected their edible fruits and used the leaves to make strong ropes. Lucas & Synge (1978) list *M. argun* among endangered species on a world scale and, together with other data, give some information on its distribution.

3. Littoral and coastal lands, and wetlands

Littoral lands in the region are usually occupied by maritime sands or rocks, in a narrow belt parallel to the Mediterranean seashore, whose width varies from several metres to 1–2 kilometres; the vegetation of these regions will be referred to as littoral vegetation. Coastal lands, however, occupy a much broader belt, bordering the littoral zone from the seaside and penetrating inland 30 km or more. They may include vegetation which is neither littoral, nor desert or montane; the vegetation of this zone will be referred to as coastal vegetation.

The littoral vegetation, although rather poor in number of species, comprises well-defined communities with species which characterize that type of vegetation, such as *Ammophila arenaria* (L.) Link (usually dominant), *Euphorbia paralias* L., *Cakile maritima* Scop., *Silene succulenta* Forssk., *Eryngium maritimum* L., *Pancratium maritimum* L., and some other species often encountered on the maritime dunes along the coast. Zohary (1973) describes the coastal psammophytic communities known from littoral sands and gives the figures 15%, 30%, 35% and 60% for the cover in four different associations representing the different habitat types along the littoral zone of the eastern and south-eastern Mediterranean. None of the above species, nor the others associated with them (see Zohary, loc. cit.) seem to be rare, threatened or endangered, unless the littoral communities are locally destroyed due to the construction of summer resorts on the sea shore. Moreover, most of these are circum-Mediterranean species and some even extend in distribution beyond the basin. However, exceptionally rare or endemic species are also known from the maritime zones. Avishai (pers. comm.) has commented on *Iris atropurpurea* Baker, an endemic species restricted to the loamy sands and littoral region north of Gaza to south of Nayanya, along the western coast of Palestine. This habitat seems to be quite exceptional for any of the irises of Sect. *Onocyclus,* which are known to grow in the coastal or inland arid hot regions of the East Mediterranean, away from the littoral zone.

In contrast to the littoral zone, the coastal vegetation is extremely rich in the number of species, as well as in the number of endemics. It comprises different types of habitats and soils which contribute to the diversity of its plant communities. Over

Fig. 1 The fan palm *Medemia argun* (Mart.) Württ. ex Wendl. in Dungul Oasis, Egypt.

90% of the Libyan flora is known from the coastal strip. The Mediterranean coastal species in Egypt comprise about 52% of the flora of the country (Täckholm, 1974; Boulos, 1975) which is a very high proportion if we take into consideration that Egypt's northern border is the Mediterranean and its eastern is the Red Sea, apart from the three rich montane regions of Sinai, Red Sea hills and Gebel Elba. In general, the coastal flora of the Eastern Mediterranean countries as a whole constitutes a very high proportion of the entire flora of the area. Moreover, coastal floristic elements can penetrate a considerable distance inland, as in some northern regions of Jordan.

Diverse human activities greatly affect the coastal flora and subject many species to different degrees of threat. The most important are grazing (by goats, sheep, camels, donkeys etc.), replacement of dry seasonal farming by perennial, irrigation canal systems, uncontrolled deforestation and the introduction of exotic arboreal species, the reclamation of salt marshes, and the establishment of industrial and tourist centres which bring problems of water and air pollution. As an example of the effect of grazing on coastal vegetation, the following experiment (Ayyad, 1980) was carried out within the framework of the 'SAMDENE' project (Systems Analysis of Mediterranean Desert Ecosystems of Northern Egypt) in the Omayed inland, non-saline depression, 83 km west of Alexandria and 10 km south of the Mediterranean Sea coast. An experimental area of 50 hectares was protected against grazing by the erection of fences in 1974. Since 1977, 20 hectares has been maintained with 50% grazing pressure, and 10 hectares with 25% grazing pressure. After three years of protection, there was a remarkable overall increase in the density and cover. Some palatable species inside the fenced area started to appear, such as *Artemisia monosperma* Del. and *Stipa lagascae* Roem. & Sch. Others increased in density, e.g. *Anabasis articulata* (Forssk.) Moq. and *Helianthemum lippii* (L.) Pers. Some showed increase of cover, such as *Echiochilon fruticosum* Desf., *Gymnocarpos decandrum* Forssk. and *Helianthemum lippii.* Only a few species exhibited a trend of decrease in density or cover.

In Israel, some vegetation noda are threatened with extinction due to the reclamation of land for agriculture. In that connection Zohary (1959) has written: 'In the Coastal Plain only small remnants of Mediterranean halophytic communities were left. Their destruction threatens some very rare species with extinction. In the same plains, the *Eragrostis bipinnata – Centaurea procurrens* association is about to disappear as a result of the rapid extension of citriculture'.

There are many other cases which can be quoted in relation to the rare and endemic species which are endangered within the coastal flora of our region. *Iris jordana* Dinsm. and *I. petrana* Dinsm. (Iridaceae) from Jordan, which have promising potential as ornamentals are, according to Lucas (1980), threatened. The Mediterranean coastal flora in general is known to be rich in bulbous species. Many of these are threatened in those regions where mechanical ploughing has been introduced. Other threats include extensive collection of some species for their ornamental or medicinal value. The following bulbous species are endangered (Lucas, 1980): *Leopoldia albiflora* Täckh. & Boulos and *L. longistyla* Tackh. & Boulos (Liliaceae) in Egypt, *Bellevalia salah-eidii* Täckh. & Boulos (Liliaceae) in Egypt and Libya, and *Allium crameri* Aschers. & Boiss. (Alliaceae) in Egypt.

At the same time, some of the commonest and most typical elements of the coastal flora of Egypt, such as *Papaver rhoeas* L. (Papaveraceae) and *Anemone coronaria* L. (Ranunculaceae), are becoming less and less common. This may be attributed to the drastic changes which are taking place in the mode of agriculture and drainage systems as well as the industrialization and urbanization of the region. Coastal forests in Latakia in northern Syria are also threatened by new, extensive tourism projects (Dr. Ibrahim Nahal, pers. comm., 1978). If this is the case with what we used to call 'common' species and vegetation-types, we must view with great concern the present and future threats that may await the numerous endemic and rare species.

In the arid Eastern and South-eastern Mediterranean region, the habitats that have suffered most are the wetlands – both coastal and inland. Zohary (1959) claims that the hydrophytic vegetation in Israel was the first to suffer from large-scale drainage activities. He adds that the Huleh swamps were until recently one of the largest representative samples of the Near-Eastern hydrophytic vegetation; similar but smaller areas have been completely destroyed in the Sharon Plain. Dafni & Agami (1976) set out a list of 26 species, which are now, according to them, extinct in Israel, together with the date when these plants were last collected or seen in the country. 76% of them occurred in wetland habitats.

In Egypt, due to the changes in agricultural and drainage systems, many species of wet-lands suffered or became endangered. The following species became quite rare: *Ranunculus rionii* Lagger, *Utricularia inflexa* Forssk., *U. gibba* L. subsp. *exoleta* (R. Br.) P. Taylor, *Ottelia alismoides* (L.)

Pers., *Najas minor* All. and *N. graminea* Del. Lake Mariut, near Alexandria, which was once a fresh-water lake, became briefly connected with the Mediterranean Sea about 180 years ago, and its water is now brackish. Lake Qarun, near Faiyum, which was known as a habitat for several interesting fresh-water flowering plants and a rich micro-flora and fauna, has recently become brackish, again due to sudden and drastic changes in the drainage system. *Myriophyllum spicatum* L., an aquatic herb, very rare in Egypt, but classically known from Lake Qarun, is no longer found there. *Ottelia alismoides,* a submerged freshwater species, typical of the flora of rice fields and irrigation canals, and according to Täckholm (1974) 'often choking the water entirely', is now becoming more and more infrequent. The case histories which follow, of *Cyperus papyrus* L. and *Ceruana pratensis* Forssk., illustrate two other examples of how changes in irrigation systems, in historical or recent times, using by means of either simple and primitive or advanced and sophisticated methods, may bring about changes in the ecology of hydrophytes which threatened their existence.

4. Hilly and mountainous regions

The elevated lands in the eastern and south-eastern Mediterranean region most often consist of low hills, usually rising to less than 1,000 m, although high mountains to over 2,000 m also exist. In Libya, the maximum elevation of Gebel Akhdar (Green Mountain) in Cyrenaica is 878 m, and that of Gebel Nefousa is c. 800 m. No conspicuous high land is encountered in the Mediterranean part of Egypt, if the plateau of Sollum (c. 250 m) near the Libyan frontier is excluded. Gebel Maghara (735 m), Gebel Halal (890 m) and Gebel Yelleg (1,090 m) are the most important areas of high ground in the Northern Desert of Sinai. The high plateau in southern Sinai has several peaks, of which Gebel Katherina (2,642 m) is the highest. The Red Sea highland, within the Eastern Desert of Egypt, extends more or less parallel and close to the coast, and is interrupted in several places by desert wadis; among the highest mountains are Gebel El-Shayib (2,184 m) and Gebel Hamata (1,978 m). In the southeast corner of Egypt, is the Gebel Elba granite massif, of which Gebel Shendib (1,912 m) is the highest summit within the Egyptian territory. In Palestine, a wide belt of highland extends from the southern foot of Mount Lebanon to the desert of Northern Sinai, with the highest point (1,208 m) in the north at Gebel Jarmak. The plateau of Jordan which extends from north to south has its highest point at Gebel Rum (1,754 m) in the southern part of the country. The main mountain chains in Lebanon and Syria are the Lebanon Mountains, with their highest peak (3,090 m) in Northern Lebanon, the Anti-Lebanon Mountains, parallel to and east of Lebanon Mountain, and Gebel El-Sheikh (Mt. Hermon), south of the Anti-Lebanon, with the highest peak at 2,700 m. Gebel Druz (1,735 m), an irregular basalt dome, is located in Southern Syria.

The most conspicuous floral element of the whole region is the montane flora, in particular the forests. Cedar of Lebanon, *Cedrus libani* A. Rich. (Pinaceae), is one of the most celebrated trees, of the Old World, and its recorded history goes back to time immemorial. It is unfortunate that the forests of our region are among those which have suffered most devastation, compared to other parts of the world, and what is now left is barely 5% of the original area. I quote Zohary (1959) in this connection: 'There are a few 'sacred forests' in the Mediterranean territory which have been preserved through the ages, testifying to the beauty of landscape in the distant past. They are also reliable indicators of climaxes in various areas, dominated today by 'man-made deserts'. These invaluable remnants are threatened with destruction, unless prompt measures are taken to give them the necessary protection.'

The mountains of Lebanon and Syria contain many interesting arboreal species which are conspicuous components of the forests. Mouterde (1966) describes *Cedrus libani* as the 'king' of them all. The best cedars that are still thriving grow between 1,500 and 1,800 m. Other coniferous trees, which accompany the cedars are *Juniperus excelsa* Bieb., *J. oxycedrus* L., and *J. drupacea* Labill. Towards the north, *Abies cilicica* (Ant. &

Ky.) Tchih. may replace the latter. *Quercus coccifera* L. and *Q. infectoria* Oliv., two coastal species, are also encountered on the mountains, together with *Q. cerris* L., *Q. pinnatifida* C.G. Gmel., *Q. cedrorum* Ky and *Q. brantii* Lindl. Other trees and shrubs typical of these mountains are *Styrax officinalis* L., which may grow to an altitude of 1,600 m or more, *Sorbus umbellata* (Desf.) Fritsch, *Cotoneaster nummularia* Fisch. & Mey, and *Lonicera nummulariifolia* Jaub. & Spach. Mouterde (loc. cit.) adds that after the greenness of cedars, and without any transition, a type of desert follows which is not inhabited by trees, and where the vegetation is mainly made up of spiny cushion-like herbs and undershrubs, which tend to grow near rocks as in other mountains of arid regions, e.g. many *Astragalus* and *Acantholimon* species. At the highest altitudes, up to 2,800 m, the arboreal vegetation is represented by *Juniperus excelsa* Bieb.

In southern Jordan, some interesting herbaceous elements are restricted to the southern arid mountains of Ras en Naqb at about 1,400 m, e.g. *Thalictrum isopyroides* C.A. Mey., *Beibersteinia multiflora* DC., *Tulipa polychroma* Stapf, *Pyrethrum santolinoides* DC. (Boulos & Al-Eisawi, 1977a). The arboreal vegetation of the southern mountains of Jordan most typically comprises, however, a *Quercus coccifera* - *Juniperus phoenicea* association (Zohary, 1962) at an altitude of 1,000 to 1,700 m. Associated species are: *Crataegus azarolus* L., *Pistacia palaestina* Boiss., *P. atlantica* Desf., *Rhamnus palaestinus* Boiss., and the endemic evergreen shrub *Daphne linearifolia* Hart, which is frequent in Petra and the highlands northwards from there.

According to Zohary (1973), the most common association found in Palestine and Jordan is that of *Pinus halepensis* - *Hypericum serpyllifolium*. Associated trees and shrubs are *Quercus coccifera* L., *Pistacia palaestina* Boiss., *P. lentiscus* L., *Rhamnus palaestinus* Boiss., *Arbutus andrachne* L., *Crataegus aronia* (L.) Bosc. ex DC. and *Ceratonia siliqua* L., as well as many shrubs and herbs. There are other associations of *Pinus halepensis* with *Cupressus sempervirens* or with *Juniperus oxycedrus,* the latter species being very rare (Zohary, 1962, 1972). In general, *P. halepensis* (Aleppo pine) occupies pale, calcareous soils at 100–1,000 m.

Juniperus phoenicea L., a circum-Mediterranean coniferous element, is known from the mountains of Northern Sinai (Gebel Maghara, Halal and Yelleg) where some old trees grow; it is the only known spontaneous conifer in Egypt. These trees seem to be remnants of a forest which was once abundant but is no longer regenerating. In the mountains of Southern Sinai, some interesting endemics are known; among these are *Cotoneaster orbicularis* Schlecht., *Rosa arabica* Crép., *Phlomis aurea* Decne., *Ballota kaiseri* Täckh., *Nepeta septemcrenata* Ehrenb., *Caralluma sinaica* (Decne.) A. Berger, *Hypericum sinaicum* Hochst. and *Primula boveana* Decne. Some other species are only known in Egypt from Sinai, e.g. *Crataegus sinaica* Boiss., *Pistacia atlantica, Astragalus palaestinus* Eig, *A. fresenii* Decne., *Epipactis consimilis* Don (the only orchid recorded from Egypt), *Alkanna orientalis* (L.) Boiss. and *Sageretia brandrethiana* Aitch.

In the maquis of Gebel Akhdar of Cyrenaica, Libya, a small endemic tree, *Arbutus pavari* Pamp. (Fig. 2), grows in association with some typical Mediterranean elements, e.g. *Ceratonia siliqua, Pistacia lentiscus, Quercus coccifera* and *Phillyrea angustifolia*. According to Bartolo *et al.* (1977), the endemic elements in Cyrenaica represent about 10% of the flora. Among these endemics are some of potential ornamental value, including *Crocus boulossi* Greuter, *Cyclamen rholfsianum* Aschers. and *Arum cyrenaicum* Hruby. The *Cyclamen* and *Arum* are rather common, in contrast to the *Crocus,* which is only known from the 1967 type collection and another, made 10 years after its discovery, at the locus classicus. This should not be understood as an appeal to exploit these interesting plants on any scale by collecting specimens from the wild, but rather to encourage some scientific institutes to keep among their living collections a few specimens from unusual localities. Some other spices are of interesting medicinal value or of culinary use as species, e.g. *Thapsia garganica* L. var. *sylphium* Ascherson, *Origanum akhdarensis* Ietswaart & Boulos and *O. cyrenaicum* Bég. & Vacc.

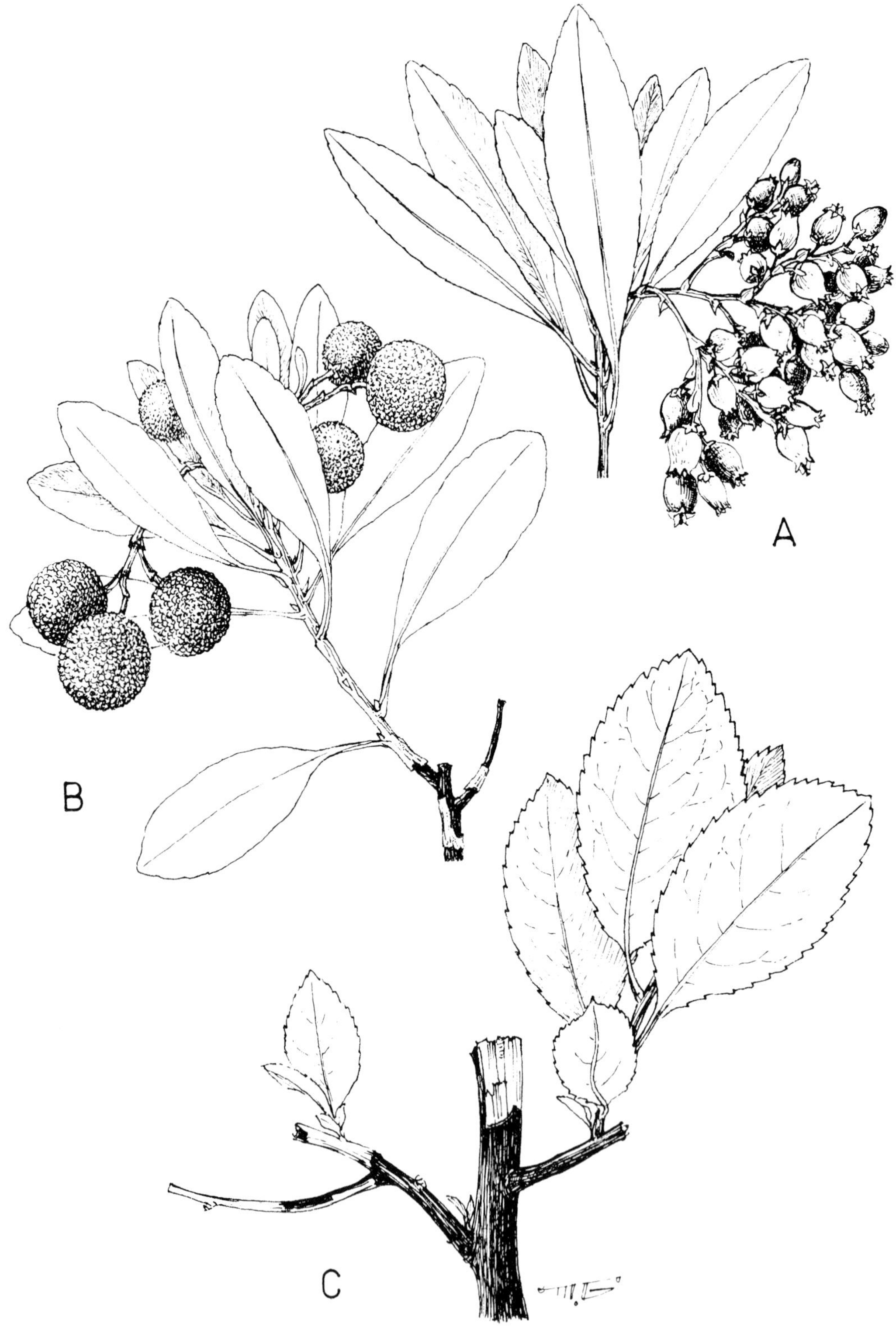

Fig. 2. Arbutus pavari Pamp., A Flowering branch; B. Fruiting branch; C. Lower leaves on vegetative branches.

In Gebel Nefousa, Tripolitania, more arid than Gebel Akhdar, patches of evergreen *Ceratonia siliqua - Pistacia atlantica* maquis are found. In some regions, e.g. Yefren and Gharian, a few old trees (probably over 600 years old) of *Pistacia atlantica* stand as relics from the past, probably representing the remnants of a once dense vegetation.

5. Some case histories of rare and threatened plants of the south-eastern arid regions

5.1 Ceruana pratensis *Forsk. (Compositae) (Fig. 3)*

The genus *Ceruana* is monotypic, comprising one species, *Ceruana pratensis,* described by Forsskål (1775) from Cairo, Egypt. The plant is a herbaceous annual with yellow discoid flower heads and stout branches which become stiff and rather woody upon maturity. According to Fayed (1979), the species is known from Egypt, Sudan and tropical north-western Africa (Chad, Nigeria, Ghana, Mali, Senegal and Gambia) (see map, Fig. 4). The distribution of this hydrophyte follows the course of the rivers Nile, Niger, Benue, Volta and Senegal. Its most characteristic habitat seems to be the muddy banks of these rivers, their tributaries, major irrigation canals and the inundated ground in their vicinities, usually on alluvial soil.

Seemingly, the plant was common both in ancient and modern Egypt. In Ancient Egypt, the plant was used for making baskets and mummy coffins. It is said that the basket in which the child Moses was put into the Nile was made of *Ceruana pratensis.* In modern Egypt, the entire plant is used to make brooms (Boulos & El-Hadidi, 1967). The Ancient Egyptians would have obviously depended on easily available and abundant raw material of *Ceruana.* The manufacture of brooms from the entire plant until the sixties of this century provides good evidence of its recent abundance in Egypt. Boulos & El-Hadidi (1967) enumerate *C. pratensis* among 150 common weeds in Egypt, with the note: 'Especially common on Nile banks in Upper Egypt' and El-Hadidi & Ghabbour (1968) listed *C. pratensis* among the plants recorded in summer 1967 from two localities in the Nile Valley region at Aswan, Egypt, with the remark: 'weed on the Nile banks'. For several other weedy species, the remark given was 'rare weed', which suggests that *C. pratensis* was not among the rare weeds of Aswan before 1967.

The plant was not recorded from the Nile region in Egyptian Nubia (Boulos, 1966), in the area from Aswan High Dam southward to the Egyptian-Sudanese frontier. All available records from herbarium specimens examined by the author or cited by Fayed (1979) show that the area of distribution of this species in Egypt lies between Aswan and Delta Barrage several kms Northwest of Cairo (see map, Fig. 5). The record from the western Mediterranean coast given by Täckholm (1974) seems to be based on a misidentified specimen.

After the construction of the Aswan High Dam in 1964, the annual water flood, with its vast amounts of alluvial silt, ceased to cover the traditional basin-irrigated land of Upper (southern) Egypt. As a result less and less silt was deposited on the banks of the Nile and irrigation canals, as well as on the hundreds of thousands of acres in Upper Egypt. Consequently, our species became increasingly less common in the new artificial habitat, and now (1980) it is extremely difficult or even impossible to collect *C. pratensis* from any of its classical Egyptian localities in the Nile Valley.

Looking through the specimens deposited in the herbaria of Cairo University (*CAI*) and the Agricultural Research Centre, Dokki, Cairo (*CAIM*), one notices the existence of abundant collections up to 1964, in contrast to the very few records from 1967 to 1971. Efforts to collect specimens during the last five years (1975–1980) from classical localities on the Nile banks from Aswan to Cairo, even in small quantities for taxonomical and cytological research, met with no success. Only two records, both from 1977, are known to me: the first from Aswan in 1977, and the second from one of the islands of the First Cataract area in Aswan. No more plants have been afterwards on that or on any other island (Mrs. Irina Springuel, pers. comm.), in spite of frequent visits to the area - almost every fortnight for the last four years (1977–1980). This suggests that the plant does not regenerate and is in

Fig. 3. Ceruana pratensis Forsk.

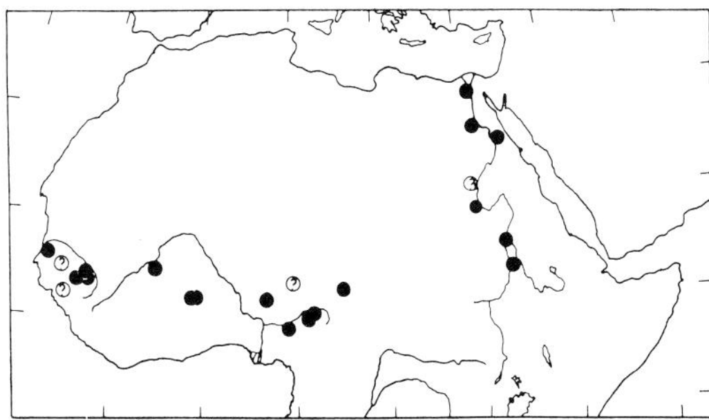

Fig. 4. General distribution of *Ceruana pratensis*, mainly after Fayed (1979) and partly after Ghabbour (pers. comm.).

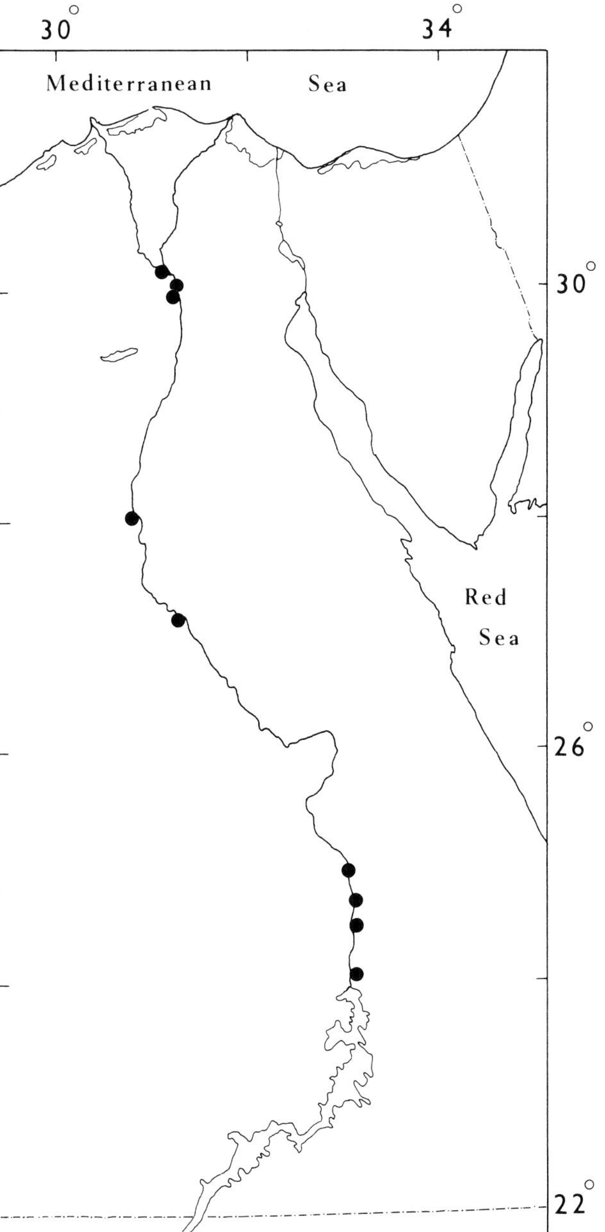

Fig. 5. Distribution of *Ceruana pratensis* in Egypt, based on herbarium specimens from CAI and CAIM.

the process of disappearing. In fact, there is evidence enough to prove that *C. pratensis* has already disappeared from throughout its range of distribution in Egypt, after a long and well-documented history that dates back several thousand years.

5.2 Diplotaxis villosa *Boulos & Jallad (Cruciferae) (Fig. 6)*

Diplotaxis villosa, was a desert annual discovered in 1975 (Boulos & Jallad, 1975) in a remote area in the southern desert of Jordan (Fig. 7). I repeat some of what we wrote in 1975 (Boulos & Jallad, loc. cit.): 'The occurrence of this remarkable new species within a vast area stretching over a few square kilometres, with thousands of individuals almost in pure stands, may draw attention to the need to carry on further floristic studies...' A few months later, I made an excursion in late spring 1976 with two of my students to the El Jafr-Bayir desert, southern Jordan (Boulos, 1977) in order to collect some seeds of this peculiar, yellow-flowered, showy desert annual. To our surprise, the rich ephemeral vegetation and the high percentage of cover of the previous year had been replaced by very scanty vegetation, obviously due to the scarcity of rain in 1976 in contrast to the exceptionally good rain in 1975. Few, small specimens of *D. villosa,* mostly with unripe fruits, denoting late and scanty rainfall, were seen scattered here and there, with much more bare ground than in the previous season. As a result, we failed to bring back even a fraction of a gram of seeds.

In the desert of southern Jordan, as in all other deserts in the southern and south-eastern Mediterranean countries, rain is most erratic, both in terms of quantity per season and per shower, as well as in the length of the intervals between different showers during the short winter rainy season. As a result, the vegetation expressed in terms of

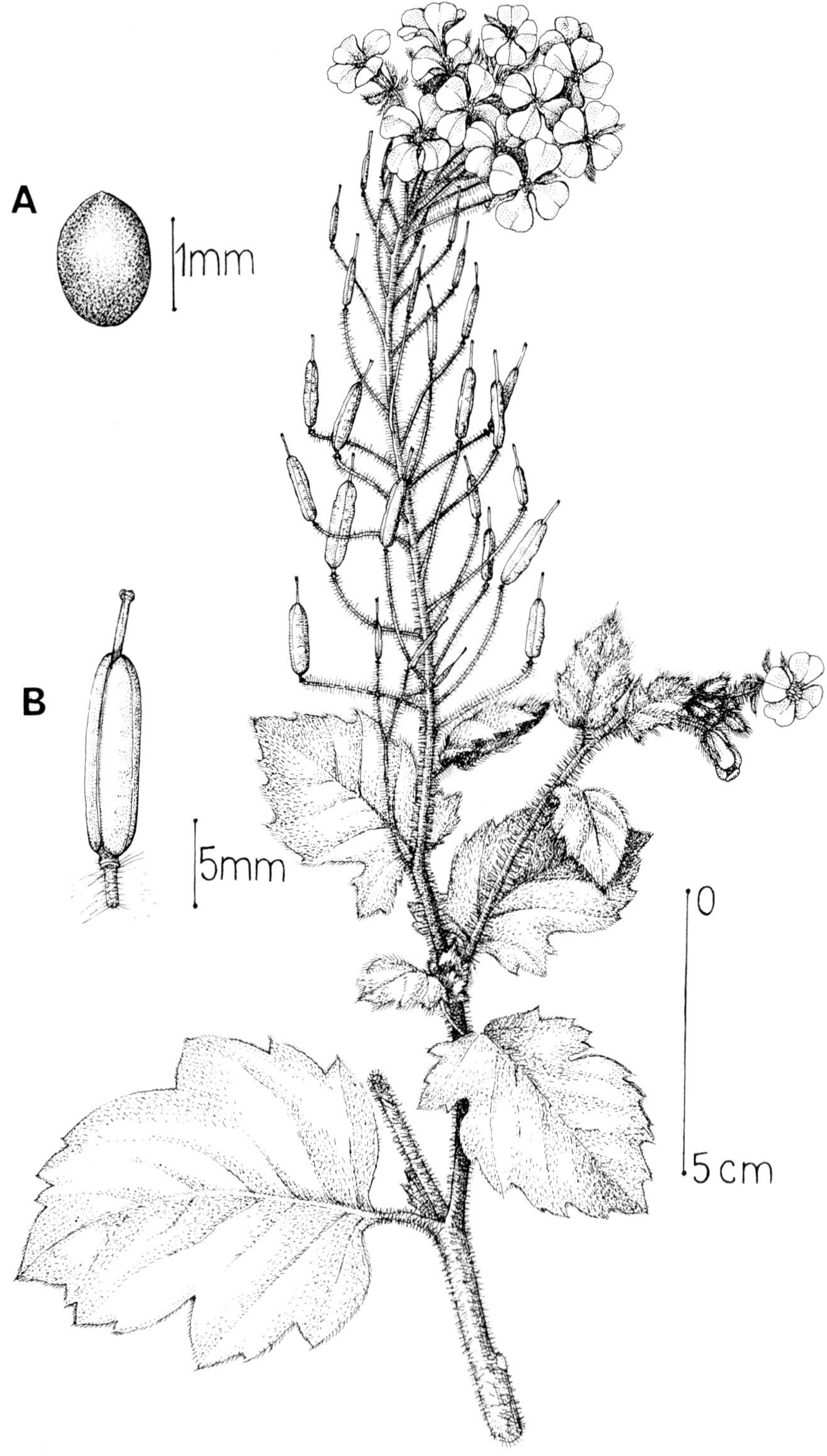

Fig. 6. Diplotaxis villosa Boulos & Jallad.

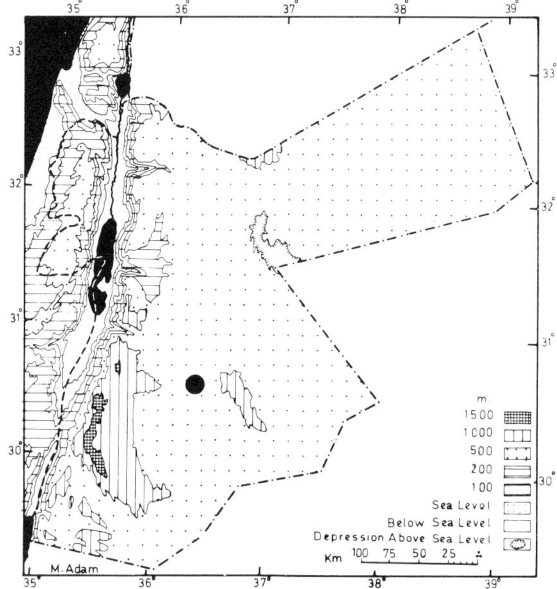

Fig. 7. Distribution of *Diplotaxis villosa* in Jordan.

percentage of cover and vigour of ephemerals, strongly reflects the scarcity or the abundance of rain.

I may add here that the highest observed rainfall in 24 hours in El-Jafr area was 29.5 mm in February 1975, while the mean annual rainfall is as low as 35.6 mm (Boulos, 1977). Again, this may clearly explain why we observed such an exceptionally rich ephemeral vegetation in 1975, when our new species was firstly collected.

D. villosa is endemic to Jordan, and represents one of those desert endemics which are restricted to one or a few limited sites of special ecological identity. However, it should be looked for in other desert wadis both in Jordan and in Northern Saudi Arabia.

D. villosa should not be classified as endangered, but would be better classified as rare (though common in one rather limited site). However, it may be worthwhile to be reminded of the possibility of several successive dry seasons occurring in the future, in connection with overgrazing, as the plant is palatable to sheep. These conditions would move this crucifer closer to the 'Vulnerable' species category.

5.3 Cyperus papyrus *L. (Cyperaceae)*

The papyrus plant, *Cyperus papyrus* L. played an important role in almost every aspect of life in ancient Egypt. Apart from the well-known paper rolls or 'papyri' which were manufactured from its long, stout, spongy culms, the plant became a symbol of the Kingdom of Lower Egypt, and was used for food, medicine, to make boats, cords, sandals, mats, baskets, boxes, chairs, beds and funeral garlands and offerings (cf. Täckholm & Drar, 1950; Laurent - Täckholm, 1951).

Cyperus papyrus is one of the giant sedges native to Central, West and South Tropical Africa. Most probably it was introduced into Egypt tens of thousands of years ago from Central Africa, as floating clumps, with the summer floods of the Nile. It became naturalized in the shallow waters of swamps, marshes and the banks of irrigation and navigation canals, which were especially abundant in Lower Egypt. This particular type of vegetation, is well documented in Ancient Egyptian history, probably because it was also a favourable habitat for large game animals which allowed water hunting, a sport practised by Ancient Egyptians. The famous relief at the tomb of Mereruka, Sakkara which dates back to the fifth dynasty, c. 2500 B.C., shows hunters in papyrus boats pursuing hippopotamus in the marshes, with crocodiles in the water and different bird species within the dense vegetation of papyrus thickets (Fig. 8).

The products of papyrus constituted one of the main economic resources of the Ancient Egyptians, especially through the manufacture of paper which was exported from Alexandria in enormous quantities during the prosperous Graeco-Roman period (Täckholm & Drar, 1950: 130). It is difficult to believe that the state-monopolized paper industry depended exclusively on the limited supply of the naturalized papyrus plants. Therefore, it is probable that papyrus was cultivated in Ancient Egypt. Dr. Hassan Ragab (pers. comm.) adds: 'In order to produce a good quality paper, the Ancient Egyptians must have depended on plants with standardized specifications, which is difficult to achieve unless the plants were cultivated under control and in huge areas or 'farms'; the cultivation

Fig. 8. Relief from the tomb of Mereruka, Sakkara, 5th dynasty (c. 2500 B.C.) showing hunters in papyrus boats pursuing hippopotamus in the marshes. Note the crocodiles and birds within the dense papyrus thickets (Photo Lehnert & Landrock, Cairo, with kind permission).

of papyrus was most probably, like the manufacture of paper, monopolized by the state'.

Papyrus cultivation and the paper industry continued to flourish in Egypt until the Arab conquest of the country in the seventh century A.D. Täckholm & Drar (1950) says: 'Papyrus was cultivated and manufactured by the Arabs in Egypt until the 8th and 9th centuries A.D.'. However, due to the introduction of new agrarian systems, marshes and swamps dried up, which caused the papyrus plants to disappear gradually (Ragab, 1980). Left without any care they faced strong competition from at least two native, troublesome, aquatic weeds; the perennial grasses *Phragmites australis* (Cav.) Trin ex Steud., (common reed) and *Vossia cuspidata* (Roxb.) Wall & Griff. (H. Ragab, pers. comm.).

P. australis possesses a well-developed root system emerging from stout creeping rhizomes, which may extend for 10 metres or more. *V. cuspidata* possesses submerged or floating, many-noded culms which produce numerous roots from the submerged nodes – 'A floating grass, which with *Saccarum spontaneum* L. makes the great grass bars of the Nile' (Willis 1973). These two grass species, with their well developed root systems, are vigorous competitors and were able to retain the ecological niche which they had shared for centuries with the adventive *Cyperus papyrus*. In this connection, Ragab (1980) points out that when the rhizomes of papyrus grow in a pond, they rise a bit closer to the water surface every year. Young branches develop over an older bed of moribund and decaying rhizomes (see Fig. 9). When there is a shortage of water supply the rhizomes dry up and the plants die.

The Arabs, moreover, introduced into Egypt in the tenth century A.D. a revolutionary method for the manufacture of paper. They used linen fibres as a raw material and abandoned papyrus pith (Täckholm & Drar, 1950; Ragab, 1980). The newly introduced method not only provided cheaper paper but its manufacture was also much easier to carry out than the obscure, or perhaps secret, formula of the Ancient Egyptians. As a result, the cultivation of papyrus was neglected. Forgotten papyrus thickets continued to propagate under rather unfavourable conditions, behaving as semi-wild plants, but quickly declined and became res-

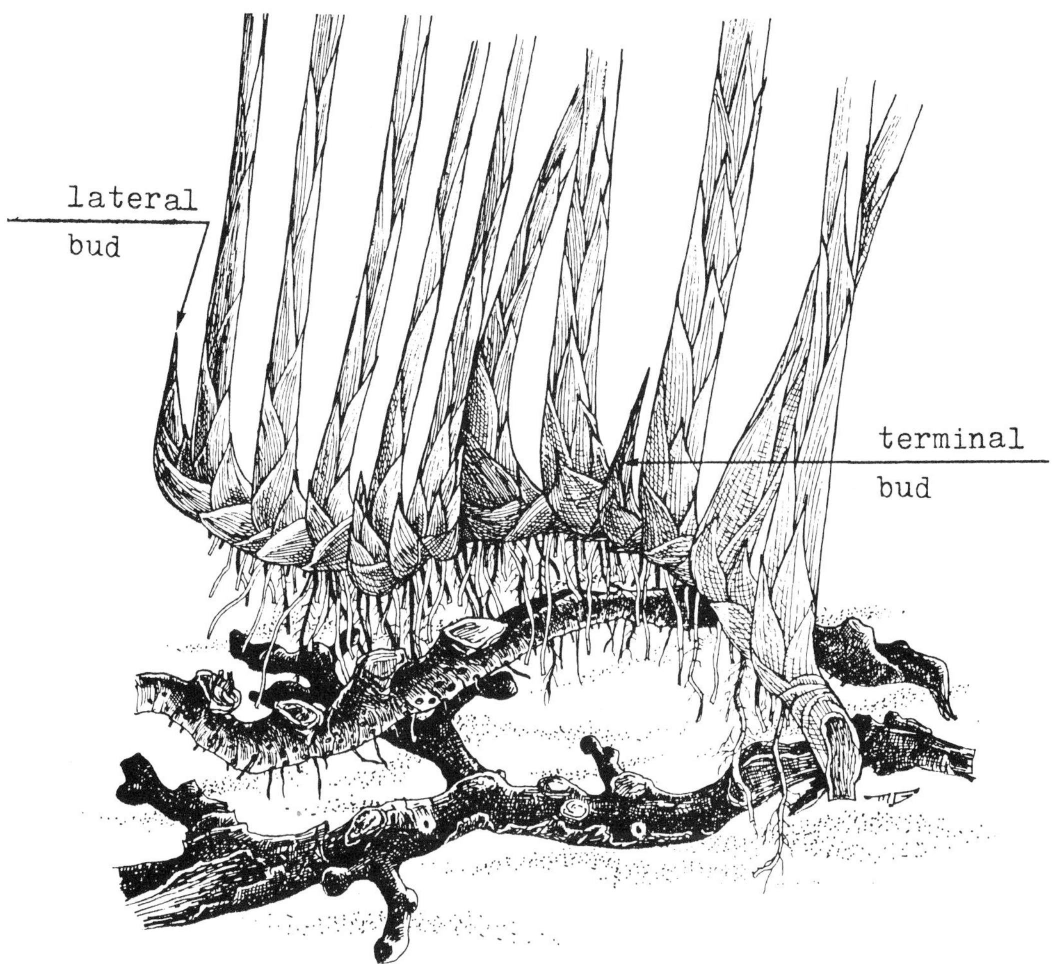

Fig. 9. Young rhizome branches of *Cyperus papyrus* L. developing over an older bed of declining and decaying rhizomes (after Ragab, 1980, with kind permission).

tricted to isolated areas due to lack of water supply and neglect by man. According to Täckholm & Drar (1950), *Cyperus papyrus* had not been recorded in any of its natural habitats in Egypt since 1820–21, when Baroness H. v. Minutoli observed it near Damietta and on the banks of Lake Manzala at the northeast corner of the Nile delta, not far from the Mediterranean seashore. The plant had been considered extinct in Egypt until it was rediscovered by Professor M.N. El-Hadidi in 1968 in the Wadi Natrum depression (west of the Nile delta) in a freshwater marsh, where some 20 plants were growing among other reeds (El-Hadidi, 1971). The papyrus plants of Wadi Natrum were later described as subsp. *hadidii* Chrtek & Slaviková (1977). According to Lucas & Synge (1978), this newly discovered subspecies of *C. papyrus* is endangered because its habitat is decreasing due to the extraction of water from the Nile and to certain changes in the irrigation pattern of the country. *C. papyrus* subsp. *hadidii,* together with other subspecies of papyrus are now in cultivation at the Papyrus Institute farms of Dr. Hassan Ragab in Giza, Egypt.

5.4 The 'Sylphium' of Antiquity

The source of the 'Sylphium' legend is Libya. The reason why it may be called a 'legend' rather than a 'story' is that the plant on which it is based is still a mystery, or at least not known to us with certainty. However, it is rooted in historical fact.

One of the most celebrated herbs of Libya, which brought great richness to the country during ancient times, was the 'Sylphium' of antiquity, a precious plant which yielded a gum-resin, the price of which was estimated by an equal weight of silver (Boulos, 1970). It was known to occur in Cyrene (now Shahhat, Cyrenaica), and was exported to neighbouring countries because of its extraordinary property of curing many diseases, and its supposed mysterious power as an aphrodisiac. The continuous over-exploitation of sylphium for several centuries led to its disappearance around the fifth century A.D. The question which therefore needs an answer is: which plant species did sylphium represent? It is difficult to give a definite answer to this question as the available documented information on that plant has come down to us only in the form of some decorative drawings, such as those on old coins. Examining the figures which represent sylphium on coins (Fig. 10), it might be possible to attribute the plant in question to the family Umbelliferae. From the information available about the present vegetation in Shahhat (old Cyrene), good evidence exists to support the view that *Thapsia garganica* L. var. *sylphium* (Viv.) Aschers. (= *Thapsia sylphium* Viv.) is the species that most nearly matches the coin figures of sylphium. This plant still grows in the vicinity of Shahhat and in several other localities in Cyrenaica (Pampanini, 1931). I collected *T. garganica* var. *sylphium* several times in Libya from Shahhat, and it is locally known as 'Drias'. I

Fig. 10. Two coins with designs representing Sylphium.

was told by some dwellers in the region that the plant is harmful to livestock, especially to camels. French (1971) gives the following information on the biological properties of *T. garganica* L. from the Mediterranean region - 'Medicine: epispastic, revulsive, for thinness, chest ailments, sterility, rheumatism, gout'. Apparently, these biological properties correspond to *T. garganica* var *garganica,* known from North Africa, west of Libya. The biological and medicinal properties of both varieties would be expected to be close to each other. The properties enumerated by French, (loc. cit.) seem to match those of Sylphium of antiquity.

The reason for the plant's regeneration after its disappearance around the fifth century A.D., could be due to the persistence of underground parts of plants for some years under unfavourable conditions. Sylphium, especially its use as a medicinal plant, was forgotten for a long period of time, and the mysterious plant and its traditional use were ignored, due to the shortage of detailed descriptions or any documented information left for following generations. The long interruption of trade between Europe and North Africa during the Middle Ages might have served to make sylphium a completely forgotten item, either as a herb with mysterious curative effects or as a plant of any historical importance.

References

Avyad, M.A. (1980). A preliminary assessment of the effect of protection on the vegetation of the Mediterranean ecosystems. Taeckolmia 9: 85-101.

Bartolo, G., Brullo, S., Gugliemo, A. & Scalia, C. (1977). Considerazioni fitogeografiche sugli endemismi della Cirenaica settentrionale. Arch. Bot. Biogeogr. Ital. 53: 131-154.

Batanouny, K.H. (1983). Human impact on desert vegetation. In: Holzner, W., Werger, M.J.A. & Ikushima, I. (Eds.), Man's Impact on Vegetation, pp 139-150. The Hague: Junk.

Boulos, L. (1966). Flora of the Nile Region in Egyptian Nubia. Feddes Repert. 73: 184-215.

Boulos, L. (1968). The discovery of *Medemia* palm in the Nubian desert of Egypt. Bot. Notiser 121: 117-120.

Boulos, L. (1970). Medicinal herbs in Lybia. Al Harad 16. Esso Standard Libya Inc. Publ.

Boulos, L. (1975). The Mediterranean element in the flora of Egypt and Libya. In: La flora du bassin méditerranéen: Essai de systématique synthétique. Colloques Intern. du C.R.N.S. 235: 119-124.

Boulos, L. (1977). Studies on the flora of Jordan: 5. On the flora of El Jafr-Bayir Desert. Candollea 32: 99-110.

Boulos, L. & Al-Eisawi, D. (1977a). Studies on the flora of Jordan: 6. On the flora of Ras en Naqb. Candollea 32: 111-120.

Boulos, L. & Al-Eisawi, D. (1977b). Studies on the flora of Jordan: 8. New and noteworthy plants. Candollea 32: 269-276.

Boulos, L. & El-Hadidi, M.N. (1967). (illustrated Magdy El-Gohary). Common weeds in Egypt. El Cairo: Dar Al-Maaref.

Boulos, L. & Jallad, W. (1975). Studies on the flora of Jordan: 1. *Diplotaxis villosa* sp. nov. (Cruciferae). Bot. Notiser 128: 365-367.

Chrtek, J. & Slavikova, Z. (1977). A new subspecies of *Cyperus papyrus* from Egypt. Preslia 49: 183-185.

Dafni, A. & Agami, M. (1976). Extinct plants of Israel. Biol. Conserv. 10: 49-52.

El-Hadidi, M.N. (1971). Distribution of *Cyperus papyrus* L. and *Nymphaea lotus* L. in inland waters of Egypt. Mitt. Bot. Staatssammlg. München 10: 470-475.

El-Hadidi, M.N. & Ghabbour, S. (1968). A floristic study of the Nile Valley at Aswan. Rev. Zool. Bot. Afr. 78: 393-407.

Fayed, A. (1979). Revision der Grangeinae (Asteraceae-Astereae). Mitt. Bot. Staatssamlg. München 15: 425-576.

Forsskal, P. (1775). Flora aegyptiaco-arabica. Copenhagen: Carsten Niebuhr.

French, D.H. (1971). Ethnobotany of the Umbelliferae. In: V.H. Heywood (Ed.), The Biology and Chemistry of the Umbelliferae, pp. 385-412. London: Academic Press.

Laurent-Tackholm, V. (1951). Faraos blomster. Stockholm: Natur och Kultur.

Lucas, G. (Ed.) (1980). First preliminary draft of the list of rare, threatened and endemic plants for the countries of North Africa and the Middle East. Kew: I.U.C.N. Threatened Plants Committee.

Lucas, G. & Synge, H. (1978). The I.U.C.N. Plant Red Data Book. Morges: I.U.C.N.

Mouterde, P. (1966). Nouvelle flore du Liban et de la Syrie. Beyrouth: Imprimerie Catholique.

Pampanini, R. (1931). Prodromo de la flora Cirenaica. Forli.

Ragab, H. (1980). Le *Papyrus*. Thése, Institut National Polytechnique, Grenoble.

Tackholm, V. (1974). Student's Flora of Egypt, 2nd. ed. Cairo University, Cairo.

Tackholm, V. & Drar, M. (1950). Flora of Egypt 2. Bull. Fac. Sci. Cairo Univ. 28: 99-146 and 296-302.

Willis, J.C. (1973). A Dictionary of the Flowering Plants and Ferns, 8th ed. Cambridge: Cambridge University Press.

Zohary, M. (1959). Wildlife protection in Israel (flora and vegetation). Proc. I.U.C.N. 7th Techn. Meeting, Brussels 5: 199-202.

Zohary, M. (1962). Plant life of Palestine. New York: The

Ronald Press.
Zohary, M. (1972). Flora Palaestina 2. Jerusalem: Israel Acad. Scienc. and Human.
Zohary, M. (1973). Geobotanical Foundations of the Middle East. Stuttgart: Gustav Fischer Verlag.

Author's address:
National Research Centre
Dokki, Cairo, Egypt

Present address:
Kuwait University
P.O. Box 5969
Kuwait

DS 451

CHAPTER 9

The Maghreb countries

J. MATHEZ, P. QUÉZEL and C. RAYNAUD

1. Introduction

The flora of the Maghreb countries is very rich and highly interesting (Maire, 1952–76; Quézel & Santa, 1962–63; Cuenod, 1964). The presence of high mountain ranges, the relative isolation of the Mediterranean biome from its European or Eastern counterparts, the long interface with the Saharo-Sindian region, the presence of Macaronesian elements in the Western part or of Boreo-Alpine relicts in the high mountains, as well as many other factors have contributed to give this flora an especial appeal for all botanists interested in the Mediterranean area. With regard to conservation, it is generally assumed that the flora and vegetation of North Africa are relatively well preserved, but this is far from true (Quézel, 1959). Since a comprehensive survey is not possible, we will mention some representative habitats or subregions so as to give at least some impression of the real situation.

2. The situation in Morocco

2.1 The High Moroccan Mountains

For many years, the problems of the conservation of the nearly one thousand species which grow in the high mountains of Morocco have been practically ignored. Now, with alarms being increasingly sounded it is advisable to adopt an attitude of caution. The highest summits of North Africa house a relatively rich flora with an important proportion of endemics. At least 160 of them are found above the forest level. Another important group of plants are southern representatives of the Boreo-Alpine stock, so that whatever their distributional status (endemic species or sub-species, vicariants, non-endemics, etc.) their presence in Morocco poses highly interesting biogeographical problems. As indicated by Emberger (1936) and Quézel (1957), many taxa are known from extremely reduced localities, and are represented by a small number of individuals. Such is the case for *Agrostis atlantica* Maire & Trab., *Agropyrum embergeri* Maire, *Ranunculus mgounicus* Quézel, *Cytisus pulvinatus* Quézel, *Alyssum flahaultianum* Emberger, *Saxifraga luizetana* Emberger, *Epilobium psilotum* Maire & Samuels., *Phagnalon iminouakense* Emberger, etc.

A fundamental reason for the vulnerability of many Atlas taxa is the increase in grazing pressure. Myriad sheep and goats roam the mountains in the summer, giving rise to intense overgrazing that prevents proper fruit-set in many species. This occurs in the spiny scrub level, where hemicryptophytes have become rarer and rarer, and also in compact meadows (pozzines) where there is a number of remarkable dwarf plants, either endemic (such as *Gentiana tornezyana* Lit. & Maire, *Plantago rhizoxylon* Emberger, *Potentilla guillermondii* Emb. & Maire, *Carum asinorum* Lit. & Maire, etc.) or of northern origine. The latter are,

Gómez-Campo, C. (ed.), Plant conservation in the Mediterranean area.
© 1985, Dr W. Junk Publishers, Dordrecht. *ISBN 90 6193 523 7.*

in fact, the most threatened due to their extreme rarity (e.g. *Botrychium lunaria* (L.) Swartz, *Polystichum lonchitis* (L.) Roth., *Polygonum bistorta* L., *Luzula multiflora* L., *Cardamine pratensis* L., *Gentianella tenella* (Rottb.) Bornes, *Meum athamanticum* Jacg.) Many species are represented by endemic variants which are particularly vulnerable; *Carex leporina* subsp. *atlasica* Lindl., *Gentiana verna* subsp. *peneti* Lit. & Maire, *Achnatherum argentea* subsp. *mesatlantica* Quézel, and *Aconitum lycoctonum* var. *atlanticum* (Coss.) Maire.

However, in the past few years the montane Atlas flora has faced a new source of danger; the increase in human population has brought about the cultivation of vast areas for summer cereals, on land that was formerly spiny scrub. Hundreds of hectares have been already tilled in the calcareous Middle Atlas mountains and the indications are that this trend will continue. Many species will then become threatened. In a similar way, the search for fuelwood has considerably reduced former stands of *Juniperus thurifera* L. whose depletion over the past thirty years in the zone of Jbel Ahansal has been dramatic. A similar fate can be predicted for several shrubby Genisteae, such as *Genista florida* var. *maroccana* Ball, *Retama dasycarpa* Cosson, and *Adenocarpus anagyrifolius* Cosson & Balansa.

In general, for a significant proportion of the Atlas flora, precise taxonomic interpretation is still needed.

2.2 The Tangier Peninsula and the Atlantic coasts

Due to its mild climate, this area has attracted many foreigners to settle here in the past hundred years. The indigenous population has also grown to a significant extent. The town of Tangier is gradually expanding towards Cape Spartel, thus replacing natural environments with privately owned properties, country residencies, gardens and the like. The list of plant species which have not been observed since the end of the 19th century or the beginning of the 20th century is already substantial and liable to grow longer.

For instance, no recent data are available about *Euphorbia bivonae* Steudel, represented on the rocks of Cape Spartel by an endemic variety (var. *tangerina* Pau), nor about *Celsia commixta* Murb., which is endemic to both sides of the Straits of Gibraltar. The same uncertainly prevails with regard to a whole set of species which, though not endemic, were found in the vicinity of Tangier, as their only locality in Morocco or even in North Africa, i.e. *Pteris palustris* Poiret, *Iris lutescens* subsp. *biflora* (Brot.) Webb & Chater, *Galega officinalis* L., *Thymelaea myrtifolia* (Poiret) Webb, *Hydrocotyle vulgaris* L., *Ammiopsis daucoides* Boiss., *Euphorbia akenocarpa* Gussone, *Mentha cervina* L., *Verbascum densiflorum* Bertol, *Verbascum laciniatum* (Poiret) O. Kunze, *Galium uliginosum* L., *G. parisiense* L. and *Adoxa moschatellina* L. The status of some of these species (*Thymelaea myrtifolia* and *Galium uliginosum*) is sometimes thought to be doubtful; others may perhaps survive in private estates to which botanists have limited access.

To the above list can be added many rare or interesting species which seem to have disappeared from the Tanger area, although they probably still occur in other Moroccan localities or elsewhere in North Africa. In particular mention may be made of *Danthonia decumbens* (L.) D.C. in Lam & DC., *Fuirena pubescens* (Poret) Kunth, *Eleocharis multicaulis* (Sm.) Desv. *Juncus bulbosus* L., *Asparagus maritimus* (L.) Mill., *Iris xiphium* L., *Quercus pyrenaica* Willd., *Thymelaea lanuginosa* (Lam.) Ceballos & Vicioso, *Euphorbia paniculata* Desf., *Teucrium scordium* subsp. *scordioides* (Schreb.) Maire & Petitmengin, *Linaria tingitana* (Boiss. & Reut., *Bartsia aspera* (Brot.) Lange, *Pulicaria sicula* (L.) Moris, *Crepis bourgeaui* Babo. ex Maire, *Crepis salzmannii* Babc., etc. Several other species are increasingly threatened, either in Tanger or in its surrounding area, because of urban development, tourist facilities, drainage of wetlands (for cultivation or for mosquito eradication) or simply because of the restricted nature of their habitats. Such is the case with *Sphagnum rufescens* (Nees & Harnsch.) Warnst, *Davallia canariensis* (L.) Sm., *Asplenium hemionitis* L., *A. marinum* L., *Osmunda regalis* L., *Equisetum telmateia* Ehrh., *Molinia caerulea* (L.) Moench, *Juncus tingitanus* Maire & Weiller, *Iris tingitana* Boiss. &

Reut. var. *tingitana* (see below), *Quercus fruticosa* Brot., *Erica ciliaris* L., *Anagallis crassifolia* Thore, *Pinguicula lusitanica* L., *Plantago macrorhiza* Poiret in its endemic variety var. *tangerina* Pau, *Tanacetum annuum* L. and *Senecio jacobaea* L.

For those who have observed the transformation of Moroccan agriculture through the introduction of modern mechanized techniques, thousands of hectares have acquired a completely new appearance. Fifty years ago, in many areas, swing ploughs used to pass between the clumps of dum palms (*Chamaerops humilis* L.) thus preserving a significant part of the flora. Then, the Rharb plains in springtime were an immense field of irises (*Iris tingitana* Boiss. & Reut. var. *tingitana*), as far as the eye could see. Mechanization has wiped out this plant, now only a curiosity outside flower gardens. *Ornithogalum reverchonii* Lange was only known from one locality in North Africa, in the clumps of dum palms on clay soils in El Hajeb; it seems now to have disappeared from that site. The following species have also gradually become rare among crops on deep clay soils: *Triguera osbeckii* (L.) Willk. and the endemics *Delphinium cossonianum* Batt. and *Lavatera maroccana* (Batt. & Trabut) Maire. The case of *Avena maroccana* Gand. is commented upon below.

On the other hand, the development of tourism all along the Atlantic shore is having a devastating effect in many parts of the Moroccan coasts. Though some endemics such as *Gaudinia maroccana* Trabut are not endangered, other species are much more threatened or vulnerable, as for example *Sesuvium portulacastrum* L., with only one locality on the Atlantic coast between Tarfaya and Portugal, *Asplenium marinum* L., *Silene reeseana* Maire, an endemic species not recently observed in one of its two recorded localities, *Silene mollisima* var. *auriculifolia* (Pomel) Pau along the rocky coasts or *Calystegia soldanella* (L.) R.Br. on sea sands. Besides, many estuaries are currently subjected to a quickly developing evolution. For instance, the dam built on Bou Regreg, a few kilometres above Rabat, regulates the flow of the river and, by supressing floods favours the silting-up of the estuary between Rabat and Salé. This new infilled land will acquire a high value for the construction of urban facilities. This means the imminent destruction of its remarkable glasswort vegetation with such uncommon plants as *Althenia filiformis* Petit or *Suaeda maritima* (L.) Dumort in an endemic variety (var. *perennans* Maire) which is unknown anywhere else.

3. The situation in Algeria and Tunisia

3.1 Algerian forests, shrublands and pastures

The flora of Mediterranean Algeria is well known (Quézel & Bounaga, 1975). The radical transformation in the methods of land-use following the French colonization, had quickly given rise to considerable modification of certain environments. Though hygrophilous ecosystems were the most deeply affected, they were not the only ones. Various inventories were drawn up by Battandier (1894) and Faurel (1959). The war of independence and the demographic explosion that followed it hardly improved matters. Grazing has notably increased in steppe areas, in forests and shrublands of the Tell and especially in the Saharian Atlas Mountains. This has had its effect in spite of the considerable efforts made in reforestation and soil management projects carried out in the past fifteen years by the Algerian government.

Forest ecosystems have undergone a very important regression over the past 25 years, and this has led to a considerable impoverishment of their floristic content. Some forest elements have suffered especially heavily. This is particularly the case of *Pinus nigra* subsp. *mauritanica* (Maire & Feyerh.) Schwarz, whose kabyle race is now represented by no more that one hundred individual trees on the southern side of the Djurdjura. The case of *Abies numidica* de Lannoy and *Cedrus atlantica* (Endl.) Carrière forests are treated with more detail later in this chapter. As for the accompanying herbaceous species, though a recent verification was not possible, we should be worried about the actual status of various endemics in the Constantinian littoral zone, such as *Digitalis atlantica* Pomel, *Pedicularis numidica* Pomel, *Epimedium perralderianum* Coss., *Silene reverchonii* Batt., *Thlaspi bulbosum* subsp. *atlanticum* (Batt.) Quézel, *Cyclamen repandum* subsp. *atlanticum* Maire, and for other species of Mediterranean or European origin such as *Asplenium hemionitis* L. and *Gennaria diphylla* (Link) Parl. on the coast of Algiers, *Galium odoratum* (L.) Scop. and *Convolvulus dryadum* Maire in the Babors region, *Pteris cretica* L. in Guerrouch forest or *Woodwardia radicans* (L.) Sm. and *Luzula campestris* subsp. *multiflora* (Retz.) Buch. on the Edough. Similarly, *Clematis vitalba* L. was never seen again at its only North African locality on the northern slopes of the Cheliah in the Aures, nor *Podanthum rigidum* Willd, subsp. *aurasiacum* (Batt. & Trab.) Dambolt in the forest of Sgag.

The regression of maquis and garrigue vegetation is less pronounced than that of the forests, but many species proper to these shrubby environments had already begun to become rare before independence. This is particularly the case of endemics such as *Adenocarpus umbellatus* Coss. of the coast of Oran and *Adenocarpus faurei* Maire from the surroundings of Tiaret. Similarly *Erica cinerea* var. *numidica* Maire from Cape Rosa is known from one locality only. On the coastal Tell, *Polygala mumbyana* Boiss. is not directly threatened, but this is unfortunately not true for *Thymelaea myrtifolia* (Poiret) Webb and *Acanthus spinosus* L. of Algiers region or of *Phlomis caballeroi* Pau and *Carthamus arborescens* L. on the coast of Oran.

The herbaceous flora of the Tell is very rich in perennials, but it is the abundance of annuals which perhaps better characterizes this environment. Having become adapted for a long time to human influences, they do not seem to have suffered particularly in the past few years. However, a few were already rare or have disappeared after the publication of the new Flora of Algeria (Quézel & Santa, 1962–63). We may mention *Themeda triandra* Forssk., *Holcus setosus* Trin., *Ophrys pallida* Rafin., *Silene glaberrima* Faurré & Maire, *S. scabrida* Soy.-Will. & Godr., *S scabriflora* Brot., *Dianthus tripunctatus* Sibth. & Sm., *Valerianella leptocarpa* Pomel., *Evax mauritanica* Pomel, and *Filago pomelii* Batt. & Trabut. The presence of *Gundelia tournefortii* and *Erucaria hispanica* (L.) Druce in the outskirts of Oran is now doubtful.

In the arid pasturelands the problem is fairly similar, though we still lack sufficient information on certain species considered among the most rare of the Algerian flora, notably *Avena breviaristata* Barratte from Oued Sahara near Zarhez Chergui, which is only known by its type specimen, *Otocarpus virgatus* Dur. from Saida-Tiaret (see below), *Mecomischus halimifolous* (Munby) Hochr. from Ain Sefra and *Lyauteya ahmedi* (Batt. & Pitard) Maire. Some other species could also be considered vulnerable, although often we do not have recent information on them, particularly *Chrysopogon aucheri* Boiss., *Trisetum loeflingianum* (L.) Presl., *Apera interrupta* (L.) Beauv., *Kochia prostrata* (L.) Schraden., *Lathyrus numidicus* Batt., *Atractylis coerulea* Batt. and various *Carduncellus*. The continental dunes of the high lands in Mediterranean Algeria also shelter many rare Saharan species such as *Scrophularia hypericifolia* Wudl., *Neurada procumbens* L., *Eremobium aegyptiacum* (Spreng.) Hochr. and the endemic genus *Saccocalyx* Coss. & Dur., (*S. satureioides* Coss. & Dur.).

The saline littoral or continental (sebkha) soils constitute an important habitat, but their flora does not seem to have suffered much under present conditions of human utilization, except perhaps in the immediate proximity of big cities. That is why *Limoniastrum monopetalum* (L.) Boiss. became very rare in Oran, as well as *Limonium ferulaceum* (L.) O. Kuntze.

A similar situation to that described for Morocco exists for many species associated with clayey substrates, as a consequence of intensive mechanical land clearance for agricultural purposes. Thus, *Convolvulus durandoi* Pomel, as well as *Mandragora autumnalis* Bertol, seem to have vanished from the Agiers region. Similarly *Triguera osbeckii* (L.) Willk. and *Linum grandiflorum* Desf. have become very rare in the proximity of Oran, and *Psoralea americana* L. or *Hammatolobium kremerianum* (Coss.) C. Mullm have never, to our knowledge, been seen again, in the Cheliff Valley and near Ghazaouet respectively.

Many of the rarest elements of the Algerian flora must be perhaps cited among the weeds of cultivation, and ruderal flora. In the proximity of

Algiers, for instance, the enigmatic *Lathyrus allardii* Batt. seems to have vanished, while the populations of *Onopordum algeriense* (Munby) Pomel are still existant but threatened. From the Constantinian coast we can mention *Sorghum annuum* Trabut and *Digitaria debilis* (Desf.) Willd, and from the Tell cultures *Ononis megalostachys* Munby, *Teucrium spinosum* wl., *Bupleurum odontites* L. or *Linaria supina* (L.) Chaz. We do not have recent information on the South Constantinian species *Prosopis steffaniana* (M.B.) Kuntze and *Leontice leontopetalum* L.

3.2 The Mediterranean relicts of the Central Saharan mountains

The existance of Mediterranean floristic relicts was one of the main surprises of the botanical explorations of the Central Saharan mountains, not only the Hoggar and the Tassili n-Ajjers (Maire, 1940; Quézel, 1964a) but also the Tibesti (Maire & Monod, 1950; Quézel, 1958). Though the significance of this flora has been recently studied (Quézel, 1965, 1979) we will comment here on its present conservation status. The presence of some endemics among the remaining relicts of recent Mediterranean climatic phases (Quézel & Martinez, 1961), indicates a relative antiquity.

Among the five endemic species of undoubted Mediterranean origin in the Hoggar and Tassili, two are extremely endangered. They are *Cupressus dupreziana* Camus whose status is commented upon in the IUCN Red Data Book (Lucas & Synge, 1978) and *Phagnalon garamantum* Maire which is only known from the specimens that Maire collected in 1928. The six endemic species of northern origin in the Tibesti do not seem to be directly endangered. Unfortunately this is not the case for the two endemic genera of this range, *Monodiella* Maire (*M. flexuosa* Maire) and *Quezelia* Scholz (*Q. tibestica* Scholz, 1966).

The whole group, whose biogeographic and historic importance is quite evident, consists of some 75 species. We should mention some additional species which are locally endangered: *Pistacia atlantica* Desf., *Clematis flammula* L., *Andryala cosyrensis* Guss., *Carex distans* L. for the Hoggar, the populations of which do not exceed a few dozen individuals, and also *Luzula atlantica* subsp. *tibestica* Quézel and *Campanula filicaulis* var. *tibestica* Quézel of Tibesti.

3.3 The Mediterranean flora of Tunisia

The flora of Tunisia would hardly differ from that of Algeria (Pottier-Alapetite, 1959, 1979–81; Cuenod, 1964) were it not for a relatively important number of species of East Mediterranean origin, which increase as the Libyan frontier is approached. A few South Italian species are also present. Strictly speaking, only three taxa are endemic to Tunisia at the specific level (Lucas, 1980); *Paronychia chabloziana* Beauverd, *Silene barrattei* Murb. and *Rumex tunetanus* Barr. & Murb. However, several other species are shared with Algeria or Libya and are still rare of vulnerable. Other more widespread taxa are locally rare or doubtfully so.

The sandy shores ane littoral meadows often possess some remarkable taxa which are in clear regression, such as *Silene barrattei* Murb. in the region of Cap Bon, *Maresia doumetiana* (Coss.) Batt. & Trabut, and also various East Mediterranean elements (*Ononis vaginalis* Vahl., *Hippocrepis cyclocarpa* Murb., *Daucus syrticue* Murb., etc.). Special mention should be of some species whose sole representation in the African continent is on Zembra Island, especially *Erodium maritimum* (L.) L'Her., *Sarcopoterium spinosum* (L.) Spach and *Iberis semperflorens* L. The present situation of these species causes concern because of the proliferation of domestic goats (Pottier-Alapetite, 1954). *Rumex tunetanus* Barr. & Murb. is an aquatic species living in temporary marshes of the Sedjenanne Lake region.

Tunisian shrublands also contain some rare or vulnerable taxa such as *Thymelaea sempervirens* Murb., *Phlomis floccosa* Don., *Origanum onites* L. or *Teucrium alopecurus* de Moé. As examples of vulnerable species of rocky habitats we can mention *Scabiosa farinosa* Coss. from Cape Bon and *Hypericum robertii* Coss. subsp. *robertii* from the calcareous lappiaz of the Thelepte region.

4. Humid environments in general

Though endemism is not very common in the flora of humid zones, the situation of these habitats in the Maghreb deserves comment because they are tending to disappear and because of their vicinity to one of the driest zones or earth.

'Dayas' are seasonal pools which are filled with water during the rainy season but become dry in summer. Intensive cultivation around them reduces their area and disturbs their margins. They are sometimes drained and tilled. In Morocco, several permant swamps (merjas) in the Rharb Plain have been fully drained and given over to cultivation. Others (such as Rhedira merja, south of Larache) have been or will soon be drained in the struggle against malaria. A few of them are the subject of protective measures (Bokka merja and Sidi-bou-Rhaba, on the advice of the Scientic Institute, Mohamed V University). Soon, due to the opportunities for irrigation, the humid lower valleys may be transformed over vast areas by major projects for agricultural development. Through this process, natural biotopes with an exceptional flora may soon disappear from the Loukkos lower valley.

A similar situation occurs in Algeria where the Rassauta marshes or those of Reghaia Forest near Algier are now practically destroyed. By referring to an inventory made by Chevassut & Quézel in 1956, it can be appreciated how much the area has been degraded since that time. The same can be said of M'Sila Forest near Oran (Daumas, Quézel & Santa, 1952). It seems that *Marsilea diffusa* Lepricur, *Glinus lotoides* L., *Alternanthera sessilis* (L.) R.Br. and *Oldenlandia capensis* L. fil. can be considered as lost for Algeria, although the last species still exists in Morocco in a few dayas around Salé. Another group of species such as *Pilularia minuta* Durieu ex A. Breun., *Airopsis tenella* (Cav.) Ascherson & Graebner, *Antinoria agrostidea* (DC.) Parl., *Vulpia obtusa* Trabut, *Illecebrum verticillatum* L., *Cardamine parviflora* L. and many species of *Sedum* ad *Elatine* have become locally extremely threatened.

Permanent aquatic environments such as those of Mitidja, Annaba and El Kala in Algeria, are probably among those most disturbed by human action. The setting up of a National Park for the last of the Lakes at the Algero-Tunisian border was a necessary safeguard operation. It seems that more than one hundred species had vanished from Algerian territories in the last century, and many others have become endangered, even if it is very difficult to compile a precise inventory. This is especially the case for various elements of African origin, mostly localized in the El Kala region: *Cyperus polystachyos* Rottb., *C. michelianus* (L.) Link, *Scirpus inclinatus* (Del.) Asch & Greb., *Fuirena rubescens* (Poiret) Kunth, *Fimbristylis squarrosa* Vahl., *F. dichotoma* (L.) Vahl., *Rhynchospora rugosa* (Vahl) S. Gale, *Rumex palustris* Sm., *Polygonum senegalense* Meus., *Hibiscus roseus* Thore ex Loisel., *Ludwigia palustris* (L.) Elliot, *Trapa bispinosa* Roxb., *Laurenbergia tetrandra* (Schott.) Kanitz., *Utricularia gibba* subsp. *exoleta* (R.Br.) P. Taylor, *Dryopteris gongyloides* (Sch.) Kuntze, and *Typha elephantina* Roxb., etc.

The same is true for numerous northern elements of Eurasian origin which share the same habitats as the previous ones: *Butomus umbellatus* L., *Nuphar lutea* (L.) Sibth. & Sm., *Nymphaea alba* L., *Frangula alnus* Miller, *Hydrocotyle vulgaris* L., *Parnassia palustris* L., *Thelypteris palustris* Schott., *Salvinia natans* (L.) All.

5. Some case histories of rare and threatened species in the Maghreb countries (Fig. 1)

5.1 Some vulnerable gymnosperms

Abies numidica de Lannoy, the Numidian fir, (Fig. 2), belongs to the group of Mediterranean firs with obtuse leaves (Trabut, 1889) which also includes *Abies nebrodensis* (Lojac.) Mattei and *Abies cilicica* (Ant. & Kotschy), Cosson. These are quite different from other Mediterranean firs with pointed leaves such as *A. pinsapo* Boiss. and *A. cephalonica* Loudon which are clearly adapted to warm and dry conditions. The growth of the Numidian fir is rapid and its wood is of excellent quality.

A. numidica is restricted to Djebel Babor and Tababort in the Petite Kabylia in Algeria. On the

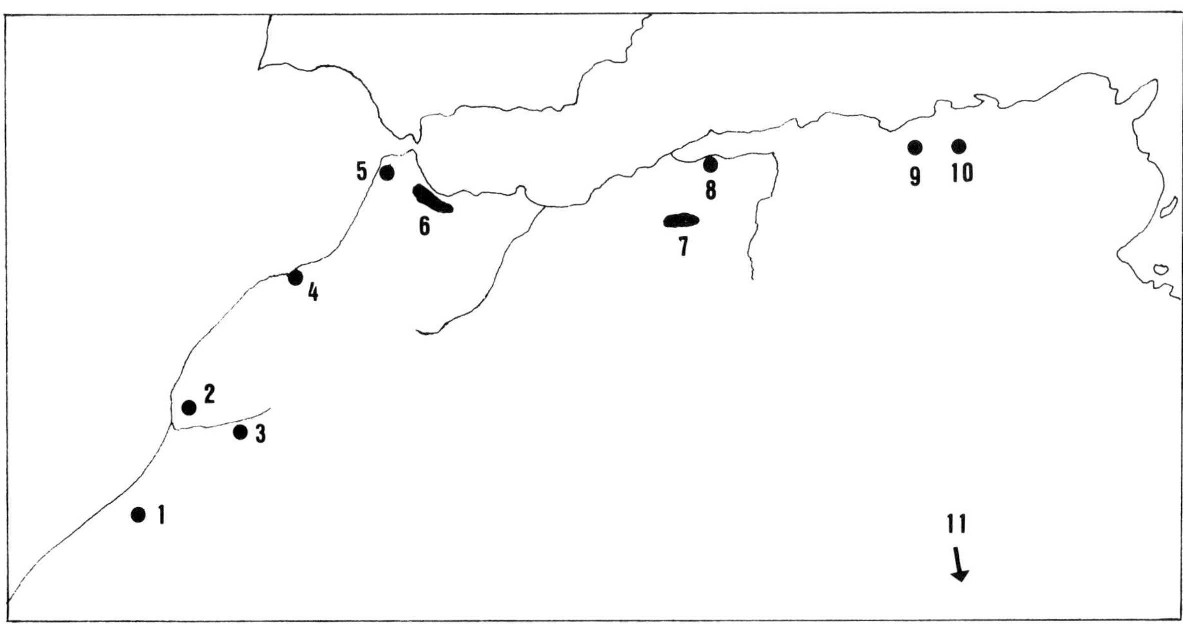

Fig. 1. Map showing the geographical positions of some of the endemic taxa discussed in the case histories of this article: 1. *Commelina rupicola*, 2. *Teucrium tananicum*, 3. *Phlomis antiatlantica*, 4. *Benedictella benoistii*, 5. *Iris tingitana* var. *tingitana*, 6. *Leuzea fontquerii Ptilostemon pseudo-hispanicum* and *Marrubium fontianum*, 7. *Otocarpus virgatus*, 8. *Teucrium santae*, 9. *Abies numidica*, 10. *Legousia juliani*, 11. *Cupressus dupreziana*.

Babor, its population covers an area of about 250 hectares which are mainly located on north-orientated slopes and ridges above 1,750–1,800 m altitude. This population was still in excellent condition in 1958, but it has been much ill-treated since then and the number of trees has decreased by half in the last twenty years. The dieturbance has mostly consisted of fires and cutting. On the Tababort, the firs are also localized on north-oriented slopes, on screes and eroded soils. They do not occupy more than a few dozen hectares, and the populations are also often much degraded. The climatic requirements of *Abies numidica* are similar to those of *Cedrus atlantica* Manetti and both trees do in fact co-exist in the mentioned area, together with *Quercus canariensis* Willd., *Taxus baccata* L., *Ilex aquifolium* L., *Acer obtusatum* Waldst. & Kit. ex Willd. and *Populus tremula* L. Among the accompanying herbaceous species, the most remarkable ones are *Neottia nidus-avis* (L.) Rich. (its only African locality), *Galium odoratum* (L.) Scop., *Senecio perralderianus* Coss. & Dur., *Ribes petraeum* Wulf. and *Corydalis solida* var. *bracteosa* Batt. & Trabut.

The populations of *A. numidica* have justified the creation of a National Park at the Babors, but unfortunately its effectiveness is not satisfactory. Seeds are collected each year and germinated for reforestation.

Cupressus atlantica was described by Gaussen in 1950 from the upper valley of Oued N'Fiss in the Central High Atlas Mountains. It also exists in other remote stations in the Western High Atlas. The species is closely related to *C. sempervirens* L. and a number of authors consider it to be a variant of the latter. Because of its hardiness, fire resistance and straightness of the trunk, the Atlas cypress is a useful conifer for reforestation providing fire protection although it is still insufficiently used.

The Atlas cypress grows on north-orientated cyrstalline slopes where it replaces *Juniperus phoenicea* L.. This habitat is between 1,100 and 2,000 m altitude, above the *Tetraclinis* associations and below the holm oak zone. The species accompanying it are those of the Phoenician juniper communities, including *Pterocephalus depressus* Cosson, *Thymelaea broussonetti* Ball, *Ormenis scar-*

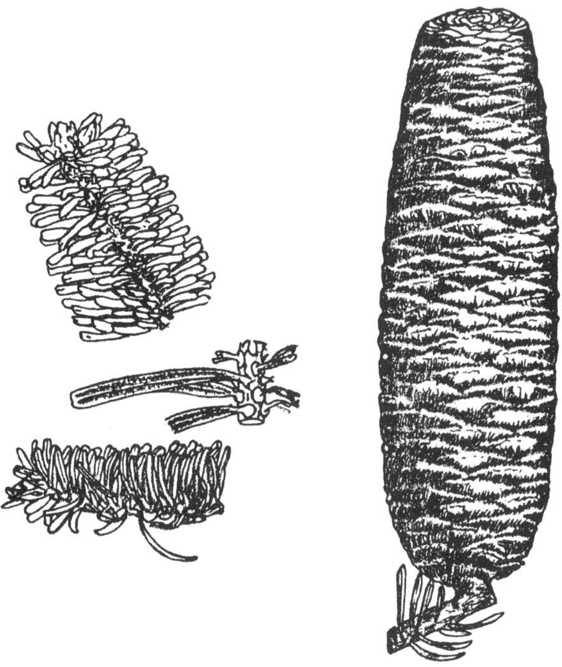

Fig. 2. Abies numidica Lanoy (Icon. Maire, 1952).

iosa (Ball) Lit. & Maire, *Cytisus purgans* subsp. *balansae* (Boiss.) Maire, and *Globularia nainii* Batt., etc.

Being aware of the value of this species for reforestation purposes, the Moroccan Forestry Commission has made a detailed record of its localities and the necessary protection measures have been undertaken.

Another highly interesting case is that of *Cupressus dupreziana* Camus of which only 153 living specimens are known in the Tassili n'Ajjer massif in the Cental sahara, Southern Algeria. Since it is included in the IUCN Red Data Book (Lucas & Synge, 1978) we refer the reader to that source, as well as to the monograph by Barry et al. (1970).

5.2 Avena maroccana *Gand. (Gramineae), a sparsely distributed ruderal*

This plant was described by Gandoger from the vicinity of Sebta (Ceuta, in the Strait of Gibraltar) and in the Chaffarin Islands (on the Mediterranean coast) as far back as 1908. It was classified as a variant of *Avena sterilis* L. by Jahandiez & Maire (1934) and this position was maintained by Maire (1953). Certainly, it is very similar in size and appearance to the much more widely distributed hexaploid *Avena sterilis* L. with which it is often associated and may be easily confused. However, *Avena maroccana* Gand. is a tetraploid (2n=28) and in the opinion of Baum (1977) it constitutes the Sect. *Pachycarpa* together with its vicariant *A. murphyi* Ladizinsky from Southern Spain. These two species are the only recorded allotetraploids of *Avena* with an AACC genome.

The species has been recently recorded inland in the Rabat area, near Maâziz, and it was described under the name *A. magna* Murphy & Terrell. A systematic search has failed to discover another locality in similar biotopes along the Atlantic margins of the Moroccan Central Plateau. Our knowledge of the geographical distribution of this species is therefore very obscure.

Its ecology is that of a weedy species, since it grows on cultivated lands, waste lands and roadsides, preferably on clay soils. In Atlantic Morocco (the area South of Tanger) it has been observed in association with the following species: *Allium nigrum* L., *Ammi visnaga* (L.) Lam., *Asparagus stipularis* Forsk., *Astragalus boeticus* L., *Atractylis gummifera* L., *Biscutella auriculata* L., *Bupleurum lancifolium* Hornem., *Centaurea eriophora* L., *C. diluta* Ait., *Convolvulus rharbensis* Batt. & Pitard (endemic), *C. tricolor* L., *Cynara cardunculus* var. *sylvestris* Fiori, *Daucus muricatus* (L.) L., *Delphinium cossonianum* Batt. (endemic), *Echinops strigosus* L., *Eryngium argyreum* Maire (endemic), *Leuzea acaulis* (L.) Holub, *Mantisalca delestrei* (Spach) Briq. & Cavill., *Moluccella laevis* L., and *Teucrium resupinatum* Desf. etc.

From the point of view of conservation, the species should be regarded as vulnerable. It is particularly liable to be adversely affected by possible future developments in weed control procedures. Any eventual damage to the survival of *Avena maroccana* is to be deplored and should be prevented. It has not only a great phylogenetic interest within the genus, but being related to such cultivated species as *A. sativa,* it might also play a role as a gene donor in breeding programmes.

5.3 Iris tingitana *Boiss. & Reut. var.* tingitana *(Iridaceae) on the Tangier plains*

Though the species as a whole is distributed over the whole northern half of Morocco and in western Algeria, the type-variety is exclusively located in the Rif Mountains and in the Atlantic Plains north of Oued Sebou. In the last few years it has fully disappeared from mechanically cultivated lands of the Rharb and Tangier Plains, only surviving outside fields or perhaps in the last few areas tilled by traditional ploughs.

Iris tingitana belongs to the section Xiphion. The type-variety, var. *tingitana,* is characterized by its many bulblets which are grouped into large clusters. Its flowers are commonly pale blue with yellow patches, but they may range from pure white to dark blue. The related variety var. *fontanesii* (Gren. & Godr.) Maire has less numerous bulblets in smaller groups, and darker blue or violet flowers. The same section includes *Iris xiphium* L. (now practically disappeared from Tanger area) as well as *Iris filifolia* Boiss. and *Iris juncea* Poiret. Some additional information about this group can be obtained from Gattefossé (1952), Maire (1959), Sauvage (1959), Ionesco and Stefanesco (1967), etc.

The type-variety of *Iris tingitana* shows a marked preference for relatively humid clay soils, which are the most fertile in northern Morocco. It belongs to essentially hygrophilic plant associations which grow on clay or calceareous soils and include few Moroccan endemics, but with some interesting taxa of the West Mediterranean basin, such as *Ammi visnaga* (L.) Lam., *Asparagus stipularis* Forssk., *Astragalus echinatus* Murray, *Echium boissieri* Steudel, *Eryngium dilatatum* Lam., *Genista clavata* Poiret, *Hedysarum aculeolatum* Munby, *H. coronarium* L., *H. spinosissimum* LW., *Lavatera trimestris* L. and *Narcissus papyraceus* Ker-Gawler.

This bulbous plant can be easily cultivated. It would therefore be advisable to suggest to the department concerned with agronomic research in Morocco that a few acres be assigned, on a local research station or on an experimental farm, to the conservation and multiplication of an *Iris* population under nearly natural conditions. As a first step, a thorough field investigation would furnish a better knowledge about the present status of the plant as well as determine some aspects of its polymorphism with a view towards preserving colour variability and possibly to suggest other suitable conservation measures.

Iris tingitana var. *fontanesii* seems to be in danger only along the tourist routes because of collecting on the roadsides. However, its distribution is much wider as it reaches Western Algeria. *Iris tingitana* var. *tingitana* is not only much more localized, but it has experienced a dramatic regression recently, so that its collection should be regulated or even forbidden.

5.4 Leuzea fontqueri *Sauvage and* Ptilostemon pseudo-hispanicus *(J. Arènes) Raynaud & Sauvage, two rare Compositae from the Rif range*

Both species are found in the western calcareous massif of the Rif mountain range in Northern Morocco, east and south of the village of Chauen. They are more or less directly associated with the flora which accompanies tree formations of *Abies pinsapo* subsp. *maroccana* (Trabut) Emb. & Maire in that area.

Leuzea fontqueri (Fig. 3) was described recently (Sauvage, 1968) and though it can be frequently encountered in the Rif limestones near Chauen above 1,700 m of altitude it is never abundant. It shows a preference for calcareous stony soils in

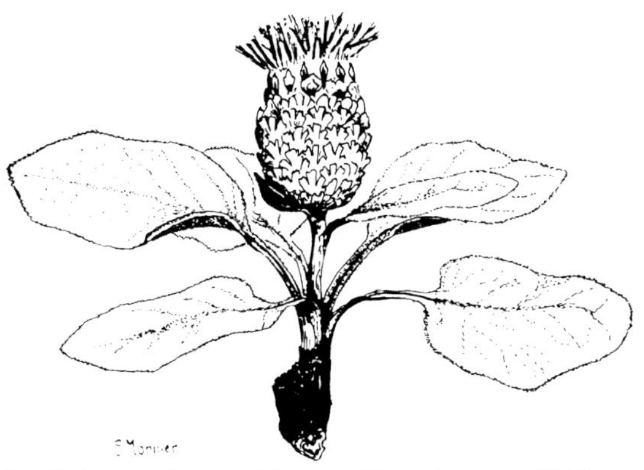

Fig. 3. Leuzea fontqueri Sauvage (Icon. Sauvage, 1968).

Abies or *Pinus* forests where it is associated with *Cephalaria mauritanica* subsp. *rifana* Maire, *Marrubium heterocladum* var. *microdontum* Sauvage, *Onosma fastigiata* subsp. *mauritanica* Maire and *Convolvulus mazicum* Emb. & Maire. It is a dwarf plant (4–5 cm in height) and is entirely covered with an araneous tomentum. The closest relative to *L. fontqueri* is probably *L. berardioides* Cosson; both species along with *L. acaulis* are the only presently described acauline or subcauline *Leuzea* of the Moroccan flora.

Ptilostemon pseudo-hispanicus is even more closely associated to *Abies* forests. Observed and described by Arènes in 1958, it has probably not been observed since that date. It is associated with other *Abies* companion species such as *Acer granatense* Boiss., *Berberis hispanica* Boiss. & Reuter, *Paeonia coriacea* var. *maroccana* Pau & Font-Quer, *Arabis josiae* Jahand. & Maire, *Isatis djurdjurae* Cosson, *Origanum grosii* Pau & Font Quer, *Sternbergia colchiciclora* Waldst. & Lit., *Ionopsidium prolongoi* (Boiss.) Batt., etc.

Through its leaf-morphology and its corymbose inflorescence, *Ptilostemon pseudo-hispanicus* differs from *P. riphaeus* (Pau & Font Quer) Greuter, whose flowers are grouped into ears of clusters. On the contrary, it is closely related to *P. hispanicus* (Lam.) Greuter by its habit, its inflorescence and the general features of the involucre.

The creation of a national park in Talassemtane region (the area where *P. pseudo-hispanicus* was found) has been contemplated. The flora of this area has been recently studied by Raynaud & Sauvage (1978). If the project proceeds, both species *Ptilostemon pseudo-hispanicus* and *Leuzea fontqueri* would be protected, together with the existing stands of *Abies pinsapo* subsp. *maroccana*. In the meanwhile, the future of these three species, and perhaps of others, will depend on the type of forest management applied by the Forestry Commission.

5.5 Teucrium santae *Quézel & Simm. and* Teucrium tananicum *Maire (Labiatae)*

Due to the characters exhibited by its calyx and inflorescence, *Teucrium santae* belongs to the Sect. *Stachyobotrys* Benth., and in particular to the Moroccan group which comprises *T. bracteatum* Desf. from Western Oran and East Morocco, *T. collinum* Cosson from the Macaronesian sector of Morocco, and *T. rupestre* Cosson, endemic to Ida ou Tanane region.

T. santae is a small perennial, localized on some calcareous cliffs and escarpments in the region of Oued Riou (Inkermann, Algeria) where it was discovered in 1959 by Quézel & Simmonneau. In that station, semi-arid climatic conditions are prevalent and the surrounding vegetation is much degraded, consisting mainly of *Pistacia atlantica* Desf., *Olea europaea* subsp. *sylvestris* Brot., *Ceratonia siliqua* L. and *Pistacia lentiscus* L.. In the rupicolous community where *T. santae* is found, *Stachys saxicola* subsp. *chelifensis* Quézel & Simmonneau is also present. The area of this subspecies coincides exactly with that of *T. santae* and as known at present covers only a few kilometres on the hills of the southern banks of Oued Cheliff. It is advisable to look for this species in the lower Valley of Cheliff region, where it may be present in other localities. Today *Teucrium santae* is much exposed to human activities and should be labelled as endangered.

A similar case is provided by *Teucrium tananicum* (Maire 1932), another rupicolous species very localized on the limestone of Ida ou Tanane region in the Western High Mountains of Morocco. It is accompanied there by *Sideritis cossoniana* Ball., *Convolvulus pitardii* var. *glaouorum* (Br. Blanquet & Maire) Sauvage & Vindt, *Malope rhodoleuca* Maire, *Polygala balansae* Cosson, and *Adenocarpus artemisiifolius* Jah., Maire & Weiller, among others. This species belongs to section *Polium* and it is somewhat similar to the sympatric *T. rupestre*, but can be clearly differentiated by its indumentum, inflorescence density and calyx and corolla-morphology. Like most chasmophytes it scarcely spreads to other areas.

Since this species has not yet been found in any other place, and is comparatively rare within its locality, conservation measures should be taken to protect it. The creation of a natural park in the Ida ou Tanane was considered as early as 1974. Such a measure would undoubtedly contribute to the efficient protection of the flora of this area of Morocco which is rich in endemics.

5.6 Some other threatened Labiatae from Morocco

The Moroccan Labiate flora is very highly diversified. To illustrate this, it may suffice to say that at least seventy-five endemics of this family grow in Morocco (Lucas, 1980). The potential value of this flora to the cosmetic industry is considerable; in particular there are four endemic species of *Lavandula* and nine of *Thymus*. In general, an important percentage of these species has restricted geographical areas and they pose possible problems of conservation. Reference has been made above to *Teucrium tananicum* Maire; in fact there are twenty-one endemic species of *Teucrium*. In continuation, two additional Labiate species, belonging to the genera *Phlomis* and *Marrubium*, are considered.

Phlomis antiatlantica Peltier (Fig. 4) has been recently described (1976) from the central section of the Anti-Atlas Range, south of Auluz. It lives on breccias and conglomerates or rhyolites, most often on slopes or along small oueds between 1450 and 1600 m in altitude. It is often associated to *Chamaerops humilis* L., *Polygala balansae* Cosson, *Thymus leptobotrys* Murbeck., *Launaea lanifera* Pau, *Convolvulus trabutianus* Schweinf. & Muschler, and *Acacia gummifera* Willd.

It is closely related to *P. italica* L. from the Balearic Islands, but it differs from it by its leaves contracted into petioles, and by the campanulate calyx which is markedly infundibuliform in its upper part. *P. antiatlantica* is also related to *P. caballeroi* Pau, but the calyx-teeth of the latter are rather longer than wide.

The very remoteness of the locality and its difficult access alone, are thought to result in an inherent protection for this endemic; additionally, cattle do not seem to pose a problem because they tend to ignore this plant.

Marrubium fontianum Maire (Fig. 5) was formerly confused with *M. echinatum* Ball by Pau & Font Quer (1930) who distributed it with that name under No. 555 in 'Iter maroccanum'. Maire decided it was a new species (Jahandiez & Maire, 1934) which he described and dedicated to one of the finders. It is a rare plant, confined to the limestone summits of the Western Rif range (Jebel Lakraa and Jebel Tissouka) between 1800 and 2100 m. In Jebel Lakraa it was found in 1955 but was not detected in 1971.

Fig. 4. Phlomis antiatlantica Peltier (Icon. Peltier, 1976).

The following plants can be collected in the same habitat: *Juniperus communis* subsp. *hemisphaerica* (Presl.) Nyman, *Festuca hystrix* Boiss., *Festuca*

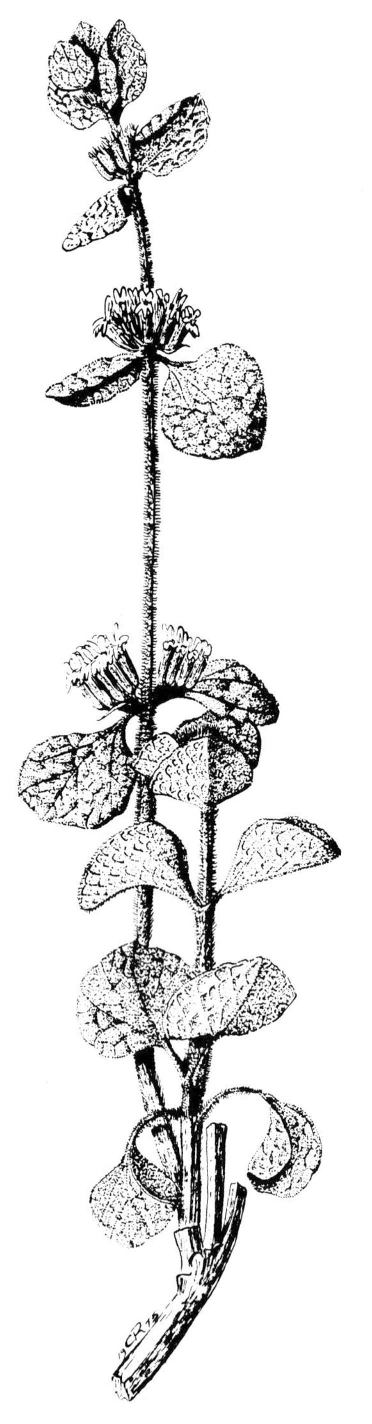

Fig. 5. Marrubium fontianum Maire (Icon. Raynaud & Sauvage, 1978).

yvesii Lit., *Festuca rifana* Lit. & Maire, *Dianthus cintranus* subsp. *cintranus* Boiss., *Ptilotrichum spinosum* (L.) Boiss., *Astragalus fontianus* Maire, and *Viola lecomtei* subsp. *embergeri* (Font Quer & Maire) Maire.

This area is not far from that of *Leuzea fontqueri* and *Ptilostemon pseudo-hispanicus,* and it would benefit from the proposed creation of a natural park. Once this project of the Moroccan Forestry Commission becomes a reality, these endemics and many other interesting plants could be effectively protected.

5.7 Benedictella benoistii *Maire (Leguminosae) on the Rabat-Casablanca area*

Benedictella is a monotypic genus strictly endemic to Morocco. It belongs to the tribe of *Loteae* and it displays many affinities with the genus *Lotus* from which it differs by its higher number of leaflets and indehiscent membranous fruits (Maire, 1924).

The strict habitat requirements of this species and its proximity to heavily populated areas make it doubly endangered. *Benedictella benoistii* is found in 'dayas' or temporary pools when these are gradually drying. It is an annual, and excessively humid winters are not favourable for its germination. Additionally, the plant cannot live under water, so it can only complete its annual cycle when rainfall has been suitable low during the winter preceding germination. These temporary biotopes contain other very interesting plants such as *Isoetes velata* Braun, *Ranunculus ophioglossifolius* Vill., *Trifolium ornithopodioides* L., *Marsilea strigosa* Willd., *Eryngium atlanticum* Batt. & Pitard, *Elatine brochonii* Clavaud, and *Oldenlandia capensis* L. fil.

Benois collected this species by 1918 in the Casablanca 'dayas' and Maire repeated the collection in 1922–23. These 'dayas' disappeared long ago, as well as several others where the plant was also found. Urban development and the extension of such cities as Casablanca and Rabat-Salé constitute a serious danger. Also, the reclamation of these pools for the purpose of mosquito eradication will sooner or later entail the extinction of this plant. However, as it appears sporadically, it cannot be observed every year and this makes monitoring it more difficult.

5.8 Commelina rupicola *Font Quer (Commelinaceae) an endangered endemic from Southwestern Morocco with tropical affinities*

This is the only species ever recorded in North Africa of the essentially tropical genus *Commelina.* It belongs to Sect. *Heteropyxis* Clarke. The plant contributes to the tropical appearance of the flora in the area, (Fig. 6). Several other taxa ranging from family to species (*Enteropogon rupestris, Striga gesnerioides,* etc.) have their only extra-tropical localities in this region of S.W. Morocco, where they may be represented by endemics. For instance, the Sapotaceae are here represented by *Argania spinosa* L., the genus *Leptochloa* by the species L. *ginae,* and the species *Kalanchoe laciniata,* by the subspecies *faustii.*

The only recorded site of *Commelina rupicola* is southwest of the city of Ifni and amongst cracks and fallen debris of a small dolomitic cliff facing south. The habitat covers no more than one hundred square metres and contains only a small population of some tens of individuals. As this habitat is located near inhabited and cultivated lands, there is a clear danger from the spread of invasive vegetation - mainly *Opuntia ficus-indica* (L.) Miller from cultivated lands.

Apart from the invading *Opuntia*, the species found with *Commelina rupicola* are *Euphorbia echinus* Hooker fil. et Coss., *Euphorbia obtusifolia* subsp. *regis-jubae* (Webb) Maire as dominants, along with, among others, *Senecio anteuphorbium* L., *Kalanchoe laciniata* subsp. *faustii* (Font Quer) Maire, *Leptochloa ginae* Maire, *Scilla latifolia* Willd., *Vagaria ollivieri* Maire, *Caralluma commutata* subsp. *hesperidum* Maire, *Enteropogon rupestris* (Schmidt) Chev., and *Echium petiolatum* Barratte & Coincy, *Striga gesnerioides* (Willd.) Vatke, mostly with Macaronesian or tropical affinities.

The high number of rare or interesting species in this area, whether endemic or not, would highly justify the creation of an integral reserve in this small section of cliffs. Proper monitoring of its evolution would be necessary, as well as the eradication of *Opuntia* if it continues to proliferate in an alarming way. These measures would be beneficial to *Commelina* but given the present situation, it seems urgent to preserve at least a small seed sample in a seed bank. Fortunately for this species, the small pale-coloured flowers it exhibits do not especially interest horticulturists so that direct collection seems unlikely. The general appearance of the plant is described in the work by Mathez (1977).

5.9 Otocarpus *Dur. (Cruciferae) and other allied genera from Norhwestern Africa*

The monospecific genus *Otocarpus* Dur. is only known from a few localities between Saida and Frenda, about 150 km to the southeast of Oran, Algeria. Among the sites where this species has been collected are those near Saida, Hachem Gharaba, Khaled Charaba, Tafaroua and Frenda. These gatherings in fact are very old and we lack accurate information concerning the present-day distribution of this genus. However, Santa gathered it in 1950 near Saida, and seeds were also collected in approximately the same place in 1971 to be stored in a seed bank (Gómez-Campo, 1972). *Otocarpus virgatus* Dur. (Fig. 7) is a plant showing strong preference for clayish soils and it occupies the zones situated between the limits of the Atlas Tellien and the highlands. The bioclimate is semiarid (annual rainfall between 400 and 600 mm) between 800 and 1100 m. altitude. It can be seen within annual grasslands near *Pinus halepensis* forests and also in local cultivations.

The genus *Otocarpus* with its sole species *Otocarpus virgatus* represents one of the rarest Algerian endemics (Quézel, 1964b). Taxonomically it is related to *Rapistrum*, within the tribe Brassiceae sub-tribe Raphaninae. There is at least one dozen of Crucifer genera endemic to the Maghreb countries, most of which belong to the tribe Brassiceae. From a practical point of view this tribe is perhaps the most interesting of the family since it includes most crop species of the genera *Brassica, Sinapis, Raphanus, Eruca* and *Crambe.*

Among these endemics, *Ceratocnemum* Coss. &

Fig. 6. Commelina rupicola Font Quer (Icon Mathez, 1974).

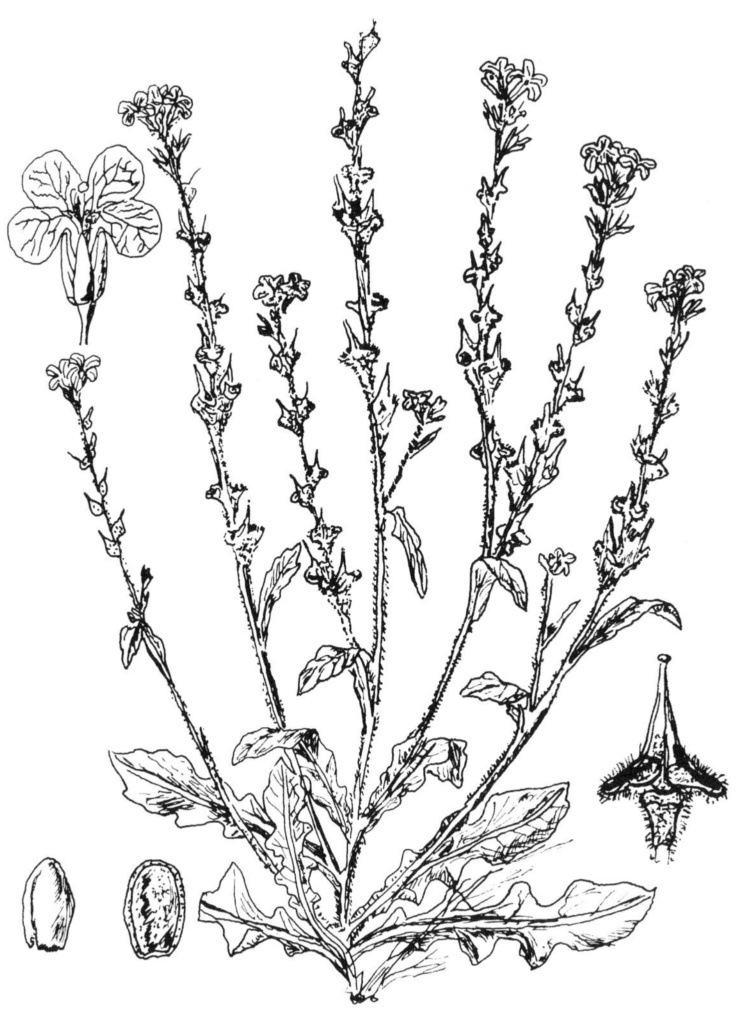

Fig. 7. Otocarpus virgatus Dur. (Icon. Maire, 1952).

Ball from central and western Morocco (also monotypic with *C. rapistroides* as unique species) is also near to *Otocarpus*. On the other hand, *Crambella* Maire, *Kremeriella* Maire and *Muricaria* Desv. are three related, also monotypic and narrowly distributed, genera. The first grows in the Moulouya Valley and the second in the Beni Snassen region. The largest area seems to correspond to *Muricaria* which grows along a sub-desertic belt between Morocco and Libya.

Two other endemic monotypic genera (*Fezia* Pitard and *Rytidocarpus* Cosson) are restricted to the region of Fes in Morocco. Both seem to occupy ruderal habitats, the former showing a preference for clayish slopes and the latter for saline depressions. Other interesting genera are *Trachystoma* O.E. Schulz with three local species in Morocco, *Raffenaldia* Godr. with two species, one of them very local in the High Atlas, *Eremophyton* Béguinot and *Foleyola* Maire, both monotypic and with sub-desertic haditat, and *Ammosperma* which extends from Algeria into Libya.

Hemicrambe Webb deserves special mention. It is found on the calcareous cliffs of the mountains around Sebta (Ceuta) and Tetuan, near the Gibraltar Strait, and was thought to be monotypic for a long time until a second species was found (Townsend, 1971 under *Fabrisinapis*) on Socotra Island in the Gulf of Aden, some 7,000 km away. The significance of this disjunction in relation to the history of the Mediterranean region (Gómez-Campo, 1978) runs parallel with other known similar cases.

Perhaps with the exception of *Hemicrambe,* these taxa are somewhat weedy and likely to be affected by the increasing use of herbicides. Seeds of all of them, together with those of many other Cruciferae are now stored in the seed-bank mentioned above.

5.10 Two endangered hygrophilous species from the Constantinian region in Algeria

Legousia juliani (Batt.) Briq. (Campanulaceae) has been recorded from Djebel Ouach near Constantine and the only known specimens are those of Julian and Giraud (1889 in herb.). To our knowledge, this plant has never been found again (Quézel, 1953) in spite of many search attempts, the last in 1977.

The exact habitat of *L. juliani* is unknown, but it could perhaps be in therophyte meadows, on the borders of marshes or in humid depressions with *Isoetes.* Djebel Ouach, a few kilometres to the north-east of Constantine, is made of Numidian sandstone and reaches 1289 m altitude. In the last century it was almost entirely deforested and only some *Quercus suber* subsisted in the vicinity. Many reforestation projects have been carried out in the past fifty years, and the mountain is now the favourite public walk for the Constantinians.

Though it is always very risky to label a species as extinct, the disappearance of *Legousia juliani* seems to be highly probable. However, some hopes of finding this plant still remain. If so, seeds should be collected and sent to specialized botanic gardens or seed banks around the Mediterranean.

Also vanished from Djebel Ouach is the species *Convolvulus durandoi* Pomel. (Convolvulaceae). This is a species linked to heavy soils and thus especially sensitive to cultivation. Battandier & Trabut (1890) expressed some concern on its conservation: 'A remarkable species on the way to disappearance. Only one land clearance destroys it forever'. The plant was relatively abundant in the Tell (coastal and sub-coastal region) between Alger and the East of Constantine and there are collections from Mitidja, Fondouk, Djebel Ouach, Djidjel and El Kala. Most of the suitable habitats from these localities seem to have vanished and *Convolvulus durandoi* has never been seen again in the vicinity of Algiers for at least 30 years. This is also the case for Djebel Ouach and El Kala. However, in 1977 we were lucky to find two localities a few kilometres west of Djidjel. The species should be looked for in the other old localities cited above.

Convolvulus durandoi is strictly bound to humid clayish substrata, even those which are partially submerged during some of the year. We can find it in wet depressions or on the margins of Cyperaceae and Juncaceae marshes. It is confined to rainy Algerian littoral zones receiving at least 800 mm of rain. Though it is bound to clayey soils, it does not seem to tolerate calcareous ones. Present populations are in an environment with *Populus alba* but always near to a *Quercus suber* forest.

C. durandoi is not the only species to be found in this situation. *Mandragora autumnalis* Bertol. for instance, has all but disappeared from all Algerian territories. This plant does not happen to be grazed by cattle, so that its disappearance may be caused by the destruction of appropriate habitats by man.

References

Arènes, J. (1958). Un *Cirsium* nouveau pour la flore du Maroc et pour la science. Bull. Mus. Hist. Nat. Paris 30: 212.

Barry, J.P. et al. (1970). Essai de monographie de *Cupressus dupreziana* A. Camus, cyprés endémique du Tassili des Ajjer (Sahara Central). Bull. Soc. Hist. Nat. Afrique Nord, Alger 61: 95–178.

Battandier, A. (1894). Considération sur les plantes refugiées rares ou en voie d'extinction de la flore algérienne. Assoc. Franc. Avancement Sciences. Congrès de Caen. 1-7.

Battandier, A. & Trabut, J. (1885-1895). Flore de l'Algérie. Paris: Jourdan. 2 vols.

Baum, B.R. (1977). Oats, wild and cultivated: A monograph of the genus *Avena* L. (Poaceae). Department of Agriculture, Ottawa.

Cuenod, A. (1964). Flore analytique et synoptique de Tunisie. S.E.F.A.N., Tunis 1: 1-287.

Chavassut, G. & Quézel, P. (1956). Contribution à l'étude des groupements végétaux de mare temporaire à *Isoetes velata* et de depressions humides à *Isoetes histrix* en Afrique du Nord. Bull. Soc. Hist. Nat. Afrique Nord, Alger 47: 59–75.

Daumas, P. Quézel, P. & Santa, S. (1952). Contribution à letude des groupements rupicoles d'Oranie. Bull. Soc. Hist. Nat. Afrique Nord, Alger 43: 186–1202.

Emberger, L. (1936). The flora and vegetation of the High

Atlas. The New Flora and Sylva, London 5: 77-82.

Faurel, L. (1959). Plantes rares et menacées d'Algerie: I.U.C.N. 7th Technical meeting, Athens 5: 140-155.

Font Quer, P. (1935). De flora occidentale adnotaciones. Cavanillesia 7: 149-150.

Gattefossé, J. (1952). Contributions a le connaissance de la flore du Maroc. Bull. Soc. Sci. Nat. Maroc 32: 53-59.

Gaussen, H. (1950). Especes nouvelles de cyprés. Le monde des plantes 270-1: 55-56.

Gómez-Campo, C. (1972). Preservation of West Mediterranean members of the cruciferous tribe *Brassiceae*. Biol. Conserv. 4: 355-360.

Gómez-Campo, C. (1978). *Hemicrambe fruticosa* (Towns.) Gz. Campo comb. nov. Lagascalia 7: 189-190.

Ionesco, T. & Stefanesco, E. (1967). La cartographie de la végétation de la region de Tanger. Al Awamia 22: 17-147.

Jahandiez, E. Maire, R. (1931-1934). Catalogue des plantes du Maroc. Alger.

Lucas, G. Ed. (1980). First preliminary draft of the list of rare, threatened and endemic plants for the countries of North Africa and the Middle East Kew: I.U.C.N.-T.P.C.

Lucas, G & Synge, H. (1987). The I.U.C.N. Plant Red Data Book. Kew: I.U.C.N. - T.P.C.

Maire, R. (1924). Contribution a l'étude de la flore de l'Afrique du Nord. Bull. Soc. Hist. Nat. Afrique Nord, Alger 15: 383-387.

Maire, R. (1932). Contribution a l'étude de la flore de l'Afrique du Nord. Bull. Soc. Hist. Nat. Afrique Nord, Alger 23: 210-211.

Maire, R. (1940). Etude sur la flora et la végétation du Sahara central. Mém. Soc. Hist. Nat. Afrique Nord, Alger 3: 1-433.

Maire, R. (1952-1976). Flore de l'Afrique du Nord. Paris: Lechevalier.

Maire, R. & Monod, J. (1950). Etudes sur la flore et la végétation du Tibesti. Mem. Inst. Fr. Afrique Noire 8: 1-141.

Mathez, J. (1977). Contribution a l'étude de la flore de la region d'Ifni. Bull. Soc. Science. Phys. Nat. Maroc 54: 20-23.

Pau, C. & Font Quer, P. (1930). Iter maroccanum, num. 555.

Peltier, J.P. (1976). Une espéce nouvelle de la flore marocaine: *Phlomis antiatlantica* (Labiatae). Candollea 31: 5-10.

Pottier-Alapetite, G. (1954). L'ile de Zembra; excursion phytosociologique. Mem. Soc. Scienc. Nat. Tunisie, Tunis 2: 34-44.

Pottier-Alapetite, G. (1959). Espèces végétales rares ou menacées de Tunisie. I.U.C.N. 7th Technical Meeting, Athens 5: 135-139.

Pottier-Alapetite, G. (1979-1981). Flore de la Tunisie. Ministère de l'Enseignement Superieure et de la Recherche Scientifique et Ministère de l'Agriculture. Tunis.

Quézel, P. (1953). Les campanulacées d'Afrique du Nord. Feddes Repert. 56: 1-65.

Quézel, P. (1957). Peuplement végétal des hautes montagnes de l'Afrique du Nord. Paris: Lechevalier.

Quézel, P. (1958). Mission botanique au Tibesti. Kem. Inst. Rech. Sahar. Univ. Alger 4: 1-357.

Quézel, P. (1959). Quelques aspects de la dégradation du paysage végetal au Sahara et en Afrique du Nord. I.U.C.N. 7th Technical Meeting, Athens, 5:

Quézel, P. (1965). La végétation du Sahara, du Tchad à la Mauritanie. Stuttgart: Gustav Fischer Verlag.

Quézel, P. (1964a). Contribution à l'étude de l'endemisme chez les phanerogames sahariens. C.R. Soc. biogeogr. 359: 89-103.

Quézel, P. (1964b). L'endemisme dans la flore algérienne. C.R. Soc. Biogeogr. 361: 137-149.

Quézel, P. (1979). Analysis of the flora of Mediterranean and Saharan Africa. Ann. Missouri Bot. Garden, 65: 479-534.

Quézel, P.l & Bounaga, D. (1975). Apercu sur la conaissance actuelle de la flore l'Algerie et de Tunisie. Colloq. Intern. .C.N.R.S., Montpellier 235: 125-130.

Quézel, P. & Martínez, C. (1961). Le dernier interpluvial au Sahara central. Lybica 6-7: 211-227.

Quézel, P. & Santa, S. (1962-1963). Nouvelle flore de lAlgerie et des régions désertiques méridionales. Paris: C.N.R.S., 2 vols.

Quézel, P. & Simmonneau, p. (1959). A propos de deux labiées nouvelles d'Algérie. Bull. Soc. Hist. Nat. Afrique Nord, Alger 50: 146-152.

Raynaud, C. & Sauvage (1978). Catalogue des végétaux vasculaires de Talessemtane (Rif calcaire). In: Etude de certains milieux du Maroc et de leur evolution recente, pp. 170-171. Paris: C.N.R.S.

Sauvage, C. (1959). Au sujet de quelques plantes rares ou menacées de la flore du Maroc. I.U.C.N. 7th Technical Meeting, Athens 5: 156-157.

Sauvage, C. (1968). Notes de floristique rifaine. Collectanea Bot. 7: 1100-1104.

Scholz, H. (1966). *Quezelia,* eine neue Gattung aus der Sahara. *(Cruciferae, Brassiceae, Vellinae).* Willdenowia 4: 205-207.

Townsend, C.C. (1971). *Fabrisinapis fruticosus.* C.C. Townsend, *Cruciferae, Brassiceae.* Hooker's Icones Plant., 7 (tab. 3673): 1-3.

Trabut, L. (1889). *L'Abies numidica.* Rev. Gen. Bot. 1: 405.

Author's address:

J. Mathez and C. Raynaud
Institut de Botanique
Université des Scienes
et Techniques du Languedoc
Montpellier, France

P. Quézel
Faculté des Scienes
et Techniques de
Saint-Jérôme
Université d'Aix
Marseille, France

CHAPTER 10

The Mediterranean islands

S. SNOGERUP

1. The insular situation

To be an insular endemic means, at the present day, to be threatened. But why should an island plant be more endangered than one on the mainland? There are several reasons for this.

The island is a limited system, which imposes more or less strict limits on the population sizes of its plants. It also has a flora and fauna different from that of the mainland, often lacking certain aggressive plant species or predative animals. Man is constantly changing the natural plant communities on islands by introducing species from different sources, including escapes from cultivation. An insular species will inevitably encounter competition or predation from species with which it has never been coadapted. Islands, especially the smaller ones are more likely to suffer changes of land-use if carried out all at once over the entire area, thus leaving no refuges for plants of the former vegetation.

For these reasons, insular endemics must be regarded as always potentially threatened, even if they occur in a comparatively safe number of localities. They can be regarded as secure only when sufficient parts of their original habitats have become effectively protected in reserves. This is why some species that occur in a hundred or so localities and contain millions of individuals have been included in this chapter. In northern Europe we have seen formerly common species becoming rapidly reduced to the verge of extinction by the introduction of modern, intensive land-use. The same mistakes should not be allowed to be repeated in the case of far more vulnerable endemics of the Mediterranean islands.

Some islands already illustrate the worst possible examples of destructive management, like St. Helena in the South Atlantic, where most of the original vegetation has been destroyed. Similar situations are also obvious on some Mediterranean islands, such as those of the central southern Aegean, where the present endemic flora grows almost exclusively in habitats that are inaccessible to man and his domestic animals. It is probable that an unknown number of endemics of more open and accessible habitats become extinct before the advent of botanical research to these islands.

There are particular aspects of island taxa that make them especially interesting for studies of evolution and phylogeny. One is the exceptionally high proportion of relict taxa among them. The barriers to the migration of new competitors from the mainland no doubt enhances the survival chances of ancient, conservative elements. Certain islands, like the New Caledonia islands of the Pacific, are well known because they have a very spectacular and taxonomically isolated relict endemic flora. But many examples are also known from the Mediterranean islands, and some will be included among the cases treated below. Not only are there relicts from the Tertiary flora which inhabited the area before the Quaternary climatic disturbances, but also some elements of undoubtedly much greater age.

Gómez-Campo, C. (ed.), Plant conservation in the Mediterranean area.
© 1985, Dr W. Junk Publishers, Dordrecht. *ISBN 90 6193 523 7.*

The islands have to some extent functioned as museums for those elements of former vegetation that have generally been replaced elsewhere. This fact has led some authors to the conclusion that the insular situation involves a decrease or halt in evolutionary processes. Such suggestions have repeatedly been expressed by Greuter (1972, 1979) and other authors. However, even if a certain species group of the Tertiary European flora has survived on some Mediterranean island, it has not necessarily done so in an unchanged state. It has, for instance, been proposed by Davis (1951) and Snogerup (1971) that Aegean cliffs have served as refuges for certain threatened elements, which, as a secondary development, have been forced towards chasmophytic specialization. In many cases there is not one single and morphologically constant relict taxon, but several closely related species, subspecies or differing populations on different islands. Such variation, indicating recent or present evolution in island populations in the Aegean, has been documented by, for example, Strid (1970) for *Nigella,* von Bothmer (1974) for *Allium,* and Snogerup (1967a, b) for *Erysimum.*

The extensive body of recent information on the flora of the Aegean area is scattered among many journals and herbaria, so that for general reference it is still necessary to consult Rechinger (1943, 1951). A balanced survey of different categories of endemism among insular endemics was presented by Cardona & Contandriopoulos (1979).

The relict element is no doubt an important part of insular endemism. Other insular taxa, however, are often cited in textbooks as examples of rapid and spectacular evolution. Isolation itself may stimulate differentiation by the imposition of barriers against gene exchange with related taxa or populations. The fact that the duration of the isolation can often be determined more or less precisely, makes island taxa specially interesting subjects for investigations concerning the rates of evolution. In the case of the Mediterranean islands the isolation has lasted, in a geological sense, for rather short periods. Wide land connections occurred in the Messinian, c. 5 million years ago, as is considered upon from a phytogeographical angle by Bocquet et al. (1978). During the Quaternary fluctuations in sea level, many now-isolated islands would also have been connected or closer to the mainland or to other islands. Thus, the results of evolution that we can see in the Mediterranean are less spectacular, in terms of taxonomic rank and morphological differences, than on old and effectively isolated oceanic islands. The study of these early stages of differentiation can readily include more refined techniques such as crossing experiments, and can thus give more details about evolutionary mechanisms than can older and more advanced examples.

Any island population of a widely distributed plant, having been isolated for a considerable time, may be regarded as a potential endemic. For the purpose of understanding the modes of differentiation under insular isolation, it is equally important to study such unfinished cases of differentiation, as the cases where the process of speciation, or even generic differentiation has already taken place. This also raises special demands for protection of the island populations of widespread species, a demand that will be more urgent the longer the isolation has lasted in a particular case.

More detailed consideration of different aspects of island plants are found in works by Carlquist (1965, 1974), McArthur & Wilson (1967), Balgooy (1969), Bramwell (1979) and Moore (1983) where references to the classical literature on the subject can also be found.

2. Some case-histories of rare and threatened species in the Mediterranean islands (Fig. 1)

2.1 Brassica hilarionis Post. (Cruciferae), on Cyprus

Brassica hilarionis was first discovered at the ruins of Ag. Hilarion castle, where it still occurs in abundance, and was described by Post in 1900. It has since been found in several more localities, but grows only on vertical cliffs of hard limestone, which limits its possible area on Cyprus rather strictly. It is only found in the Kyrenia range of the northern part of the island, from Kornos in the west to the Yaila Peaks in the east. It is locally

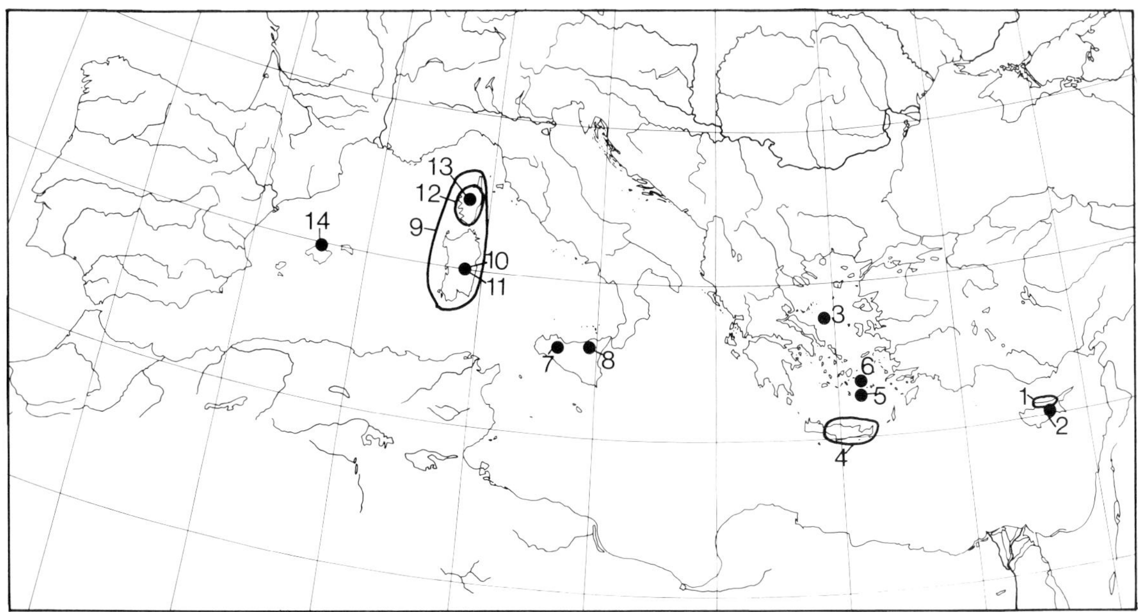

Fig. 1. Map showing the locations of the species discussed in this chapter. 1. *Brassica hilarionis,* 2. *Bupleurum sintenisii,* 3. *Aubrietia scyria,* 4. *Phoenix theophrastii,* 5. The island of Anaphi, Kikladhes, 6. The island of Amorgos, Kikladhes, 7. *Valantia deltoidea,* 8. *Petagnia saniculifolia,* 9. *Stachys corsica,* 10 and 11. *Rhamnus persicifolius* and *Lamyropsis microcephala.* 12. *Juncus requienii,* 13. *Draba loiseleurii,* 14. *Naufraga balearica.*

common, being one of the characteristic species of the limestome cliffs.

B. hilarionis is a small, perennial shrub that belongs to Sect. Brassica in a limited sense, which includes the closest relatives of the cultivated coleworts and cabbages. It shows morphological resemblances to *B. cretica* Lam. of Greece and southwestern Anatolia, as well as to *B. macrocarpa* Guss. of the Isole Egadi off the west coast of Sicily. It is quite characteristic in its combination of fleshy, glabrous leaves, broad, smooth to faintly reticulate pods and white to pinkish flowers with purplish sepals. It must be regarded as a separate species, equivalent to the more western members, of the group, which are usually less locally distributed. These species are, according to their distributions, probably more ancient in origin than the Messinian period. *B. hilarionis* may therefore safely be interpreted as a relict endemic of considerable antiquity.

This is an example of an old endemic, rather strictly local and with narrow ecological preferences, but still competing successfully and even dominant in some of the communities in which it occurs. The inaccessibility of these cliffs will contribute to securing its future survival. The only severe threat that might occur would be extensive quarrying of the limestone.

Although *B. hilarionis* can be said to be an endemic which is not particularly threatened, it is also one that may have future economic importance. Because of the weak sterility barriers in this group of *Brassica* species, it may be regarded as a potential source of genetic variability for future breeding of various *Brassica* crops. In its isolated occurrence, it has undoubtedly preserved and developed properties that are not found in other species of the group. It should therefore be regarded as a definite subject for protection, and not only the species as such should be saved, but also if possible the entire range of genetic variability present. This means that any heavy reduction of the chasmophytic communities of the Kyrenia range should be regarded as disastrous.

The *Brassica oleracea* group, to which *B. hilarionis* belongs, is almost entirely Mediterranean. Its other members are equally important subjects for protection, as they occur in a comparatively small number of localities. They have been further commented upon by Onno (1933) and Snogerup (1980), among others.

There is a special conservation problem for the members of this, group, because crossing can occur between its taxa and even with members of other groups and genera. Cultivated coleworts and cabbages in fields near the cliffs might cause introgressive populations to form. We have no indications that this has happened to any population of *B. hilarionis,* but there are suspicions concerning some western populations of the *B. oleracea* group. This problem is not unique to *Brassica,* but has to be considered for many wild relatives of cultivated plants.

2.2 Bupleurum sintenisii *Huter (Umbelliferae), on Cyprus*

Bupleurum sintenisii was discovered by Sintenis and Rigo, who collected it in the hills between Nicosia and Kythraea at the end of May 1880. The description was, however, not published until 1905 by Huter. Since its discovery it has only been observed a few times, and only within a limited area of east Cyprus, on plains and hillsides at 100–250 m. Other known localities are at Potami, Athalassa near Nicosia, Lymbia near Arkangelos Monastery, Pergamos and Troulli.

B. sintenisii is morphologically very different from the other annual *Bupleurum* species. Its fruit carries rows of hooked bristles, a feature not found in any other member of the genus. Such a fruit character would have merited generic rank in other groups of the family Umbelliferae. Its closest relationship is probably with the pre-dominantly coastal species of the *B. tenuissimum* group. *B. sintenisii* is a small plant, 5 cm or less in height, with leaves 2 cm or less. Apart from the fruit form it is also characterized by its petals, which are 0.5 mm, rounded and very pronouncedly cucullate (hooded). The distinguishing characters between annual *Bupleurum* species are usually rather small, so *B. sintenisii* is definitely an old and taxonomically isolated endemic.

B. sintenisii occurs in the rather common-place vegetation of dry, more or less rocky hillsides and plains in a grazed lowland area. It may seem astonishing to those not familiar with the genus, that it has remained within a very limited area in spite of the rather wide and continuous occurrence of its habitat. This is, however, known to be the case with several other local endemic annual species of the genus *Bupleurum.* Apparently they disperse their seeds only very close to the parental plant. The hooked glochidia of the fruits of *B. sintenisii* might be presumed to be an adaption towards more effective dispersal, but it has not enabled the spread of the species from its present situation as a local endemic.

B. sintenisii is not even common within its rather small area of distribution. Annual *Bupleurum* species frequently occur in great numbers of individuals at a locality, but in this case the number of known localities is only half a dozen. The occurence in an open, grazed habitat puts it at a great risk in relation to future land-uses. Aforestation or cultivation of its few localities might rapidly bring it towards extinction, despite a great number of individuals. This is an example of how, on an isolated island, even apparently trivial habitats may contain taxonomically isolated, narrow endemics. It is for that reason important in such areas that carefully selected reserves of such commonplace habitats are established to protect them from changes in land-use. To plan such measures, a detailed, modern local flora is necessary. In the case of Cyrpus the publication of such a work has been started by Meikle (1977).

2.3 Aubrieta scyria *Hal. (Cruciferae), on Skiros, central Aegean*

This species was discovered by Dr. Tuntas in 1901 and received its name from Halácsy in 1910. It has since been observed and collected several times at the type locality on the island of Skiros. It occurs on the north-east-facing cliffs on the north-eastern part of Mt. Kochilas, south-east of the chapel of Ag. Artemios. Its taxonomic relations were discussed in the revision by Phitos (1970).

Skiros and the surrounding islands have been investigated during the last ten years in the course of specialized floristic research, without the discovery of any further localities of *A. scyria*. It can now at least be safely stated that it does not occur in any of the neighbouring islands or islets nor in the northern part of Skiros. It was also looked for in vain on the cliffs further to the southeast on Mt. Kochilas. Therefore, it seems most probable that *A. scyria* is a truly narrow endemic, which grows only on that single cliff, i.e. in an area of 0.5×2 km.

A. scyria is most easily separated from the other members of the genus by its fruits. These are small, only c. 2.5×15 mm, with very hard an thick valves and an indumentum of very small stellate hairs only. The plant forms small, dense mats with several dense, c.10-flowered inflorescences of small, light-violet flowers.

By its preference for limestone, *A. scyria* resembles the other Greek species of the genus. But unlike them, it occurs only on almost vertical cliffs, rather than on the flatter rocks or on the cliff top area. This strict, chasmophytic specialization is very common amongst other Aegean endemics. It no doubt reflects the fact that this type of locality was most suited for survival during particular periods of the past.

The same cliff flora contains also some other restricted species such, as *Aethionema retsina* Phitos & Snog, *Scorzonera scyria* Gust & Snog, *Campanula merxmuelleri* Phitos, *Centaurea rechingeri* Phitos and *Galium reiseri* Hal. Such species are also represented in a varying number of other localities on Skiros and the surrounding islands. This cliff is the richest of a system of cliffs on and around Skiros, but *A. scyria* is the only species limited to this locality. It is, accordingly, a key species for the determination of the best place to establish a reserve to save the endemic chasmophytic elements of this area. This pattern, with a series of concentric distribution, is typical for such groups of insular endemics. The common central locality is probably as a rule also the real main refugium, where the element survived during the most severe periods of the past. It is probably also the place best suited in which to secure the future survival of these species and of the entire communities which they form.

A. scyria is not all threatened at present, because most of its habitat is quite inaccessible to grazing animals as well as to human collectors. There might be potential threats in the future through limestone quarrying or the dumping of waste in connection with tourism. These risks can however be easily eliminated by making this cliff a reserve and by concentrating such activities, if any, to other parts of the mountain.

The existence of this species in one comparatively small population, with perhaps only a few hundred individuals, might be thought to involve a risk of genetic deterioration. However, any plant or plant group living exclusively as a chasmophyte must have been forced to occur in small populations during long periods. Only by genetic adaption to this mode of life have they been able to survive, and they will therefore not run any great risk by continuing to exist in a situation that is, for them, quite normal. We should distinguish between a species that normally occurs in large populations and is suddenly forced to exist in a small isolated population, and a species that has been adapted to such an existence during the course of geological periods.

2.4 Phoenix theophrastii *W. Greuter (Palmae), on Crete (Fig. 2)*

The occurrence of a wild date palm on Crete (Phoenix) has been known in botanical literature since the days of Theophrastos. It was long regarded as representing the introduced *Ph. dactylifera* L. in spite of the fact that Theophrastos had stated that it was different to cultivated plants. Greuter (1967, 1968) published the results of a careful examination, that lead him to the opinion that the palms of Crete belong to a separate species, which he named after the man who had first noted its existence. It is distinguished from *Ph. dactylifera* by its low, basally branching habit and by its fruit which is only c. 15×9 mm with a thin, non-edible layer around the seed.

Since its description, *Ph. theophrastii* has been found to grow in several places on Crete. Barclay (1974) mapped 3 localities on the south coast, the type locality at Vaï on the north-eastern cost and another at Gazi on the north coast near Iraklion (Heracleon). I have also observed it in the small

Fig. 2. A large specimen of *Phoenix theophrastii* Greuter showing vegetative propagation in a stream course west of Vaï, Kriti.

valley north west of the monastery of Toplou not far from Vaï and along the Stram bed inside Vaï. There are some more unpublished localities in rarely visited parts of the southern coast of Crete, where the land is rocky, dry and sparsely inhabited; and the species has recently been reported from the sout-west coast of Turkey.

The localities of *Ph. theophrastii* are all in places with good access to ground water, i.e. along rivulets, water courses and in depressions. It is often associated with other species regarded as indications of underground water such as *Nerium oleander* L. and *Juncus heldreichianus* Marsson ex Parl. Such places with better water supply than their surroundings have usually been subjected to cultivation or heavy grazing. The wild palms have probably been able to survive because of their habit of growing along the margins of the stream beds.

By reference to old literature, place-name studies and traditions, Greuter gives reliable evidence that this species still occurred in other parts of Greece during historic times. The genus *Phoenix* was widespread in Europe until the Miocene, and in the Aegean area there exist an almost continuous series of records up to the present time. Thus *Ph. theophrastii* might be a relict of the Tertiary European *Phoenix* which has persisted in some places with a favourable climate and where human activity has been low.

The well-known locality at Vaï is now much visited by tourists and the palm trees have suffered some destruction in the beach area, but those growing higher up along the stream bed are growing vigorously. Other populations run rather little risk of destruction at present, because cultivation is gradually becoming less intense and the localities on the southern side of the island are not easily accessible. Grazing will hardly harm the palms, which are effectively protected by their spines. There is, however, another danger from the habit of setting fire to grazed areas. In some cases fires have even been lit around the trees in an attempt to destroy them. The palms may have become regarded as a hindrance and a weed in the best grazing plots, and this has perhaps caused the reduction of the species to its present vulnerable status.

As land-use in the more inaccessible parts of south Crete is today much less intense than in the past, *Ph. theophrastii* is not really threatened at present.

However, if the planting of forest trees on a large scale is ever carried out, then the *Phoenix* might become endangered and perhaps even intentionally destroyed. At present the total number of palms is only a few thousand, but as the formation of new shoots from the bases is vigorous the species may be regarded as safe as long as it is not directly destroyed by man.

Ph. theophrastii is an example of a comparatively rare insular endemic occurring in the mesophytic habitats. By this habitat preference it is exposed to threats from land-use practised by the human population. Its future existance is entirely dependent upon the attitude to conservation of local and national autorities.

2.5 Relict occurrences of Quercus ilex *L. (Fagaceae), in the central and southern Kikladhes*

This and the following case are intendend to illustrate two different and partly opposing problems for conservation. They are taken from an island where human activity has been very intense for at least 5,000 years and has already changed the entire vegetation. This first case refers to the situation for a forest species that probably covered large parts of the island before overgrazing and cultivation changed the landscape. At present there is no forest in the Kikladhes, only scattered trees within the cultivated areas.

Quercus ilex is common on both the mainland areas and most larger islands of the north-eastern Mediterranean. But in the central and southern parts of the Kikladhes (Central Aegean), this tree is today only found in groups of a few individuals. On the island of Naxos it occurs only on a few cliffs and rocky localities, but seems to be not really threatened. On the island of Amorgos, the situation is much worse for *Q. ilex*. The only known locality is on the south-facing cliffs of Panagia Chozoviotissa on Mt Ag. Elias. On a visit in 1963, I could observe only two individuals. Both were healthy and grew in quite inaccessible sites, where they will probably survive for a long time. However, the possibility of the establishment of a new generation before they die is certainly slight. The cliff is not a normal locality for this species, which is a forest tree. Its relict occurence there only reflects the fact that it has no chance of survival on level ground during the present heavy grazing and hunt for every single piece of wood.

It may be quite safely stated that *Quercus ilex* is not threatened in most other parts of its distribution and, therefore, considerations concerning its conservation as a species might not be thought to make much sense. Nevertheless, each island population does represent the result of an often long-term process of isolation and it may be regarded as a local variant, even if not meriting taxonomic recognition. Its difficult past history on an island like Amorgos may have even forced it into rather rapid ecological specialization. If the continued existence of this oak on the island is to be secured, this will no doubt require that some forest is allowed to develop somewhere on Mt. Ag. Elias. The few individuals on the cliff will otherwise risk becoming the last, The cliff holds abundant, well-adapted obligate chasmophytes, which will have a much better chance, to take over the few empty niches that occur at intervals on the cliff. As other species of the former forest are under similar stress, arguments for a forest reserve might easily be put forward.

A situation similar to that of Amorgos occurs on Anaphi, where I observed one small inaccessible specimen of *Q. ilex* in 1964. In this case the chances of its survival must be regarded as rather small.

2.6 Erysimum senoneri *(Heldr. & Sart.) Wettst. subsp.* amorginum *Snog. (Cruciferae), on Amorgos, central Aegean (Fig. 3)*

Erysimum Sect. *Cheiranthus* is a group of species limited to the Aegean area and southern Greece. They occur mainly on limestone cliffs, several of them in a small number of localities, according to the revision by Snogerup (1967a, b). *Erysimum senoneri* has ecological preferences that deviate a little from those of the other *Erysimum* species in Sect. *Cheiranthus*. The other species of the section are obligate chasmophytes, while *E. senoneri* has a tendency to occur in scrub thickets in large crevices or on the margins of the cliffs. In most cases, however, individuals occurring outside the cliffs are few and dependent upon lack of grazing.

On the island of Amorgos, separate ecological specialization has developed. The variant called subsp. *amorginum* grows within the dense and often spiny shrubs of the garrigue. As an obvious adaption to this mode of life it has a profuse branching habit. The usual leafy short shoots of *Erysimum* Sect. *Cheiranthus* are here modified into thin branches. The inflorescences are many but each is few-flowered, and the flowers and leaves are small. The plant thus only exposes a number of small leaf rosettes and inflorescenses between the twigs and spines of the protecting shrub.

E. senoneri subsp. *amorginum* has developed an adaptation to the secundary vegetation that followed the overexploitation of the island. This has allowed it to increase its numbers enormously in comparison with the chasmophytic variants. It constitutes a garrigue ecotype, which differs in morphology from subsp. *senoneri* but is not genetically isolated from it. With its greater number of individuals, it is apparently crossing with the remaining chasmophytic populations to such an extent that it threatens to destabilize their adaptation to the cliffs. The entire process suggests that *Erysimum* might migrate from the cliffs to invade the man-made vegetation of the island. The continued existance and final stabilization of this process is dependent on a continuation of present land-use.

The taxonomic delimitation of this ecotype reflects an interest in the study of its diverging line of specialization and the evolutionary consequences. Of this subsp. *amorginum* may be considered worth preserving for future studies, as are other parallel cases. In a conservation-oriented project this case may suggest the value of allowing large areas of secondary vegetation to continue under the present management regimes of heavy grazing and burning at intervals. There will probably be several other cases where the evolutionary response of certain plant species to the man-made vegetation has led to adaptations that are of great

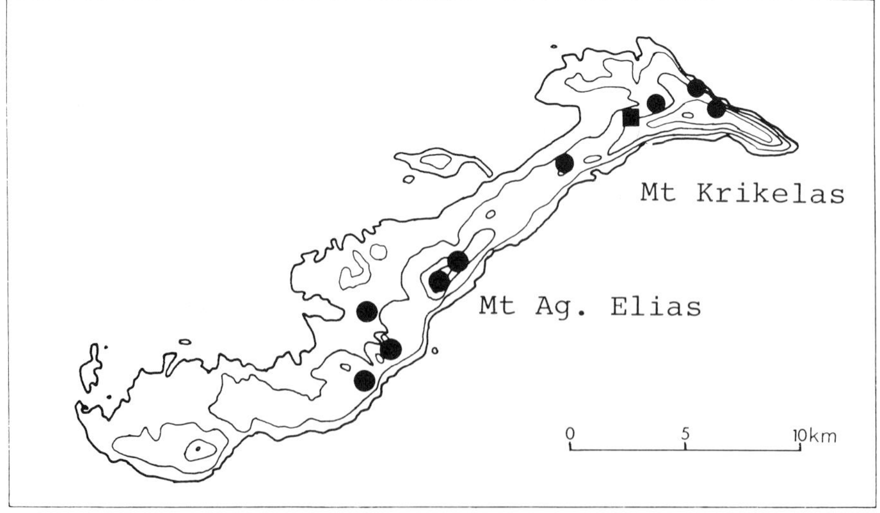

Fig. 3. Map of the island of Amorgos, Kikladhes, showing the distribution of *Erysimum senoneri* subsp. *amorginum* (dots) and the only observed locality of pure subsp. *senoneri* on the island.

interest for further study. These arguments, however, are in conflict with those set out in the case of *Quercus ilex* above concerning primary vegetation forms and taxonomic units.

2.7 Valantia deltoidea *Brullo (Rubiaceae), on Sicily (Fig. 4)*

Valantia deltoidea, a member of a genus otherwise dominated by widespread species, was not described until March 1980. It is clearly a distinct species, and can be separated from its closest relative *V. muralis* by differently-formed hairy leaves and by several differences in the fruiting corpus (Brullo 1980).

Valantia deltoidea is only known from the limestome area of Mt Rocca Bussambra, south of Palermo in north-west Sicily. It is very common in meadows at the type locality, where it was discovered by Dr. S.L. Brullo in 1976. The mountain is only 1,600 m high, but it was previously known to contain two other local species, *viz. Gagea busambarensis* (Tin.) Parl. and *Armeria gussonei* Boiss.

This case clearly illustrates how local endemics still remain to be discovered, even in comparatively well investigated parts of the Mediterranean islands. Every locality known to contain some local species must also be presumed to house undescribed taxa, until it has been very thoroughly investigated. This is a further argument to protect such places against changes in management which might destroy some species before we have even learned of their existence.

2.8 Petagnia saniculifolia *Guss. (Umbelliferae), on Sicily*

This plant represents a monotypic, endemic genus of the family Umbelliferae that is very isolated taxonomically. It has long-petiolate, 5-palmate basal leaves and high stems up to 50 cm with comparatively small inflorescences. In habit, *Petagnia* resembles rather closely the genus *Sanicula,* as it does in some details of the fruit, and some affinity to *Sanicula* can indeed be presumed. However, the inflorescence of *Petagnia* is quite different from that which is normal in the family. The partial inflorescences consist of a central female or hermaphrodite flower surrounded by 2–4 adnate male flowers. The ovary is also unique, being unilocular with one ovule, and giving rise to a one-seeded nut.

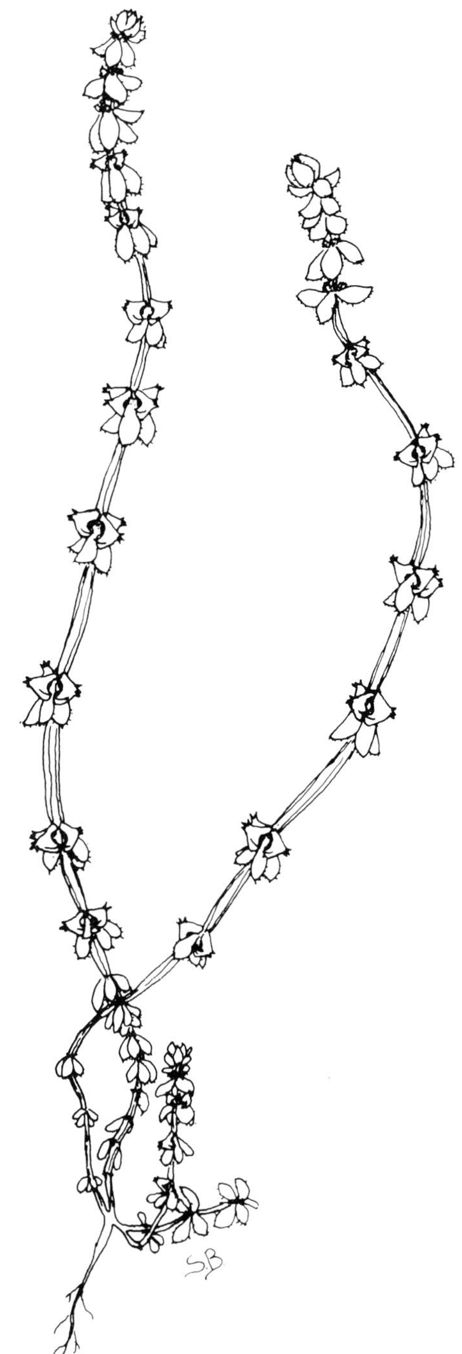

Fig. 4. Valantia deldoidea Brullo (from Brullo, 1980).

P. saniculifolia was described by J. Gussone, the great pioneer of Sicilian botany (Gussone 1827). Despite its interest as one of the outstanding endemic genera of Europe, it has only been reported from a small area of northern Sicily, in the north-eastern part of the Monti Nebrodi. As far as is known, all observed localities are situated within an area no more than 15 km wide, down to about 500 m altitude.

P. saniculifolia is vulnerable because of its dependence on a rather special ecological niche. It grows on the margins of cold mountain streams in more or less narrow valleys, with deciduous forest, often dominated by *Quercus pubescens* Willd. and *Corylys avellana* L. *P. saniculifolia* propagates both by seeds and by creeping rhizomes, and will have no difficulty in persisting at such sites, as long as they fulfil its rather strict requirements. The distribution lies, however, within an intensely cultivated area, and the mountain streams on which it depends are partly used for irrigation of surrounding farmland. There is also a risk that deforestation will threaten its existence. Sicilian botanists have repeatedly argued for the setting aside of reserves to save this especially interesting endemic (Pignatti 1971; Brullo et al., 1976). It is but one selected example among the many unique, often local species of this island, which occur in a range of environments. There is an urgent need for a new Flora of Sicily to replace that of Lojacono-Pojero (1909). Only with full and precise documentation will it be possible to argue with full and precise documentation will it be possible to argue with full strength for the preservation of the most important botanical treasures of Sicilia.

2.9 Stachys corsica *Pers. (Labiatae), on Corsica and Sardinia (Fig. 5)*

Stachys corsica is a characteristic local species related to *S. arvensis* (L.) L. *S. marrubifolia* Viv., etc. It can be readily distinguished from the other species of the group, especially by its short-lived perennial habit, broad leaves and reduced, 1 on 2-flowered verticillasters.

Stachys corsica was described by Person (1807) and has since been reported from a large number

Fig. 5. Stachys corsica Pers. (from Camarda, 1978).

of localities in Sardinia and Corsica. Camarda (1978) gave no less than 30 localities on Sardinia based on revised herbarium material. Arrigoni (1977, 1978) with several other members of a team from Firenze and Sassari, published a series of surveys of the endemics of Sardinia. Such concentrated accounts of distribution, ecology and taxonomic status will form a sound background for future conservation measures on the island.

S. corsica occurs mainly at 500–1,000 m altitude in the mountains. It is usually found at the base of cliffs and in seasonally wet parts of rocky localities, but sometimes also in open fields. It is not limited to any special substrate and shows a variability reminiscent of a widespread species adapted to different habitats. Hence it is a good example of a species limited by the effective and probably long-lasting isolation of the sarco-corsican massif from the other parts of the central Mediterranean. Such isolated species (which are not uncommon and apparently capable of active expansion) have been forced to remain as endemics in this area.

S. corsica is not at all threatened as a species. Its main scientific interest may, however, lie in the amount and causes of its intraspecific variation. It will no doubt be possible to reconstruct the geological history of the area with some precision, so that a time-scale can be given for the various stages of differentiation. This is an aspect that makes many island endemics especially important subjects for the study of evolution. However, a species like *S. corsica* will retain its full value for botanical research of this kind only if the types of vegetation containing its different variants are preserved. This case is an example of the most widespread of the sardo-corsican endemics. The following cases will illustrate some of the gradually more restricted species of this category.

2.10 Rhamnus persicifolius *Moris (Rhamnaceae), on Sardinia*

Rhamnus persicifolius has the habit of a small tree or shrub. It is most closely related to the widespread *R. catharticus* L., but differs from it by its large, lanceolate leaves and reddish drupes. It occurs at 700–1,100 m in mountains, forests, Mediterranean shrub vegetation or degraded communities.

R. persicifolius was described by Moris (1827), but he did not give a more precise locality. It has been observed in different parts of the mountain called Massichio del Gennargentu, but never in the (now rather well investigated) surrounding mountains. Therefore it seems to be a very restricted species, only occurring in an area of about 20×10 km. Even within this area it is not at all common, only occurring as scattered groups or single trees or shrubs. Arrigoni (1977) presumes that its distribution has become restricted in recent times by the degradation of the forest vegetation on Sardinia. The effects of human influence have become more severe because of the present change in the climate towards more arid conditions.

R. persicifolius is just one example of the woodland species that have suffered from human activities in Sardinia. Continued degradation of the remnants of forest and native Mediterranean shrub vegetation may bring it to extinction. Reforestation by modern methods and perhaps new tree species represent an equally serious risk. The east central Sardinian area contains more local species than other parts of the island and most of the more widespread endemics also occurs there. *R. persicifolius* is no doubt one of the useful indicator species for the selection of suitable reserves in this important area.

2.11 Lamyropsis microcephala *(Moris) Dittr. & Greut (Compisitae), on Sardinia (Fig. 6)*

Lamyropsis microcephala is an extremely thorny thistle with a basally woody stem, large leaves up to the usually solitary heads and whitish-pink flowers. It was first discovered by Moris (1827), but not recognized by its discoverer as a separate species until 1840. He had first regarded it as *Ptilostemon (Cirsium) strictus* of Italy and the Balkans. It was

Fig. 6. Lamyropsis microcephala (Moris) Dittr. of Greut. (from Diana-Corrias, 1977).

later collected around the type locality by Thomas and Gennari, but not being observed after 1862 it was believed to be distinct for a long time. However, it was rediscovered by P.V. Arrigoni in 1968 and its distribution has since been studied in more detail by several botanists. Its nomenclature and taxonomic position has been treated by Dittrich & Greuter (1972, 1973).

L. microcephala only occurs in a very restricted area in the Monte del Gennargentu, or more precisely on the western and southern slopes of Mt. Bruncu Spina at 1500–1700 m, above the river Riu Aratu. It grows on a schistose substrate. The locality is subjected to heavy erosion due to grazing and grubbing by semi-feral pigs. The normal vegetation in similar localities is said to be low mountain scrub.

L. microcephala is mainly, or perhaps exclusively, propagated vegetatively in its only known population. Its seed-set is low and the germination of the few seeds produced is also low, according to the observations of Diana-Corrias (1977). Because of these disturbances in combination with the poor management of the locality this species must be regarded as one of the most threatened of the Sardinian endemics. It occurs in the same floristically rich area as the last-mentioned species, and its case calls for inmediate protective action.

2.12 Juncus requienii *Parl. (Juncaceae), on Corsica*

Juncus requienii was described by Parlatore (1852) in his Flora Italiana, but was then for al long time overlooked. The collections by Burnat, Briquet and collaborators in 1906–1913 showed it to be comparatively common in the mountains of Corse.

J. requienii is a strictly calcifuge species, which occurs at the margins of watercourses and in wet places at 1,000–2,300 m. It propagates vegetatively by means of long, thin runners which produce culms at various intervals. The culms are only 2–10 cm and the total number of flowers is usually between 10 and 15. Seed-set is good, so that propagation is not only vegetative but also by seed.

This species is particularly interesting because it is a rare case of a local endemic in *Juncus* subgenus *Septati,* a group in which most of the species are widespread. It is one of several species which indicate that the flora of the high altitudes of Corse is the result of long-term isolation and has rather weak connections with that of the Alps or Appenini. *J. requienii* is very different from the other European species of the group. Its closest relative is *J. brachyanthus* Trab. of the Atlas Mountains. It is one of the species that may contribute to our knowledge of the history of the Corsican montane endemic element. The members of this category are typically not local but occur in many populations and are spread throughout the island. In the case of *J. requienii* there are more than 20 documented localities from different mountains.

Hence the task for the conservationist is not so much a question of saving endangered local species. Instead the most species-rich and well-preserved examples of each habitat have to be identified. In this case the montane to subalpine wetlands would be involved.

Like many other less showy species, *J. requienii* has been overlooked and misunderstood by many collectors and authors. Such species often risk becoming neglected when plans for conservation are established.

2.13 Draba loiseleurii *Boiss. (Cruciferae), on Corsica*

The presence of this *Draba* species on Corsica was known since the beginning of the nineteenth century, but it was first described as a separate species by Boissier (1853). *Draba loiseleurii* belongs to section *Aizopsis*, which comprises about on 20 more or less well-defined species in the mountains of the Mediterranean and Central Europe.

They are all small, rosulate-pulvinate plants with stiff leaves and usually yellow flowers. The differences between the species of the group are rather small and the entire section must be regarded as critical. *Draba loiseleurii* is recognized by its flat siliqua with short and often few hairs, style 1 mm long and rather large, yellow petals. It is morphologically rather uniform except for the indumentum, since subglabrous individuals occur mixed with more hairy ones.

Draba loiseleurii occurs only in the mountains of Mt. Cinto, Mt. Rotondo and Mt. d'Oro, situated within an area of about 30 km. It is limited to rocks of the summit areas above 2,300 m, and thus only exists in these three limited populations.

The species of section *Aizopsis* are no doubt a series of relicts from an older more widespread, species, whose origin must probably be sought far back in the Tertiary. The present isolated groups are taxonomically problematic and require a thorough biosystematic study. They seem to be promising subjects for a study of the effects of isolation in the Mediterranean mountain flora.

Draba loiseleurii is just one example of the rich but spatially limited alpine flora of Corsica, which contains both a considerable number of endemics and old isolates of more widespread species. Much information on these and other Corsican endemics was contained in an extensive account by Contandriopoulos (1962), which will provide a very useful background for the establishment of future conservation measures on Corsica.

Fig. 7. Naufraga balearica Constance & Cannon. Redrawn from these authors.

2.14 Naufraga balearica *Constance & Cannon (Umbelliferae), on Islas Baleares (Fig. 7)*

Naufraga is a monotypic endemic genus, belonging to the subfamily Hydrocotyloideae. It has no close relatives in Europe but is probably related to some genera in the southern hemisphere. Hence it must be considered a relict endemic of considerable age. *Naufraga balearica* is a dwarf plant, less than 5 cm high. It has rosette of basal leaves, 5–10 mm long with a petiole up to 25 mm, 3-foliolate or pinnate with two to three pairs of leaflets. The few-flowered inflorescence looks like a compound umbel with trifoliate bracts, but its true homologies are controversial. This plant was not discovered until 1962 and was described by Constance & Cannon (1967) during the preparatory work for *Flora Europaea*.

N. balearica is ecologically specialized, occurring in wet fissures of limestone cliffs near the sea. Such localities are few and limited in area, so that even if more stations should remain to be discovered, this genus is still very restricted in space and number of individuals. In Mallorca only the type locality is known, situated at Cala San Vicente on the north coast. *Naufraga* has also been apparently reported from Corsica.

This is a good example of a restricted endemic, which will run very little risk from natural changes or normal human activity because of its isolated and rather inaccessible localities. But as it is now known to many plant collectors and Mallorca is a centre for tourism at least the type locality has now become threatened by over-collecting. With only 1–2 small populations remaining, unnecessary collecting may reduce the amount of genetic variation to a degree that affects the ability of the plant to compete and establish new individuals in its own localities. Hence the problem in this case is to stop collection, not only by formally forbidding it but by wholy preventing it. It will probably help if localities other than the well-known type locality are kept as secret as possible, by papers being published without exact locations and by being communicated to very few persons. It may also contribute to the avoidance of unnecessary collecting if any garden posessing living material of *Naufraga* could propagate and distribute it extensively, since the species grows readily in cultivation.

References

Arrigoni, P.V. (1977–1978). Le piante endemiche della Sardegna. Boll. Soc. Sarda. Si. Nat. 16: 259–264 and 17: 177–328.

Balgooy, M.M.J.V. (1969). A study on the diversity of island floras. Blumea 17: 137–178.

Barclay, C. (1974). A new locality of wild *Phoenix* in Crete. Ann. Mus. Goulandris 2: 23–29.

Bocquet, G. Widler, B. & Kiefer, H. (1978). The Messinian model: A new outlook for the floristics and systematics of the Mediterranean area. Candollea 33: 269–287.

Boissier, E. (1853). Diagnoses Plantarum Orientalium Novarum, 3. Geneva: Academische Druck. u. Verlag.

Bothmer, R.V. (1974). Studies in the Aegean flora 21. Biosystematic studies in the *Allium ampeloprasum* complex. Opera Botanica 34: 1–104.

Bramwell, D. (Ed.) (1979). Plants and Islands. London: Academic Press.

Brullo, S. (1980). *Valantia deltoidea* Brullo, sp. nov. from Sicily. Bot. Notiser 133: 63–66.

Brullo, S., Grillo, M. & Cugielmo, A. (1976). Osservazioni ecologiche preliminari su *Petagnia saniculifolia* Guss., raro endemismo siculo. Giorn. Bot. Ital. 110: 293–296.

Camarda, I. (1978). Le piante endemiche della Sardegna, 21–23. Bull. Soc. Sarda Sci. Nat. 17: 227–241.

Cardona, M.A. & Contandiopoulos, J. (1977). L'endémisme dans les flores insulaires ḿediterranéennes. Mediterranea (Valencia) 2: 49–77.

Carlquist, S. (1965). Island Life. New York: Natural History Press.

Carlquist, S. (1974). Island Biology. New York and London: Columbia University Press.

Constance, L. & Cannon, F.M. (1967). *Naufraga,* a new genus of Umbelliferae. Feddes Repert. 74: 1–4.

Contandriopoulos, J. (1962). Recherches sur la flore endemique de la Corse et sur ses origines. Ann. Fac. Sci. Marseille 32: 1–354.

Davis, P.H. (1951). Cliff vegetation in the eastern Mediterranean. J. Ecol. 39: 63–93.

Diana-Corrias, S. (1977). Le piante endemiche della Sardegna, 6–7. Boll. Soc. Sarda Sci. Nat. 16: 287–294.

Dittrich, M. & Greuter, W. (1972). *Lamyropsis microcephala* (Moris) Dittr. and Greuter. In Jull 47 (Plant label only).

Dittrich, M. & Greuter, W. (1973). Neuer Beitrag zur Kenntnis der *Gattung Lamyropsis* (Compositae): die Identität von *Cirsium microcephalum* Moris. Ann. Mus. Goulandris 1: 85–98.

Greuter, W. (1967). Beiträge zur Flora des Südägäis. 8. *Phoenix theophrastii,* die wilde Dattelpalme Kretas. Bauhinia 3: 243–250.

Greuter, W. (1968). Le dattier de Théophraste, spécialité crétoise. Ann. Mus. Genève 81: 14–16.

Greuter, W. (1972). The relict element of the flora of Crete and its evolutionary significance. In: Valentine, D.H. (Ed.), Taxonomy, Phytogeography and Evolution, pp. 161-177. London: Academic Press.

Greuter, W. (1979). The origin and evolution of island floras as exemplified by the Aegean Archipielago. In: Bramwell, D. (Ed.), Plants and Islands, pp. 87-106. London: Academic Press.

Gussone, J. (1827). Florae Siculae Prodomus, Napoli.

Halacsy, E.V. (1910). Aufzählung der von Dr. B. Tuntas auf der Insel Scyros der Nördlichen Sporaden im Juni 1908 gesammelten Arten. Österr. Bot. Zeitschr. 60: 114-118.

Huter, R. (1905). Herbar-Studien (Fortsetzung). Österr. Bot. Zeitschr. 55: 358-362.

Lojacono-Pojero, M. (1909). Flora Sìcula. Tip. Boccone del Povero. Palermo.

McArthur, R.H. & Wilson, E.O. (1967). The theory of Island Biogeography. Princeton: Princeton University Press.

Meikle, R.D. (1977). Flora of Cyprus. Kew: Royal Botanic Gardens.

Moore, D.M. (1983). Human impact on island vegetation. In: Holzner, W. Werger M.J.A. & Ikusima, I. (Eds.), Man's Impact on Vegetation, pp. 237-246. The Hague: Junk.

Moris, G.G. (1827). Stirpium Sardoarum Elenchus 1. Reg. Typ. Carali.

Moris, G.G. (1840). Flora Sardoa 2. Reg. Typ. Taurini.

Onno, M. (1933). Die Wildformen aus dem Verwandtschaftskreis *Brassica oleracea* L. Österr. Bot. Zeitschr. 82: 309-334.

Parlatore, PH. (1852). Flora Italiana, II. Firenze: Le Monier.

Person, C.H. (1807). Synopsis Planatrum 2. Paris and Tübingen: Carol Fried Cramerum.

Phitos, D. (1970). Die Gattung *Aubrietia* in Griechenland. Candollea 25: 69-W87.

Pignatti, S. (1971). Vallone Calagna sopra Tortorici Censimento dei biotopi di rilevante interesse vegetazionale meritevoli di converzione in Italia. 19: 10. Sicilia, Camerino.

Post, G.E. (1900). Plantae Postianae fasc. X. Mem. Herb. Boiss. 18: 89-102.

Rechinger, K.H. (1943). Flora Aegea. Denkschr. Akad. Wiss. Wien Mat. Nat. Kl. 105, 1: 1-924.

Rechinger, K.H. (1951). Phytogeographia Aegea. Denkschr. Akad. Wiss Wien. Mat. Nat. Kl. 105, 1: 1-208.

Snogerup, S. (1967a). Studies in the Aegean Flora VIII. *Erysimum* Sect. *Cheiranthus*. A. Taxonomy. Opera Botanica 13: 1-70.

Snogerup, S. (1967b). Studies in the Aegean Flora IX. *Erysimum* Sect. *Cheiranthus*. B. Variation and Evolution in the small population system. Opera Botanica 14: 1-86.

Snogerup, S. (1971). Evolutionary and plant geographical aspects of chasmophytic communities. In: Plant Life in South West Asia, Davis, P.H., (Ed.), pp. 157-170. Edinburgh: Edinburgh University Press.

Snogerup, S. (1980). The wild forms of the *Brassica oleracea* group (2n = 18) and their possible relations to the cultivated ones.In: S. Tsunoda, K. Hinata and C. Gómez-Campo (Eds.), *Brassica* Crops and Wild Allies, pp. 121-132. Tokyo: Japan Scientific Societies Press.

Strid, A. (1970). Studies in the Aegean Flora 16. Biosystematic Studies of the *Nigella arvensis* complex. With special references to the problem of non-adaptive radiation. Opera Botanica 28: 1-169.

The Botanical Museum
Ö. Vallgatan 18
S-223 61 Lund, Sweden

Part III

Actions and solutions

CHAPTER 11

The value of information in saving threatened Mediterranean plants

C. LEON, G. LUCAS and H. SYNGE

1. Introduction

We live at a time when information is valued more than ever before. We borrow the misleading term 'data' to describe it. We ask 'information scientists' to arrange it in structures so it can be held in electronic form and then apply to it a technology of microchips with power to sort, shuffle and select, which half a century ago no-one would have imagined possible. The result is the modern 'data base'. In this chapter we want to discuss the relevance of a 'data base' to saving endangered plants and ask how valuable it is to nature conservation in Mediterranean countries.

The question is important and pressing. There is a growing feeling that time is running out for major parts of the plant kingdom; we need to be able to convince decision-makers now of the need to maintain plant diversity for tomorrow. The best way to convince somebody is to provide facts that he or she cannot refute. When decision-makers are convinced of our case, then we need to provide them with the additional information to do the job.

Another reason why this issue needs discussing is that the volume of information is growing every day and so, paradoxically, it is becoming more difficult rather than easier to find the information needed for specific tasks. Every day the number of papers and the volume of potentially useful information generated by taxonomists increases. With the rapidly developing interest in plant conservation over the last decade, the volume of the more specialist literature on threatened plants is also expanding fast. So we have to ask whether the new techniques can help us find our way through this paper jungle.

A recent bibliography covering only endangered plant lists (Jarvis, Leon & Oldfield, 1981) comprises hundreds of titles. A more comprehensive reference work entitled 'Plants in Danger: What do we know?' (IUCN, in prep.), aims to provide references to all lists and books about threatened plants world-wide, as well as listing voluntary organizations, outlining protected species legislation and briefly describing national plant data-bases. Written as an information referral system, it is designed to answer the question, 'Where can I find out about the plants of any country, which of them are known to be in danger, and whom then can I get in touch with?'

For nearly ten years, many authors, ourselves included, have emphasized the need for information on threatened plants. Some of the results have been impressive; as we outline later and as the 'Plants in Danger' book will show, threatened plant lists have been produced by IUCN for all Mediterranean countries, many countries have produced their own lists, and there are some national plant conservation programmes developing. Governments have passed laws to protect plant species. The European Economic Community and the Council of Europe have made growing commitments to conservation. The science of monitoring and managing rare species populations on scienti-

Gómez-Campo, C. (ed.), Plant conservation in the Mediterranean area.
© 1985, Dr W. Junk Publishers, Dordrecht. *ISBN 90 6193 523 7.*

fic principles is also in a phase of rapid development (see the papers in Synge, 1981). And lastly, there is a network of active botanic gardens collaborating together to grow threatened Mediterranean plants in cultivation.

Yet the crucial step is still to come: translating the information already held into action in the field to protect rare plant sites. It is hard, if not impossible, to find examples of reserves declared or planning controls agreed to protect Mediterranean species listed as Endangered. This highlights the most difficult and yet most important step in the process; that of taking an endangered species list and developing and implementing protection plans for the species it contains. To undertake this step we must first assess exactly what information is available and how it was assembled. This is the purpose of the present article, which we hope will help point the way ahead and assist those who can use the data. This book should itself serve as a major stimulus for the action that is needed.

2. The taxonomic base

An adequate classification is a prerequisite for any threatened plant list to be comprehensive. Obviously we cannot decide which botanical entities are in danger and where they still grow until we have a working system for classifying them and a series of names to which they can be referred. Nor can we be sure that the threatened plant list is complete until we have screened all the plants growing in the area covered to assess which of them are rare or in danger. This in itself requires a Flora or Checklist.

A good floristic treatment greatly facilitates the search for those species which may be rare or in danger on a world scale. Some species turn out to have more extensive distributions than previously thought and others to be no more than geographical variants of widespread species. The easiest route to a comprehensive threatened plant list is to take from the Flora all species that could conceivably be rare or threatened and then screen each one individually for its exact status in the wild. In the absence of a good systematic basis, threatened plant lists can only be based on random observations.

Conservation benefits from taxonomic research in another, equally important, way: taxonomists with the knowledge to identify plants reliably and the time to study them in the field often know better than anyone else which species are rare and where they can still be found. In many cases this knowledge is the result of a lifetime's work. All those who wish to see threatened plants brought out of danger and conserved owe a great deal to those scientists who have freely made available their life's work and experience to the conservation bodies.

Over the last 25 years there has been tremendous progress in Mediterranean plant taxonomy – the revision of critical groups, the unravelling of nomenclatural problems and the description of new species. *Flora Europaea* (Tutin et al., 1964–1980) is now complete, providing a basic systematic treatment for the countries of Europe. Many countries covered by *Flora Europaea* are also developing their own national Floras (e.g. Jordanov et al., 1963 –, for Bulgaria). On the southern and eastern side of the Mediterranean it is pleasing to record the rapid progress of the *Flora of Libya* (Ali et al., 1976–), the launching of a new Flora of Egypt (N. el-Hadidi, in prep.), a comprehensive *Flora of Cyprus* (Meikle, 1977–1985) and the steady drawing to conclusion of the massive *Flora of Turkey* (Davis, 1965–).

Of particular relevance to this book is the 'Med-Checklist' of the Organization for the Phytotaxonomic Investigation of the Mediterranean Area (OPTIMA), described by Greuter (1979). This will provide a plant checklist for all countries of the Mediterranean and should help to greatly reduce the disparities between accounts for groups in North Africa and in Europe.

As the knowledge from taxonomy and plant distributions becomes more precise, the conservation needs in terms of species and areas for protection become more refined and better reflect the situation on the ground. Obviously the first threatened plant lists and subsequent recommendations cannot be comprehensive, but as the classifications become more accurate and refined, so too can the conservation data be improved and made more precise.

3. Mediterranean threatened plant lists

Of the 2879 species identified as endemic to individual Mediterranean countries (excluding Syria, Lebanon and Turkey), 1529 are believed to be rare or threatened, using IUCN criteria (described below). Of the remaining endemics, nearly 300 are considered to be too insufficiently known to be given a category. The figures become 3583 endemics, of which 1968 are rare and threatened, if the rich floras of the Azores, Madeira and the Canary Islands are included. Considerable weight should be given to the status of endemic species because they tend to include the majority of the threatened species and good political arguments can be made for their protection.

These figures are taken from the latest versions of the lists compiled by the Threatened Plants Unit of IUCN's Conservation Monitoring Centre, as part of a continuing programme of data-gathering on threatened plants world-wide. The overall aim of this programme is to determine which species are threatened, their distribution, the causes for their decline and, wherever possible, to recommend protective action.

TPU's programme started in 1974 with a far-sighted contract from the Council of Europe to produce a list of threatened plants for the region covered by the Council. The advent of *Flora Europaea* provided a basic floristic account from which lists of endemics and other candidate species could be extracted. Extensive and detailed collaboration with botanists all over the continent provided information on the degree of threat to most of the species initially selected for screening.

A report was produced for the Council of Europe in 1976 and published by them in 1977 (IUCN Threatened Plants Committee, 1977). The gathering and analysis of the information was carried out by Hugh Synge until 1980, when Christine Leon took over as TPU Research Assistant for Europe. To reflect continuing changes in the size of field populations and to incorporate newly discovered rare or threatened taxa, a second edition was produced in 1982 and published in 1983. Each edition contains lists of endemic taxa annotated with conservation categories for France, Spain, Portugal, Greece, Italy, Albania, Yugoslavia and Malta. Lists for the Azores, Madeira and Canary Islands were added to the second edition. For Turkey, only a partial list of endemic species was included, without conservation ratings.

The threat to each listed species is indicated by one of the IUCN 'Red Data Book' categories. The criteria adopted are rigidly applied, following definitions laid down by IUCN's Species Survival Commission. These are reproduced below; a short booklet explaining their application to plants in more detail, with examples, is sent to contributors and is available free on request from the TPU.

Extinct (Ex)

Endangered (E)
Taxa in danger of extinction and whose survival is unlikely if the causal factors continue operating.
Included are taxa whose numbers have been reduced to a critical level or whose habitats have been so drastically reduced that they are deemed to be in immediate danger of extinction.

Vulnerable (V)
Taxa believed likely to move into the Endangered category in the near future if the causal factors continue operating.
Included are taxa of which most or all the populations are *decreasing* because of over-exploitation, extensive destruction of habitat or other environmental disturbance; taxa with populations that have been seriously *depleted* and whose ultimate security is not yet assured; and taxa with populations that are still abundant but are *under threat* from serious adverse factors throughout their range.

Rare (R)
Taxa with small world populations that are not at present Endangered or Vulnerable, but are at risk.

These taxa are usually localized within restricted geographical areas or habitats or are thinly scattered over a more extensive range.

Indeterminate (I)
Taxa *known* to be Extinct, Endangered, Vulnerable or Rare but where there is not enough information to say which of the four categories is appropriate.

Out of Danger (O)
Taxa formerly included in one of the above categories, but which are now considered relatively secure because effective conservation measures have been taken or the previous threat to their survival has been removed.

Insufficiently Known (K)
Taxa that are suspected but not definitely known to belong to any of the above categories, because of the lack of information.

N.B. In practice, Endangered and Vulnerable categories may include, temporarily, taxa whose populations are beginning to recover as a result of remedial action, but whose recovery is insufficient to justify their transfer to another category.

For species which are neither rare nor threatened, the symbol 'nt' is used.

Both editions comprise a full list of taxa threatened on a European scale, covering ferns, conifers and angiosperms. Taxa whose Red Data Book category for Europe *as a whole* is Extinct, Endangered, Vulnerable, Rare or Indeterminate are included but the report excluded species rare or threatened in one European country but common in another. The second part of the reports comprised a series of lists, one for each country, covering in each case all the endemics of that country, whether or not rare or threatened, and also those other indigenous species which are threatened on a European scale. A list of species whose conservation status is unknown (category Insufficiently Known) concludes each report. One of the objectives of the second edition was to establish the status of 1003 'Insufficiently Known' taxa listed in the first edition. As a result of closer examination of these little known taxa, this number has now been reduced by half.

An introduction to the reports includes a short summary of the laws protecting rare species in each country and reproduces the resolution on threatened plants agreed by the Council of Ministers of the Council of Europe. This is reproduced below:

The Committee of Ministers:

Referring to Resolution No. 2 of the European Ministerial Conferences on the Environment (Vienna 1973, Brussels 1976) on the protection of wildlife;
Having regard to the list of rare, threatened and endemic plants in Europe commissioned by the European Committee for the Conservation of Nature and Natural Resources;
Noting that man and all animals are dependent for their survival on the plant kingdom;
Recognising that plants (species, sub-species, varieties, etc.) form a genetic resource of immeasurable value to mankind and that the economic potential of the plant kingdom is as yet only partly realised;

Recognising the scientific, educational, recreational, aesthetic, cultural and ethical value of plants to mankind;

Noting that the list includes some 1,400* species as rare and/or threatened in Europe, of which more than 100 are in imminent danger of extinction and that the figure of 1,400 represents approximately one tenth of the total European flora;

Realising that once a species becomes extinct, it cannot be recreated by man, and hence that it is of the utmost importance to ensure the conservation of as many species as possible for the economic, scientific and cultural benefit of mankind;

Recommends that the government of member States of the Council of Europe be guided in their policy in this matter by the principles set out below;

1. Ensure appropriate legal protection for all plants indentified as endangered in the Euro-

* Revised in second edition to approximately 2,300.

pean list with provision for licences to be issued for approved collection purposes;

2. Provide minimum legal protection for all plants against depredations not yet covered by law;
3. Institute or complete national surveys of plants that are rare or threatened within their boundaries for appropriate dissemination and publication. Such surveys should:
   (a) Include plants that are rare or threatened only in particular countries and therefore not included in the European lists;
   (b) Identify the principal threats to the plants so listed;
   (c) Specify the action needed to ensure their survival;
4. Establish nature reserves and designate areas in which vegetation and flora are protected by law and stimulate the setting up of nature reserves by private bodies, with the long-term aim of ensuring that all species on the European lists can be found in such areas and in so doing contribute to the establishment of the European network of biogenetic reserves which was the subject of Resolution (87) 17;
5. Incorporate safeguards in future planning strategies to protect all species on the European list, as the major threat to many plants is created by changing patterns of land use;
6. Stimulate, undertake and co-ordinate through competent organisations multidsisciplinary research at national or international level, with particular emphasis on bringing together information on plants found in more than one country with a view to:
   (a) Extending and improving knowledge about the flora of those areas in Europe that are still insufficiently known botanically, so as to be able to make constructive proposals for conservation and planning purposes;
   (b) Promoting studies on the habitat, autoecology and population biology of each plant on the European list to provide the information needed from which integrated conservation management plans can be formulated;
   (c) Promoting studies on the dynamics and ecology of the vegetation types in which the plants on the European list occur;
7. Give appropriate support to scientifically based botanical gardens so that they have the facilities they need to propagate and grow the plants on the European list and to distribute propagating material to other institutions and where appropriate reintroduce plants to the wild, with the aim of reducing the pressure on wild plant populations and at the same time drawing attention to the aesthetic, cultural and scientific importance of these plants;
8. Ratify for their States, if they have not already done so, the Convention on International Trade in Endangered Species of Wild Fauna and Flora, opened for signature in Washington on 3 March 1973;
9. Acknowledge that the plant kingdom is a dynamic system and needs to be monitored at stated intervals so that the list can be revised regularly;
10. Prepare and disseminate codes of conduct on rare and threatened plants;
11. Disseminate general information on the need to protect plants and on the protective measures set out in the European Committee's list.

Ideally the Council of Europe list should become the overview of a national list or Red Data Book for each country, produced not by IUCN but by botanists and conservationists in that country. These national accounts would go rather further than IUCN lists and reports, by, for example, covering nationally threatened as well as world-threatened taxa, by including far more detailed information on each species, and maybe by including rather more lower taxa. Yet few Mediterranean countries have national Red Data Books so far.

A good model may emerge from Spain where very recently the Director General of the Environment of the Spanish Government, Dr. Saénz Lain, convened a meeting of Spanish botanists to produce a list of threatened plants for Spain, including the floristically rich Canary and Balearic Islands. The Spanish botanists will use the IUCN Red Data Book categories as criteria for the degree of threat and are taking the latest version of the IUCN list as the starting point which they will update and enlarge. The meeting followed a suggestion made by IUCN, in particular through the newly formed SSC European Specialist Group under the chairmanship of Professor V.H. Heywood. It was organised in association with Professor C. Gómez-Campo of the Universidad Politécnica in Madrid, who is also co-ordinating the production of the Spanish plant Red Data Book.

For the Mediterranean countries of North Africa and Asia, a threatened species list was issued in 1980. This was arranged in a similar way to the European list and contained lists for Morocco, Algeria, Tunisia, Libya, Egypt, Israel and Cyprus, as well as for the Azores, Madeira and the Canary Islands. The lack of a regional Flora proved a great stumbling block, as all national Floras and Checklists had to be compared and screened one by one and many differences in taxonomic treatments could not be resolved.

Table 1 shows the species in Mediterranean countries at present listed by IUCN as Endangered on a world scale. Table 2 is a similar list for species listed as Extinct; the TPU defines an 'Extinct' species as one that has not been found after repeated searches of known and likely areas. It is of course virtually impossible to say any plant is extinct, i.e. it now longer survives anywhere on earth, and indeed a number of species listed as 'Extinct' by IUCN have, to our great delight, re-appeared, either in the wild or in cultivation. This shows the value of the IUCN 'Extinct' category; by focusing attention on these often obscure species, efforts can be made to relocate them before they succumb as a result of loss of their restricted habitats. Table 3 is a statistical summary of the numbers of endemic species in each of the IUCN Red Data Book categories, according to the latest data held by TPU. All these tables include information for the floras of the Atlantic Islands of the Azores, Madeira and the Canaries.

It is to the botanists in the field, in each of these Mediterranean countries, that we owe our thanks for their co-operation and expert knowledge. Close collaboration with the Organization for the Phytotaxonomic Investigation of the Mediterranean Area (OPTIMA), through its energetic Secretariat under Professor W. Greuter, greatly in-

Table. 1 List of endangered plants in the Mediterranean countries and Macaronesian islands.

Gymnospermae		
Cupressacea		
Cupressus dupreziana A. Camus	Algeria	E
Juniperus cedrus Webb & Berthel	Canary Is. (E); Madeira (E)	E
Pinaceae		
Abies nebrodensis (Lojac.) Mattei	Italy	E
Angiospermae		
Aizoaceae		
Mesembryanthemum gaussenii Leredde	Algeria	E
Aquifoliaceae		
Ilex platyphylla Webb & Berthel.	Canary Is.	E

List of endangered plants in Mediterranean countries (continued)

Boraginaceae		
Anchusa crispa Viv.	Corsica (E); Sardinia (E)	E
Echium auberianum Webb & Berthel.	Canary Is.	E
Echium gentianoides Webb ex Coincy	Canary Is.	E
Echium handiense Svent.	Canary Is.	E
Echium pininana Webb & Berthel.	Canary Is.	E
Myosotis rehsteineri Wartm.	Italy (E)	E
Omphalodes littoralis Lehm.	France	E
Onosma elegantissima Rech. f. & Goulimy	Greece	E
Symphytum cycladense Pawl.	Greece	E
Campanulaceae		
Campanula baborensis Quézel	Algeria	E
Campanula sabatia De Not.	Italy	E
Campanula secundiflora Vis. & Pancic	Yugoslavia	E
Musschia wollastonii Lowe	Madeira	E
Wahlenbergia bernardi Leredde	Algeria	E
Caprifoliaceae		
Sambucus palmensis Link	Canary Is.	E
Caryophyllaceae		
Arenaria lithops Heywood ex McNeill	Spain	E
Dianthus guessfeldtianus Muschler	Egypt	E
Gypsophila papillosa P. Porta	Italy	E
Loeflingia tavaresiana G. Samp.	Portugal	E
Silene cirtensis Pomel	Algeria	E
Silene orphanidis Boiss.	Greece	E
Silene physalodes Boiss.	Syria (?)	E
Silene pseudovestita Battand.	Algeria	E
Silene reeseana Maire	Morocco	E
Silene rothmaleri Pinto da Silva	Portugal	E
Silene sessionis Battand.	Algeria	E
Silene velutina Pourret ex Loisel.	Corsica	E
Chenopodiaceae		
Kochia saxicola Guss.	Italy	E
Salicornia veneta Pignatti & Lausi	Italy	E
Cistaceae		
Cistus osbeckiifolius Webb ex Christ	Canary Is.	E
Helianthemum bystropogophyllum Svent.	Canary Is.	E
Helianthemum sphaerocalyx Gauba & Janchen	Egypt	E
Tuberaria major (Willk.) Pinto da Silva et al.	Portugal	E
Compositae		
Anacyclus alboranensis Esteve Chueca & Varo	Spain	E
Anthemis glaberrima (Rech.f.) Greuter	Greece	E

List of endangered plants in Mediterranean countries (continued)

Argyranthemum lidii Humphries	Canary Is.	E
Argyranthemum pinnatifidum (L.f.) Lowe ssp. *succulentum* (Lowe) Humphries	Madeira	E
Argyranthemum thalassophilum Svent.) Humphries	Salvage Is.	E
Argyranthemum winteri (Svent.) Humphries	Canary Is.	E
Artemisia granatensis Boiss.	Spain	E
Aster pyrenaeus Desf. ex DC.	France	E
Asteriscus schultzii (Bolle) Pitard & Proust	Canary Is.	E
Atractylis arbuscula Svent. & Michaelis	Canary Is.	E
Atractylis preauxiana Schultz Bip.	Canary Is.	E
Calendula suffruticosa Vahl ssp. *maritima* (Guss.) Meikle	Italy	E
Carduncellus ilicifolius Pomel	Algeria	E
Centaurea balearica J.D. Rodriguez	Balearic Is.	E
Centaurea heldreichii Halacsy	Greece	E
Centaurea horrida Badaro	Sardinia	E
Centaurea kalambakensis Freyn & Sint.	Greece	E
Centaurea lactiflora Halacsy	Greece	E
Centaurea linaresii Lazaro	Spain	E
Centaurea megarensis Halacsy & Hayek	Greece	E
Centaurea niederi Heldr.	Greece	E
Centaurea peucedanifolia Boiss. & Orph.	Greece	E
Centaurea princeps Boiss. & Heldr.	Greece	E
Cheirolophus arboreus (Webb) Holub	Canary Is.	E
Cheirolophus duranii (Burchard) Holub	Canary Is.	E
Cheirolophus junonianus (Svent.) Holub	Canary Is.	E
Cheirolophus massonianus (Lowe) Hansen & Sunding	Madeira	E
Cheirolophus tagananensis (Svent.) Holub	Canary Is.	E
Crepis claryi Battand.	Algeria	E
Crepis crocifolia Boiss. & Heldr.	Greece	E
Crepis faureliana Maire	Algeria	E
Helichrysum monogynum B.L. Burtt & Sunding	Canary Is.	E
Hypochoeris oligocephala (Svent. & D. Bramwell) Lack	Canary Is.	E
Lamyropsis microcephala (Moris) Dittrich & Greuter	Sardinia	E
Leontodon siculus (Guss.) Finch &Sell	Italy	E
Leuzea cynaroides (Link) Font Quer	Canary Is.	E
Logfia neglecta (Soy.-Will.) Holub	Corsica (E) France (E)	E
Lugoa revoluta DC.	Canary Is.	E
Mecomischus pedunçulatus (Coss. & Durieu) Maire	Algeria	E
Onopordum algeriense (Munby) Pomel	Algeria	E
Onopordum cyrenaicum Maire & M. Weiller	Libya	E
Pulicaria burchardii Hutch.	Canary Is.	E
Pulicaria canariensis Bolle	Canary Is.	E
Pulicaria filaginoides Pomel	Algeria	E
Scorzonera drarii Tackh.	Egypt	E
Senecio alboranicus Maire	Spain	E
Senecio hadrosomus Svent.	Canary Is.	E
Senecio hermosae Pitard	Canary Is.	E
Sonchus bornmuelleri Pitard	Canary Is.	E
Tanacetum ptarmiciflorum (Webb) Schultz Bip.	Canary Is.	E

List of endangered plants in Mediterranean countries (continued)

Convolvulaceae		
Convolvulus argyrothamnos Greuter	Greece	E
Convolvulus lopez-socasi Svent.	Canary Is.	E
Convolvulus massonii A. Dietr.	Madeira	E
Ipomoea sinaica Tackh. & Boulos	Egypt	E
Crassulaceae		
Aeonium saundersii Bolle	Canary Is.	E
Monanthes adenoscepes Svent.	Canary Is.	E
Cruciferae		
Alyssum akamasicum B.L. Burtt	Cyprus	E
Alyssum fastigiatum Heywood	Spain	E
Arabis kennedyae Meikle	Cyprus	E
Biscutella neustriaca Bonnet	France	E
Brassica bourgeaui (Webb ex Christ) Kuntze	Canary Is.	E
Brassica macrocarpa Guss.	Italy	E
Coronopus navasii Pau	Spain	E
Crambe arborea Webb ex Christ	Canary Is.	E
Crambe sventenii B. Petters. ex Bramw. & Sundin	Canary Is.	E
Diplotaxis siettiana Maire	Spain	E
Hormathophylla pyrenaica (Lapeycr.) Cullen & Dudley	France	E
Hutera rupestris P. Porta	Spain	E
Iberis arbuscula Runemark	Greece	E
Ionopsidium acaule (Desf.) Reichenb.	Portugal	E
Otocarpus virgatus Durieu	Algeria	E
Hutera johnstonii (G. Samp.) G. Campo	Portugal	E
Sisymbrium cavanillesianum Valdés & Castrov	Spain	E
Cyperaceae		
Cyperus papyrus L. ssp. *hadidii* Chrtek & Slavikova	Egypt	E
Dipsacaceae		
Pterocephalus virens Berthel.	Canary Is.	E
Euphorbiaceae		
Euphorbia anachoreta Svent.	Salvage Is.	E
Euphorbia handiensis Burchard	Canary Is.	E
Euphorbia ruscinonensis Boiss.	France	E
Geraniaceae		
Geranium humbertii Beauverd	Greece	E
Geranium maderense Yeo	Madeira	E
Gramineae		
Deschampsia maderensis (Hackel & Bornm.) Buschm.	Madeira	E

List of endangered plants in Mediterranean countries (continued)

Phalaris maderensis (Menezes) Menezes	Madeira	E
Stipa austroitalica Martinovsky	Italy	E
Stipagrostis drarii (Tackh.) de Winter	Egypt	E
Vulpia obtusa Trabut	Algeria	E
Grossulariaceae		
Ribes sardoum Martelli	Sardinia	E
Hypericaceae		
Hypericum aciferum (Greuter) N.K.B. Robson	Greece	E
Iridaceae		
Crocus cyprius Boiss. & Kotschy	Cyprus	E
Crocus hartmannianus Holmboe	Cyprus	E
Iris lortetii Barbey	Israel (E); Lebanon (I)	E
Labiatae		
Amaracus cordifolius Mont. & Aucher-Eley ex Benth.	Cyprus	E
Micromeria taygetea P.H. Davis	Greece	E
Nepeta sphaciotica P.H. Davis	Greece	E
Sideritis cystosiphon Svent.	Canary Is.	E
Sideritis discolor (Webb ex de Noe) Bolle	Canary Is.	E
Sideritis infernalis Bolle	Canary Is.	E
Sideritis nervosa (Christ) Lid	Canary Is.	E
Thymus camphoratus Hoffmanns. & Link	Portugal	E
Thymus carnosus Boiss.	Portugal	E
Thymus cephalotos L.	Portugal	E
Leguminosae		
Adenocarpus faurei Maire	Algeria	E
Adenocarpus umbellatus Coss.	Algeria	E
Anagyris latifolia Brouss. ex Willd.	Canary Is.	E
Anthyllis lemanniana Lowe	Maderia	E
Astragalus algarbiensis Coss. ex Bunge	Portugal (E); Spain (E)	E
Astragalus aquilanus Anzalone	Italy	E
Astragalus macrocarpus DC. ssp. *lefkarensis* Agerer-Kirchoff & Meikle	Cyprus	E
Astragalus maritimus Moris	Sardinia	E
Astragalus verrucosus Moris	Sardinia	E
Crotalaria vialattei Battand.	Algeria	E
Cytisus aeolicus Guss. ex Lindl.	Italy	E
Genista spinulosa Pomel	Algeria	E
Lathyrus lentiformis Plitm.	Israel	E
Lotus berthelotii Masferrer	Canary Is.	E
Lotus callis-viridis D. Bramwell & D.H. Davis	Canary Is.	E
Lotus kunkelii (E. Chueca) D. Bramwell et al.	Canary Is.	E

List of endangered plants in Mediterranean countries (continued)

Lotus maculatus Breitfeld	Canary Is.	E
Ononis megalostachys Munby	Algeria	E
Teline benehoavensis (Bolle ex Svent.) Santos	Canary Is.	E
Teline linifolia (L.) Webb & Berthel. ssp. *teneriffae* P.E. Gibbs & Dingwall	Canary Is.	E
Trigonella balachowskyi Leredde	Algeria	E
Liliaceae		
Allium crameri Aschers. & Boiss.	Egypt	E
Allium seirotrichum Ducell. & Maire	Algeria	E
Allium trichocnemis Gay	Algeria	E
Asparagus fallax Svent.	Canary Is.	E
Bellevalia salah-eidii Tackh. & Boulos	Egypt (E); Libya (E)	E
Chionodoxa lochiae Meikle	Cyprus	E
Colchicum arenarium Waldst. & Kit.	Yugoslavia	E
Leopoldia albiflora Tackh. & Boulos	Egypt	E
Leopoldia longistyla Tackh. & Boulos	Egypt	E
Muscari gussonei (Parl.) Tod	Italy	E
Scilla morrisii Meikle	Cyprus	E
Malvaceae		
Lavatera phoenicea Vent.	Canary Is.	E
Onagraceae		
Epilobium numidicum Battand.	Algeria	E
Orchidaceae		
Cephalanthera cucullata Boiss. & Heldr.	Greece	E
Goodyera macrophylla Lowe	Madeira	E
Orchis scopulorum Summerh.	Madeira	E
Palmae		
Medemia argun (Martius) Wurtt. ex HA. Wendl	Egypt (E)	E
Pittosporaceae		
Pittosporum coriaceum Dryander ex Aiton	Madeira	E
Plantaginaceae		
Plantago lybyca Beguinot & Vaccari	Libya	E
Plantago malato-belizii Lawalree	Madeira	E
Plumbaginaceae		
Armeria rouyana Daveau	Portugal	E
Armeria soleirolii (Duby) Godron	Corsica	E
Limonium arborescens (Brouss.) Kuntze	Canary Is.	E

List of endangered plants in Mediterranean countries (continued)

Limonium dendroides Svent.	Canary Is.	E
Limonium fruticans (Webb) Kuntze	Canary Is.	E
Limonium imbricatum (Webb & Berthel.) Hubbard	Canary Is.	E
Limonium macrophyllum (Brouss.) Kuntze	Canary Is.	E
Limonium preauxii (Webb & Berthel.) Kuntze	Canary Is.	E
Limonium spectabile (Svent.) Kunkel & Sunding	Canary Is.	E
Polygonaceae		
Calligonum calvescens Maire	Algeria	E
Rumex rothschildianus Aarons. ex Evenari	Israel	E
Rumex tunetanus G. Barratte & Murb.	Tunisia	E
Potamogetonaceae		
Potamogeton hoggarensis Dandy	Algeria	E
Primulaceae		
Primula apennina Widmer	Italy	E
Ranunculaceae		
Aquilegia cazorlensis Heywood	Spain	E
Consolida samia P.H. Davis	Greece	E
Delphinium caseyi B.L. Burtt.	Cyprus	E
Ranunculus kykkoensis Meikle	Cyprus	E
Ranunculus weyleri Mares	Balearic Is.	E
Rosaceae		
Bencomia brachystachya Svent.	Canary Is.	E
Bencomia exstipulata Svent.	Canary Is.	E
Chamaemeles coriacea Lindl.	Madeira	E
Marcetella maderensis (Bornm.) Svent.	Madeira	E
Sorbus maderensis Dode	Maderia	E
Rubiaceae		
Galium litorale Guss.	Italy	E
Galium numidicum Pomel	Algeria	E
Santalaceae		
Kunkeliella canariensis Stearn	Canary Is.	E
Kunkeliella psilotoclada (Svent.) Stearn	Canary Is.	E
Scrophulariaceae		
Antirrhinum charidemi Lange	Spain	E
Digitalis atlantica Pomel	Algeria	E
Isoplexis chalcantha Svent. & O'Shanahan	Canary Is.	E
Linaria algarviana Chav.	Portugal	E
Linaria burceziana Maire	Algeria	E
Linaria ficalhoana Rouy	Portugal	E

List of endangered plants in Mediterranean countries (continued)

Linaria hellenica Turrill	Greece	E
Odontites discolor Pomel	Algeria	E
Pedicularis numidica Pomel	Algeria	E
Verbascum cylleneum (Boiss. & Heldr.) Kuntze	Greece	E
Veronica kaiseri Tackh.	Egypt	E
Veronica musa Tackh. & Hadidi	Egypt	E
Selaginaceae		
Globularia ascanii D. Bramwell & Kunkel	Canary Is.	E
Globularia sarcophylla Svent.	Canary Is.	E
Globularia stygia Oprh. ex Boiss.	Greece	E
Solanaceae		
Solanum lidii Sunding	Canary Is.	E
Solanum trisectum Dunal	Madeira	E
Withania obtusifolia Tackh.	Egypt	E
Thymelaeaceae		
Daphne rodriguezii Texidor	Balearic Is.	E
Umbelliferae		
Angelica heterocarpa Lloyd	France	E
Bunium brevifolium Lowe	Madeira	E
Bupleurum capillare Boiss. & Heldr.	Greece	E
Bupleurum kakiskalae Greuter	Greece	E
Laserpitium longiradium Boiss.	Spain	E
Ligusticum albanicum S. Javorka	Albania	E
Monizia edulis Lowe	Madeira (E); Salvage Is. (E)	E
Valerianaceae		
Valeriana longiflora Willk.	Spain	E
Violaceae		
Viola hispida Lam.	France	E
Viola jaubertiana Mares & Vigineix	Balearic Is.	E
Viola palmensis Webb & Berthel.	Canary Is.	E

creased consistency between the different floristic accounts; their Med-Checklist will help to reduce the number of disparities in the lists and fill gaps in the coverage.

There has been little analysis of the data contained in the lists (but see Sukopp & Trautmann, 1981; Synge, 1980), especially for Mediterranean countries. However, the high number of listed species reflects the many pressures, past and present, on the vegetation of the Mediterranean region. In recent times industrialization has been rapid with accompanying urban development and the widespread over-exploitation of natural resources. Perhaps the two most dramatic effects have been the degradation of the evergreen forests and maquis, most noticeably on the northern shore

Table 2. List of extinct plants in Mediterranean countries.

Angiospermae		
Boraginaceae		
Myosotis ruscinonensis Rouy	France	Ex
Caryophyllaceae		
Minuartia olonensis (Bonnier) P. Fourn.	France	Ex
Crassulaceae		
Monanthes dasyphylla Svent.	Canary Is.	Ex
Graminaea		
Avenula hackelii (Henriq.) Holub	Portugal	Ex
Labiatae		
Thymus oehmianus Ronn. & Soska	Yugoslavia	Ex
Leguminosae		
Genista melia Boiss.	Greece	Ex
Onobrychis aliacmonia Rech.f.	Greece	Ex
Tephrosia kassasi Boulos	Egypt	Ex
Tetragonolobus wiedemannii Boiss.	Greece	Ex
Vicia dennesiana H.C. Watson	Azores	Ex
Liliaceae		
Allium rouyi Gaut.	Spain	Ex
Plumbaginaceae		
Armeria arcuata Welw. ex Boiss. & Reuter	Portugal	Ex
Primulaceae		
Lysimachia minoricensis J.D. Rodriguez	Balearic Is.	Ex
Umbelliferae		
Geocaryum bornmuelleri (Wolff) Engstr.	Greece	Ex
Geocaryum divaricatum (Boiss. & Orph.) Engstr.	Greece	Ex
Violaceae		
Viola cryana Gillot	France	Ex
Zygophyllaceae		
Fagonia taeckholmiana Hadidi	Egypt	Ex

Table 3. A statistical summary of the numbers of rare, threatened and endemic plants in the Mediterranean.

	Endemic species								Non-endemic species					
Country	Ex	E	V	R	I	K	nt		Ex	E	V	R	I	
Albania	0	1	2	11	6	2	2	24	0	0	9	54	3	66
Algeria	0	31	22	65	6	9	39	172	0	2	4	8	11	25
Azores	1	0	5	18	6	11	14	55	0	0	1	3	0	4
Balearic Is.	1	4	7	18	0	0	22	52	0	0	5	1	1	7
Canary Is.	1	64	124	121	9	43	152	514	0	2	17	6	1	26
Corsica	0	2	0	11	0	4	13	30	0	2	15	6	0	23
Cyprus	0	10	9	22	5	23	46	115	0	0	0	0	0	0
Egypt	2	12	6	38	6	4	2	70	0	3	11	17	5	36
France	3	7	10	23	3	16	11	73	0	3	63	27	13	106
Gibraltar	0	0	0	1	0	0	0	1	0	1	0	2	0	3
Greece	5	25	36	355	40	58	223	742	3	6	33	143	7	192
Israel	0	2	1	5	1	4	1	14	0	1	5	7	3	16
Italy	0	14	24	66	8	39	77	228	0	2	78	47	7	134
Libya	0	2	18	18	4	20	21	83	0	1	11	7	0	19
Maderia	0	17	30	39	0	22	23	131	0	3	18	7	0	28
Malta	0	0	0	1	0	0	1	2	1	1	3	5	1	11
Morocco	0	1	3	162	23	53	294	536	0	0	1	0	1	2
Salvage Is.	0	2	1	1	0	0	0	4	0	1	2	2	1	6
Portugal	2	11	11	33	0	1	12	70	0	2	15	25	2	44
Sardinia	0	5	3	9	0	1	8	26	0	1	12	8	0	21
Spain	1	14	18	159	7	37	265	501	0	12	46	64	20	142
Tunisia	0	1	0	1	0	0	1	3	0	0	0	4	10	14
Yugoslavia	1	1	6	85	3	21	20	137	0	1	56	84	5	146
Totals	17	226	336	1262	127	368	1247	3583						
Totals (excluding Atlantic Is.)	15	143	176	1082	112	292	1058	2879						

(see Tomaselli, 1977), and desertification (see Le Houèrou, 1977). The original and dominant vegetation cover of sclerophyllous evergreen forest has long since disappeared following conversion to agricultural land and cutting for fuel and timber. Today only fragmentary pockets of these forests remain. These threats are explored in other chapters of this book.

More recently the great popularity of the Mediterranean to visitors poses a new threat. About one third of all international tourism is in the Mediterranean (Tangi, 1977). This growth in tourism has been rapid and for the most part insufficiently planned. The deteriorating condition of the coast is all too familiar. Apart from the much publicized pollution of the Mediterranean Sea, coastal areas have suffered similar abuse. Although important as a source of foreign revenue, coastal areas are suffering loss of wildlife, pollution, trampling and increased fire risks. It is most unlikely that tourism will decline in the future, and solutions to the problems created are urgently needed.

4. Monitoring

The list of threatened species is only the beginning. A wide range of different studies is needed on the listed species. This was the topic of a very successful conference held at Cambridge, England, in 1980, whose proceedings have been published (Synge, 1981). It was remarkable to find how much all the different aspects of biological science could contribute to conserving rare plants.

Much emphasis was placed on monitoring rare plant populations in the field, but there are several steps that must be taken first. Lists of localities from the literature and from herbarium specimens must be made. Many of the records are old ones, some imprecise, while others may turn out to be erroneous. These records then need to be verified in the field and likely habitats searched for new localities. A good example of work of this kind is given by Crompton (1981) for Eastern England. These activities are very time-consuming and best done on a local scale, but are absolutely essential to conservation success. The TPU attempted to undertake these steps for the 83 species Endangered in the EEC countries, finding localities from the literature and herbarium specimens at Kew, and through correspondence with botanists in the EEC. The results (Leon, in press) were mixed, in part due to the excessively short timescale imposed by the EEC; although many localities could be found from the literature and herbaria, few could be verified without field visits.

There has been much recent discussion of the techniques for monitoring rare plant populations in the field, using methods of 'actuarial monitoring' pioneered by Harper and described by him (Harper, 1977), Davy & Jefferies (1981), White & Bratton (1981) and Williams (1981) provide overviews of the techniques available to the researcher on rare plants. The great advantage of 'actuarial monitoring' over annual population counts is that it shows precisely at which stage in the life cycle losses and gains are occurring, and so provides a crucial pointer to how the plant can be successfully managed.

5. International initiatives

In many Mediterranean countries, large disparities occur between existing wildlife legislation and between their enforcement and effectiveness. Some of these laws do not even incorporate a list of protected species; a broad statement proposing that due regard should be taken for wildlife survival in the event of development programmes is often considered adequate! Where species lists do exist, however, these have sometimes originated as a result of concern about species attractive to the horticulturist and collector, and it is often difficult to direct similar attention to less spectacular plants which are declining from other causes, in particular loss of their habitat. It is these species that require our most urgent protection and close monitoring.

In the international arena, the interest in conservation shown by the Council of Europe and, more recently, the Commission of the European Economic Community (EEC) has been particularly welcome. The council of Europe have an ambitious ecological programme (described by Baum, 1981); perhaps their most important achievement so far has been the Berne Convention of 1979, 'The Convention on the Conservation of European Wildlife and Natural Habitats', which came into force in 1982. This convention was drawn up to encourage co-operation between states in conserving their wildlife, with emphasis on the species identified as threatened internationally. The two most important articles concerning the protection of habitats and species read:

> *Article 4.*
> 'Each Contracting Party shall take appropriate and necessary legislative and administrative measures to ensure the conservation of the habitats of the wild flora and fauna species, especially those specified in the Appendices I and II, and the conservation of endangered natural habitats.'

> *Article 5.*
> 'Each Contracting Party shall take appropriate and necessary legislative and administrative measures to ensure the special protection of the wild flora species specified in Appendix I. Deliberate picking, collecting, cutting or uprooting of such plants shall be prohibited. Each Contracting Party shall, as appropriate, prohibit the possession or sale of these species.'

Appendix I contains 119 plant species Endangered in the region covered by the Council of Europe. (Appendix 2 refers to animals.)

Of the 119 species listed, over 100 occur in Mediterranean countries. Although these represent only a small fraction of those species in need of protection, they have been selected as those most seriously threatened, and set a precedent for action. Twenty countries and the EEC have signed. So far 14 have ratified including the EEC. These are: Austria, Denmark, Greece, Ireland, Italy, Liechtenstein, Luxembourg, Netherlands, Portugal, Sweden, Switzerland, Turkey and the United Kingdom.

The Berne Convention can become a most valuable tool to initiate large-scale practical conservation programmes, but detailed habitat studies and an understanding of plant population dynamics is essential before real successes can be achieved. This is the aim behind the Council of Europe's Ecological Action Plan (Baum, 1981), which comprises three main projects. First is a 'Biogenetic Reserve Scheme' to identify and select representative European ecosystems and to provide protection for their flora and fauna. The Mediterranean maquis is one of the ecosystems under study. Secondly, a project to map vegetation features is also underway; this will show the structure of potential vegetation, dominant ecosystem types and land use, as well as priority areas for the conservation of flora, fauna, ecotypes, endemic species and ecosystem types. Thirdly, a phytosociological system is under review to facilitate the identification of 'standard' habitat types throughout Europe.

Proper use of this information should greatly improve the efficiency and value of the Berne Convention, especially in bringing about protection for critical areas, supported by substantial scientific evidence. Detailed information on individual threatened species is unbelievably scarce. For example, 'The IUCN Plant Red Data Book' (Lucas & Synge, 1978) provided case histories on 29 Mediterranean threatened plants, from a total of 250 species world-wide. Only nine of these 29 species, however, receive any formal protection and in many cases, their conservation requirements are still inadequately known. A project initiated in 1980 by the European Economic Community also supports this conclusion. In an attempt to co-ordinate detailed information about all Endangered plants within EEC countries, TPU was asked to produce information on each. 73 out of the 83 species described occur in Mediterranean countries (Leon, in press). For each species TPU compiled a sheet similar to those in the Red Data Book; the sheets describe a plant's regional status, the threats to its survival and its habitat, together with conservation recommendations wherever possible. Despite much help from botanists, for which the TPU is most grateful, this short study highlighted one of the central problems of data-gathering for threatened plants: the lack of sufficiently detailed field knowledge. It is frequently because of their rarity that these plants are so poorly known in the wild. Special attention is needed to focus upon their habitat requirements and this of course is vital to any conservation programme.

The approval of the widely familiar UNEP Mediterranean Action Plan (Thacher, 1977) at Barcelona in 1975 was a landmark in the struggle to protect the Mediterranean Sea. Not only has the Plan encouraged a large-scale clean-up of the sea involving management programmes for the control of water pollution, the establishment of marine nature reserves and shared resource utilization, it has also set a precedent for co-operation and understanding between Mediterranean countries in overcoming political obstacles through straightforward determination.

By 1980, three treaties, products of the M.A.P., had been ratified by 15 Mediterranean countries and the EEC, committing the parties to 'take all appropriate measures to prevent, abate, and combat pollution and to protect and enhance the marine environment'. IUCN and WWF have meanwhile been instrumental in sponsoring projects throughout the Mediterranean Basin and it is intended that a network of protected areas be established as a result. This was the central theme of the 1980 M.A.P. Intergovernmental Meeting in Athens (see Anon, 1980). A document is already available to help governments select, manage and establish protected areas, and recently an inventory has been compiled of such protected areas and potentially protected ones.

Intergovernmental collaboration appears to be succeeding here. Now is the time to direct similar attention towards terrestrial areas, where a co-ordinated protection programme is badly needed. Hopefully the Council of Europe's Ecological Action Plan will gain momentum in the European Mediterranean, while it is hoped that the Mediterranean Action Plan will continue to provide stimulus and help to the North African countries.

More recently, IUCN and World Wildlife Fund (WWF) have started a major new initiative to save plants. In March 1984 WWF launched its Plants Campaign which will run for 18 months and be the major focus for many of its 24 National Organizations. To ensure best use of the funds generated, IUCN designed a Plant Conservation Programme; this is a set of nearly 100 conservation projects around the world, designed and selected using the data-base of the IUCN Conservation Monitoring Centre and following advice from many specialists. The projects vary from developing networks of protected areas in lowland Cameroon, Central America and Borneo, to promoting the conservation of medicinal plants and to training and institution building. Details are available from IUCN.

At the same time, WWF's in individual countries are being encouraged to develop their own national plant programmesJ Among the Mediterranean countries, a campaign has already been launched by WWF-Spain. Under the IUCN/WWF international programme, two projects have been proposed in Greece: one to compile a national Red Data Book, inclusive of recommendations for protective action, and another to promote an education programme on plant conservation. These are part of an overall IUCN/WWF set of initiatives with the Hellenic Society for the Protection of Nature.

In the Plants Programme, IUCN and WWF are anxious to encourage botanic gardens to cultivate and re-introduce threatened plants. Furthermore botanic gardens have tremendous potential to become ideal centres for informing the public about plant conservation, and in so doing illustrate a major practical use of information-gathering for conservation. The development of this concept within Mediterranean gardens could create a spring board for other more practical conservation activities, and thereby strengthen other national initiatives.

6. Conclusion

Despite the many international activities described, it is difficult if not impossible to find one Mediterranean plant which has been saved as a result of this information base. There are no or virtually no reserves declared for plant species identified as Endangered (The Cretan Date Palm

(*Phoenix theophrasti* is a notable exception). This is, of course, partly a question of time, but it does highlight the most difficult step in the process: taking an endangered species list and ensuring protection on the ground for each one. It has to be admitted that plant conservationists, in contrast perhaps to their zoological colleagues, have been more successful in building towards the long-term goals, through international agreements and sophisticated data-gathering, than in short-term action to save critically endangered species. It could be argued that the balance should be modified a little and that some urgent 'search and rescue' exercises should not only save species but would attract welcome publicity for the urgency of our common task.

This does not, however, impair the long-term importance and value of a common data-base. With the large number of species rare or threatened, organized information is essential in the long-term. As conservation steadily becomes a more important political issue, conservationists must be able to provide decision-makers with hard and accurate data to support their concern and to enable plant conservation to be integrated into land use planning. Information on the localities of rare and threatened species is perhaps the most important ingredient.

There are perhaps two ways of looking at the usefulness of the data base. The first is that data-gathering is simply the first step in a linear process towards saving species. This is a very natural and logical viewpoint but it does present the problem of when to stop gathering data and start implementing what you have got. It is very tempting to go on revising and updating and enlarging what one has, so that the second phase can easily get postponed. This viewpoint is surely only valid if the data-gathering can be done quickly.

Another way of looking at the problem is that the value of the data is to provide supporting evidence for conservation plans and projects that already exist. Under this hypothesis, there will always be field botanists who will know which sites to protect without need of endangered species lists. The lists, once accepted formally by governments, support the political arguments to achieve the task. It is much more convincing, the argument goes, to tell a politician that by protecting a piece of grassland he will save certain attractive plants from extinction than that he must protect a good example of one habitat type, which means nothing to him. Species data too are needed to refute the question politicians are sometimes tempted to ask: 'You have got one nature reserve, why do you need another?' The data base is only one part of the process; the secret is to find a balance between gathering more data, using the data in the political and intergovernmental arena to build concern for conservation as a political goal, and practical steps on the ground to save individual threatened species.

References

Ali, S.I., Jafri, S.M.H. & El-Gadi, A. (Eds.) 1976–). Flora of Libya. Tripoli: Al Faateh University.

Anonymous (1980). Protect the Mediterranean. IUCN Bulletin 11(9/10): 81–82.

Baum, P. (1981). An ecological action plan for nature conservation in western Europe. Threatened Plants Committee – Newsletter No. 8: 16–18.

Crompton, G. (1981). Surveying rare plants in Eastern England. In: Synge, H. (Ed.), The Biological Aspects of Rare Plant Conservation, 117–124, Wiley.

Davis, P.H. (1965-1984). Flora of Turkey and the East Aegean Islands. Vols. 1-8. Edinburgh University Press.

Davy, A.J. & Jefferies, R.L. (1981). Approaches to the monitoring of rare plant populations. In: Synge, H. (Ed.), The Biological Aspects of Rare Plant Conservation, 219–232. Wiley.

El-Hadidi, M.N. (Ed.) (In prep.) Flora of Egypt.

Greuter, W. (1979). Med-Checklist: Apercu du Projet d'un Catalogue de la flore du Bassin Méditerranéen. OPTIMA Newletter 8/9: 18–26.

Harper, J.L. (1977). Population Biology of Plants. Academic Press.

IUCN Conservation Monitoring Centre (1984). The IUCN/ WWF Plants Conservation Programme 1984–85. IUCN and WWF, Gland.

IUCN Threatened Plants Committee (1983). List of Rare, Threatened and Endemic Plants in Europe. Nature and Environment Series No. 27, 2nd. edition. Council of Europe, Strasbourg.

IUCN Threatened Plants Committee Secretariat (1980). First Preliminary Draft of the List of Rare, Threatened and Endemic Plants for the Countries of North Africa and the Middle East. IUCN, Kew.

Jarvis, C., Leon, C. & Oldfield, S. (1981). Bibliography of Red

Data Books and threatened plant lists. In Synge, H. (Ed.), The Biological Aspects of Rare Plant Conservation, 513–529. Wiley.
Jordanov, D. et al. (Eds) (1963–). Flora Reipublicae Popularis Bulgaricae. Sofia. 8 vols.
Le Houérou, H.N. (1977). Man and desertization in the Mediterranean region. Ambio 6: 363–5.
Leon, C. (In press). Endangered Plants. In A Draft Community List of Threatened Species of Wild Flora and Vertebrate Fauna, Vol. 1. Prepared for the Environment and Consumer Protection Service of the Commission of the European Communities, by the Nature Conservancy Council of Great Britain, November 1980.
Lucas, G. & Synge, H. (1978). The IUCN Plant Red Data Book. IUCN, Switzerland.
Meikle, R.D. (1977-1985). Flora of Cyprus. Kew: Bentham-Moxoa Trust. 2 vols.
Sukopp, H. & Trautmann, W. (1981). Causes for the decline of threatened plants in the Federal Republic of Germany. In: Synge, H. (Ed.). The Biological Aspects of Rare Plant Conservation, 113–116. Wiley.
Synge, H. (1980). Endangered monocotyledons in Europe and South West Asia. In Brickell, C.D., Cutler, D.F. and Gregory, M. (Eds.), Petaloid Monocotyledons: Horticultural and Botanical Research, 199–206. Academic Press.
Synge, H. (Ed.) (1981). The Biological Aspects of Rare Plant Conservation. Wiley. 558 pp.
Tangi, M. (1977). Tourism and the environment. Ambio 6: 336–341.
Thacher, P.S. (1977). The Mediterranean Action Plan. Ambio 6: 308–312
Threatened Plants Unit (IUCN Conservation Monitoring Centre) (1983). List of Rare, Threatened and Endemic Plants in Europe, 2nd ed. Nature and Environment Series No. 27, Council of Europe, Strasbourg.
Tomaselli, R. (1977). The degradation of the Mediterranean maquis. Ambio 6: 356–362.
Tutin, T.G., Heywood, V.H., Burges, N.A., Valentine, D.H., Walters, S.M. & Webb, D.A. (Eds.), (1964–1980). Flora Europaea. Cambridge University Press. 5 vols.
White, P.S. & Bratton, S.P. (1981). Monitoring vegetation and rare plant populations in US national parks and preserves. In: Synge, H. (Ed.), The Biological Aspects of Rare Plant Conservation, 265–278. Wiley.
Williams, O.B. (1981). Monitoring changes in populations of desert plants. In: Synge, H. (Ed.), The Biological Aspects of Rare Plant Conservation, 233–240. Wiley.

The Herbarium and Threatened Plants Unit
(Conservation Monitoring Centre, IUCN)
Royal Botanic Gardens, Kew, England

CHAPTER 12

Conservation of plant species within their native ecosystems

J. RUIZ DE LA TORRE

1. Introduction

As has been repeatedly stressed (Anonymous, 1973; Greuter, 1979) the protection of ecosystems is the easiest, safest, most natural and most economic way to protect individual species of plants, and should therefore be considered as the definitive form of defensive action. Any other forms of conservation, however essential or urgent, should only be considered as inital, emergency or supplementary measures. In the selection of ecosystems to be protected, several criteria such as landscape, vegetation and fauna are currently taken into account (Dasmann, 1972; Clapham et al., 1980). The need to protect individual plant species is usually given too little weight, so we need to emphasize their importance and adapt conservation strategies in order to fulfil this objective with the same efficiency as the others.

Equally, some flexibility should be introduced, in order to accommodate additional or future goals in the conservation of ecosystems, since it would be a rather rare case in which the sole purpose would be to protect a group of endemic plant taxa. There are, for instance, 'endemic' landscapes and those non-endemic plant species which provide distinctive support in such ecosystems should be given proper recognition in conservation programme. For example, the ecosystem might contain certain plant species that provide necessary food for endemic animals, as very often happens in the case of the insect fauna.

Whenever possible, conservation efforts should include the habitats of local or regional relict populations, even if these taxa are abundant in other countries. Their protection may generate much interest, especially when they are populations which are found at a great distance from other locations of the same taxon and where it is probable that a differentiation at the genetic level has occurred as a consequence of the distance and isolation. The problems and techniques that apply the conservation of relict populations of non-endemic taxa are identical to the conservation of endemic taxa, so that it will be sufficient to discuss the latter.

Conservation of plants within their natural ecosystems falls within the framework of conservation of the ecosystems themselves. Fundamental in the conservation of ecosystems is the maintenance of their definitive physical and biological conditions. One of the first priorities is to begin a careful study of the conditions existing in the ecosystem. For the Mediterranean area, perhaps a second premise should be added: the maintenance of a given level of exploitation, grazing or harvesting is necessary. Unlike the tropical rain forest, where most of the indigenous species can be conserved within climax formations under conditions of maximum stability, the Mediterranean region has been severely influenced by man and various other factors and is still very rich in species. Very few of these species are known to be a part of Mediterranean climax vegetation. Most of them correspond to

Gómez-Campo, C. (ed.), Plant conservation in the Mediterranean area.
© 1985, Dr W. Junk Publishers, Dordrecht. *ISBN 90 6193 523 7.*

successional stages affected by either natural or artificial exploitation, and they should be conserved under the prevailing conditions of relative instability (Downes, 1969).

The ecosystem chosen for protection may have survived incomplete or suffer fluctuations and modifications in its component taxa. Existence of as complete a system as possible must be ensured, and particularly, the endemic or unusual and interesting taxa it contains must be preserved in safe conditions. Consequently, conservation will very often need to include specific measures and activities aimed at assuring the perpetuation of such taxa. This requires a previously acquired detailed knowledge of the conditions necessary for each taxon and the significance of the ecological niche it occupies, from which the potential area required can be deduced, as well as any special regeneration or renewal requirements of its populations. In short, the autoecology of each taxon of interest should be sufficiently known so that one can deal with its conservation under optimal conditions.

Uniqueness and threat of extinction are the most commonly encountered reasons for the inclusion of a given taxon in conservation programmes (Preston, 1962). Both are often difficult to measure with exactitude, thus the term 'endemic' is frequently used for practical reasons - but these should at least be estimated so as to be in a position to make reasonable decisions.

Uniqueness could in theory be judged by the systematic rank, thus a species would be considered more important than a subspecies or variety. Among species, relatively greater attention would be paid to those which are monotypic within a genus. In plurispecific genera, it is not difficult to distinguish certain species which have high genetic differentiation, or systematic distance, with respect to others. But the measurement of uniqueness is not only a matter of phenetic distance to nearest relatives, but might also involve the degree of physiological differentiation, difference in function or in ecological adaptation, reduction of the area over which is found, the length of separation from ancestors, the geopgraphic distance to other relatives, etc.

In the case of relicts, vulnerability is often linked with rarity and/or isolation. The different kinds of rarity have been discussed by Rabinowitz (1981).

A danger can be shown to exist where intense reduction or tranformation has taken place for reasons that are expected to continue (Melville, 1970). The short-term disappearance of the endemics or of the whole system in which it is basically integrated thus become predictible. Also the prediction of threats in the near future is a certain indication of danger to a given taxon. The conservation of narrow endemics, which are associated with just one basic type of habitat, should be considered to be most urgent cases, especially if their populations are reduced and distant from one another.

There can be also intrinsic characteristics of fragility for an endemic, such as combustibility, increased susceptibility to plagues or diseases, lowered regeneration capacity and difficulty of natural regeneration or artificial reconstruction of its former habitat. Independently from its historical age, a taxon may show clear signs of 'decrepitude', a concept that is not easily measurable but which is intuitive, and certainly, observable. Fostered in most cases by human action, decrepitude may lead to extinction, even in cases where the taxon is present in widely represented ecosystems of a favourable nature. But even in the case of a young vigorous species, the survival of the endemic will be largely dependent upon continued existence of the ecosystem or group of ecosystems to which it belongs.

Some endemic taxa are found abundantly over relatively large areas where they grow vigorously, reproduce and compete well. In such cases it is obviously unnecessary to adopt special conservation measures and effort can be concentrated on other endangered species. However, all shorts of gradations are found between the two extremes of vigour and decrepitude, and since the nature of the dangers may change with time, much caution should be exercised. An upper limit should be always set for any criteria used. As an example we can cite the Ibero-Mauritanian species *Ononis speciosa* Lag. which is today expanding in the south and southeast of Spain, following the sharp drop in grazing by goats that has taken place during the past two decades. We must not forget that these changes of situation are reversible. Species that were formerly endangered but are not so today, should also be monitored and included in conservation programmes, even if given a low priority.

2. The structure and function of the ecosystems to be protected

For each kind of ecosystem to be protected, several preliminary studies will be necessary in order to acquire a basic knowledge of as many parameters as possible, both biotic and abiotic. This is an always difficult task, given the enormous complexity of any ecological system and the large number of factors involved. Below we shall refer to a simplified model of the structure of an ecosystem which is closely interrelated with a substrate and atmosphere, and made up of a set of living organisms that constitute the biocoenosis. This in turn could be divided into macrofauna, macrovegetation, microbiota and man. In terrestrial systems, soil results from the interaction between biocoenosis, atmosphere and substrate.

The non-biotic environmental conditions which are fundamental to the ecosystem itself are the availability of space, energy, water, nutrients contributed by the mineral substrate or received from outside, the presence of toxic substances, etc. The availability of space and the capacity to support living organisms are affected by the relief and the geological substrate. The availability of energy and water and their cycles are a result of the climate. Water might also depend on the climate of other regions through lateral contribution. The availability of nutrients is basically controlled by the underlying rock, and is affected by the maturity of the system. Climate and relief (as in the deposition of nutrient-laden silt by floods) as well as other factors, such as vulcanism, wind erosion and sedimentation, can also have an influence.

As far as the soil is concerned (Witkamp, 1971), the ecosystem may be associated with a single soil type, but most often it will be linked to a mosaic of several soil types, each needing identification.

Among the living organisms, we can distinguish some with a conspicuous share in the biomass, which are necessary for the very existence of a specific type of ecosystem. The endemic elements that we intend to conserve as an ultimate goal rarely belong to the group of essential taxa, but they can be considered as a second group of 'special' ones.

The general structure of the ecosystem fundamentally depends on the macrovegetation component or is closely correlated with it. In particular, it depends on the dominant life-form or on a combination of dominant life-forms in the macrovegetation, whether it be trees, shrubs, dwarf shrubs, and so on.

There are several classifications which can be used to describe the existing life-forms (Raunkiaer, 1934; Du Rietz, 1931; Cain, 1950; Emberger et al. 1968). They combine characters, such as the size or nature of the shoot, branching patterns, life-cycle, seasonal distribution of growt periods, tape and duration of the assimilation organs, presence or absence of spiny habit, type of root system and mode of regeneration.

There are also several classifications to designate the general structural types of macrovegetation. That of Schmithüsen (1968) is among the most complete, while that of Dansereau (1958, 1961) is among the most schematic. As far as we are concerned, purely structural classifications can and should be used, as they are more closely correlated with function. That of Fosberg (1967) is recommended. To be useful, the simplest classification should at least distinguish:

Forests: dominated by trees.
High scrub: dominated by tall shrubs.
Low scrub: dominated by low shrubs.
Parks: open or scattered upper stratum and a lower stratum formed by low shrubs, grasses or a mixture of both.
Mixtures of shrubs and grasses, mesophyllous or xerophyllous.
Hygrophyllous (wet) pastures.
Meadows and other mesophyll pastures.
Xerophyllous pastures, perennial or annual.
Ephemeral pastures.
Open and desert communities.

It is important to determine what position the elements to be conserved occupy in the ecosystem's macrovegetation: this will largely depend on their life-form. We can distinguish, for instance, those elements of the uppermost vegetation strata, such as lianes or epiphytes in tropical ecosystems, from those trees integral to the upper or intermediary strata, or from those shrubs and herbs include in the lower strata. Most endemics of the Mediterranean region are small shrubs or herbs.

The ecosystem will be generally characterized by the life-forms of the dominant species. The specific type will be usually linked to a group of sub-dominant taxa which may be exclusive, alternative, co-dominant of vicariant. Sometimes it is not possible to define truly dominant species. The specific type will be usually linked to a group of sub-dominant taxa which may be exclusive, alternative, co-dominant or vicariant. Sometimes it is not possible to define truly dominant species. The specific type of ecosystem will also have a characteristic group of species. This is used by the phytosociological shool of Braun-Blanquet (1965) to establish vegetation units on the basis of floristics. Environmental factors: and the reasons why these fundamental eco-floristic groups have developed should be identified so as to permit an appropriate choice of conservation measures.

To study the function of the ecosystems, the fundamental flowpaths of energy, water, nutrients and toxins (including allelopathic substances) should be identified, and their annual cycles investigated. The ecological niches within these flows, particularly those of the species to be protected should also be identified. Finally, the existing forms of harvesting or grazing should be carefully studied along with the bioproductivity to determine how much exploitation is compatible with the unimpaired maintenance of the ecosystem. One must emphasize the importance of both the study and the application of controlled exploitation methods, since a large proportion of endemics of the Mediterranean region do belong to immature systems or to systems with blocked succession due to natural or artificial exploitation.

Ecological succession (Odum, 1969; Bazzaz, 1979) begins with pioneer or colonizing structures and progresses to a climax vegetation where natural exploitation is minimal. The position of the system within succession needs to be defined, and the processes discovered whereby the present composition of the ecosystem has been reached; this process may have been either progressive, regressive or a simple lateral drift within similar maturity levels. Equally, possible modes of change, forms of recovery, conditioning factors for stability (in the sense expressed in Section 4), etc., are interesting subjects for study. In fact, these dynamic aspects have been little studied, due to intrinsic complexity and the amount of time needed, and thus remain largely unexplored. These complexities are usually replaced by oversimplifications that reduce the evolution of systems to a simple linear series or succession. As for the Mediterranean climax, it is the opinion of the author that it usually comprises a far more complex macrovegetation than is commonly thought.

The effects of the outside climate are modified within the ecosystem itself, where space, energy and water cycles are concerned. In the substrate, the cycling of nutrients is modified by the action of microorganisms, viruses and even macroorganisms. Toxic substances and selective inhibitors may be present in the soil (Muller, 1966: Rice, 1974). The microbiota includes those organisms which are responsible for the decomposition of debris and wastes, for the formation, evolution and conservation of the soil, and also for diseases which could result in alterations to the system or to cyclic changes in its structure.

Apart from the exploitation of the vegetation by the grazing species, the macrofauna influences the economy of nutrients with their excreta and remains. This influence may be important in the case of oligotrophic substrates. The mesofauna plays an important role, as it contains not only pests but also their controls. Sometimes they contain specialized consumers, such as those that feed on certain fruits and seeds, that may in some cases lead to a sharp decrease in the numbers of a species; for example, a small coleopterous beetle attacks *Lavatea oblongifolia* Boiss., an endemic shrub of the limestone mountains of Almería and Granada which shows very poor regeneration. In such cases,

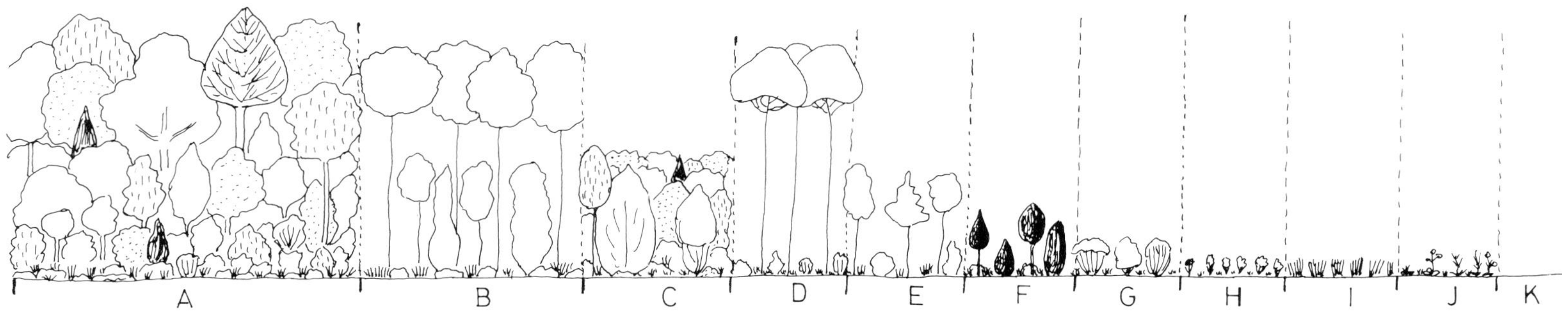

Fig. 1. Different Mediterranean vegetation assemblages ordered according to increasing disturbance and human influence. Agricultural assemblages are excluded. Structures B and H present variable disturbance which may be greater than C and I respectively. A. Heterogeneous mixed forest without differentiation of strata. – B. Homogeneous oligospecific or monospecific forest with three strata. – C. Dense maquis. – D. Xerophilous pine forest. – E. Arbustive oligospecific shrub. – F. Juniper scrub. – G. Medium size oligospecific scrub. – H. Dwarf scrub (*Thymus* etc.). – I. Xerophilous perennial grassland (*Stipa*, etc.). – J. Annual grassland. – K. Anthropogenous desert.

possible controlling mechanisms can probably be found within the same system.

3. Exploitation in ecosystems

Exploitation is currently and intuitively seen as the extraction of biomass from an ecosystem by man. But the concept is much broader, as exploitation also occurs by such natural means as fire, floods, vulcanism, etc. Any action or circumstance that tends to cause the system to regress towards more immature stages, or to maintain it as such, may be seen as an exploiting factor. Compactness of rock, relief and limiting climatic conditions, are seen as exploiting factors from this point of view. We can also speak of exploitation by other external ecosystems. For the vegetation, this will generally take the form of exploitation by grazing animals from among the larger fauna of neighbouring systems (Elton, 1958). The same may occur among the mesofauna, as in the case of the migrant locust, which originates in very distant ecosystems.

The existence of exploitation which benefits neighbouring systems should always be considered because it constitutes an important permanent factor for the ecosystem. In such cases, isolation would lead to a drop in the level of exploitation and thus to a progressive evolution which would cause a change in the ecosystem: this could lead to the disappearance of heliophyte endemics needing abundant space and with little competitive ability. In the Mediterranean region, the most important cause of exploitation is man. Such kinds of man-made or man-induced exploitation together constitute so-called artificial exploitation.

Today, most ecosystems in the Mediterranean region have been deeply altered and disturbed since prehistoric times (Fig. 1). The wide array of different forms of natural exploitation, in some cases probably very intense, formerly preserved many endemic species and made certain regions important sources of endemism. Man's influence has led to the disappearance of some species, but has provided many others with additional areas to inhabit, by rejuvenating systems where annuals, heliophytes and certain other specialized species could not previously survive.

This is why several authors have often considered and emphasized the positive aspects of human action. Westhoff (1970, 1971) says that 'man has a positive influence on diversity as long as his action aims at stabilization; on the other hand, through induced disturbances, he is impoverishing nature by increasing the amount of instability'. But the meaning of human influence will depend on the intensity and duration of the action. When considered as a whole, Westhoff's opinion is that 'human influence was much more positive than negative; it was enriching and beneficial, not impoverishing'. The transformation and removal of biomass experienced by vegetation has been such a common occurrence that as Naveh (1971) points out 'in the Mediterranean region, fire, man, his axe and livestock have become inseparable and integral parts of these semi-natural ecosystems. Therefore, perhaps their exclusion cannot be regarded as creating a 'natural' situation which will lead to the re-establishment of a hypothetical climax. On the contrary, it might be viewed as leading to a less natural situation, at least from the point of view of plant diversity'.

The above concepts are controversial and will be so for some time. They may be closer to the truth when the vegetation or the flora are considered, while they become more dubious if an integral ecological approach towards the structure and dynamics of the ecosystems is used. The concepts become philosophical when trying to define what is 'natural' and what is not. But certainly, they should be taken very seriously when dealing, as we are, with the problem of conservation of plant species in the Mediterranean area. Over large areas, conservation policies based upon the exclusion of all artificial exploitation will eventually lead, we can be sure, to the extinction of many endemic species.

Man's effect on vegetation sometimes takes place through occasional or isolated actions. Among these we can cite the clearance of vegetation for various purposes, the construction of public works and other permanent installations, the occasional destructive leaks of toxic or corrosive industrial contaminants, and so on. Fires could also enter this group (Trabaud, 1981; Komarek, 1983), although

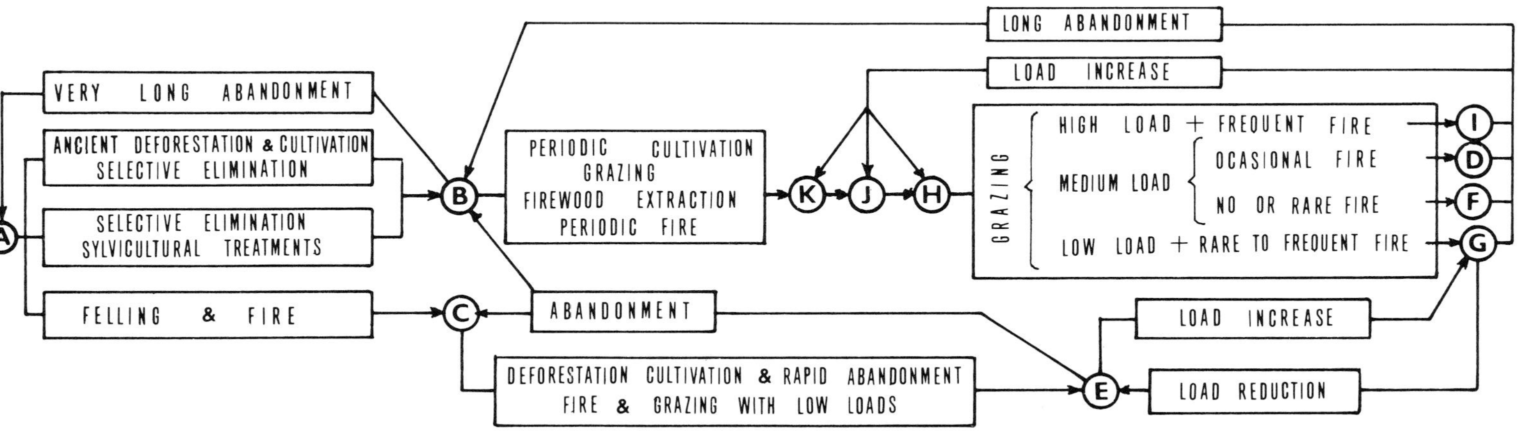

Fig. 2. A simplified dynamic model of succession in relation to possible treatments (letters as in Fig. 1.).

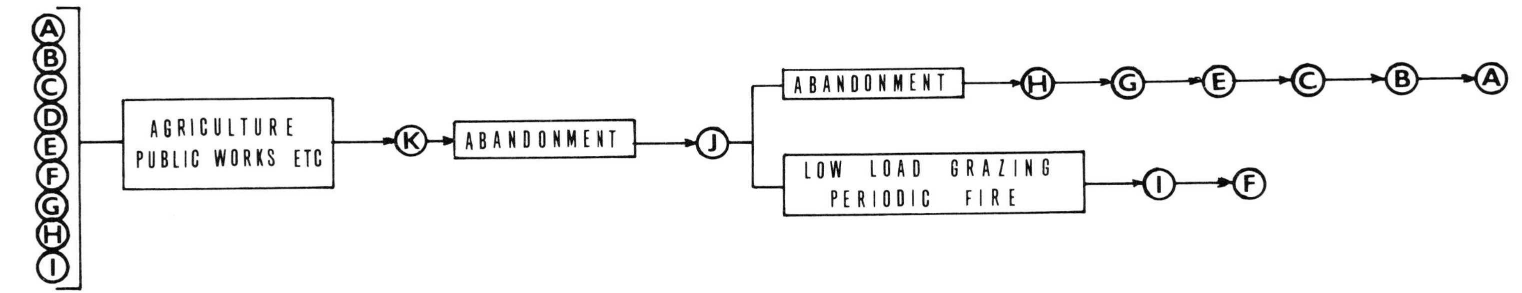

Fig. 3. A simplified model of rapid regression and continuous subsequent progressive succession (letters as in Fig. 1.).

for political and economic reasons they have occurred exceptionally frequently and intensively over the past few years in some countries. Among these isolated actions, those which are avoidable or those which have preferable alternatives should be prevented from taking place.

Other continuous, intermittent or periodic actions can be encompassed under the term 'land-use'. For the prevailing land-uses in the Mediterranean region, the classification of Ruiz de la Torre and Ruiz del Castillo (1977) may be used with little addition. The main groups in this classification are:

Agriculture: includes irrigated or non-irrigated fields located either on open lands or in terraces, tilled lands with windbreaks, floodable fields, paddy fields, fallow lands and old fields which have been abandoned.

Silviculture: provides wood or timber, fuelwood, charcoal, resins, cork and other wood products, or provides artificial reforestation;

Pastures: for a single kind of livestock or mixed grazing;

Mixed: Parkland-like combinations of the above three groups with dry land agriculture, with a mixture of different types of exploitation;

Miscellaneous: Hunting, fishing, bee-keeping, camping, sports and recreation, wildlife parks, housing estates, etc.

Reserves: Conservation oriented spaces of different types with minimal degrees of exploitation.

When studying the stability conditions of the ecosystems, land-use should be identified along with the periodicities and intensities of the cycles involved.

The above land-uses which are positive benefits to the quality of human life and to human survival, often include a number of clearly negative aspects from the conservational point of view. Apart from the occupation of land for housing, industry, or other completely urban uses and the encroachment of human activity in more open lands, we should mention the pollution of air, water an soil with very diverse types of wastes. A preliminary but useful systematization of the possible types of threat to rare and endemic species was included in the *List of rare, threatened and endemic plants* (TPU-IUCN-CMC) Anon. 1983). Indirect but important effects such as climatic or edaphic changes (variation in temperature or in the humidity of the substrate) are often brought about by the clearance of trees on a wide scale (as in the Alpujarra of southern Spain) or changes from dryland to irrigated agriculture or the reverse. The reduction or supression of certain older forms of exploitation – pasturage, wood or foliage collection, hunting, etc. – may also have a negative effect on the conservation of many established ecosystems and their components. (Figs. 2 and 3).

When considering the use of these concepts of exploitation, the greatest dangers from man's actions, as far as conservation is concerned, could be summarized as follows: (1) a disruption of the equilibrium of a system through an increase in the pressure of exploitation; (2) an arbitrary change in the nature and intensity of exploitation; and (3) a total or partial suppression of exploitation.

The extension and intensification of agriculture, silviculture and urban and recreational use falls within the first of these categories. Some types of exploitation may be negligible with regard to use of biomass, but still be extremely dangereous if they affect the genetic diversity or have a markedly selective character. Such is the collection or rare taxa as herbarium specimens, the removal of living plants for transplantation and cultivation, and the collection of medicinal or aromatic plants.

Many situations which fall into the second category may be cited, where changes in the type of activity take place with often neglected effects on the flora. As an example, the introduction of new breeds of cattle or even a new livestock species may lead to readjustments in the prevailing plant species in pasturelands, thus changing the structure of the system and in some cases causing the decline of the associated endemic plants.

The third case, the total or partial suppression of exploitation, is of greater importance these days when agriculture tends to be more intensive and to occupy less space, and pasturelands are often abandoned as cattle or sheep become increasingly reared in stockyards. If land suddenly stops being exploited, ecological succession will eventually produce shrubby or forest vegetation. Once again, it is emphasized that Mediterranean endemics only rarely belong to advanced successional stages, and thus such a transformation will hardly be beneficial to the conservation of rare species. Many small heliophillous species will be harmed, and among them are likely to be some valuable endemic species.

If we take the case of goats (French, 1965), the deleterious role being often assigned to this species is the result of inappropriate loads. From a merely qualitive point of view, goats contribute to a beneficial diversification of the macrofauna. Their dissapearance brings profound changes in the structure of vegetation with tall woody species invading. The effects of wild grazing animals as rabbits are often underrated as well.

Modern times have brought profound changes in man's attitudes and traditional interaction with Nature. These changes have proliferated in the past few decades and their increasing impact poses the most important threat to natural ecosystems and to species therein. Below we summarize those which are the most obvious by dividing them into two groups: those which derive from too intensive exploitation and those derived from too low exploitation.

Human actions involving excessive exploitation include:

- Rapid development of vast areas.
- Use of powerful mechanical tilling and earth-moving equipment.
- Extensive urban building which degrades or destroys vegetation.
- Selective building in semi-natural areas to establish second homes and holiday residences.
- Intensive use of pesticides, herbicides, and other chemicals.
- Massive emissions of pollutants of various forms.
- Deliberately or accidentally causing large fires in forested areas.
- Overpopulation that affects arid, semi-arid and desert areas.
- Increase in excursions, hunting and other open-air activities.
- Increase in the collection of rare and interesting plant races for various purposes.
- Proliferation of game reserves for large grazing animals, often with the introduction of exotic species.
- Reserves of unmanaged herbivorous wild animals.

Actions involving a reduction of elimination of artificial exploitation include:

- Depopulation of rural areas.
- Abandonment of hill agriculture.
- Abandonment of dryland agriculture in marginal areas.
- Progressive abandonment of free pasturage in general, and, particularly, by wood-eating species.
- Extension of protected areas of low exploitation.

Most of the above mentioned actions are of recent origin, and it may be expected that additional types will perhaps enter the scene in the near future. All of them affect the survival of rare plant specie as they have direct effects on the local ecosystems. As they are very often inescapable, alternative treatment programs should be studied and tested in the areas selected for protection, always giving priority to the natural components of the ecosystem.

4. Conditions of stability for immature systems

In conservation generally, immature stages of succession are of special interest, since they are necessarily present in the semi-natural or transitional zones surrounding protected areas of mature climax vegetation. Additionally, they contain a variety of information and greatly contribute to the diversity of the biosphere and its landscapes. Immature stages assume immense importance in the Mediterranean region as very few of the Mediterranean plants belong to climax or subclimax ecosystems. The greatest number are to be found in immature systems of either natural or artificial origin.

If conservation plans are simply restricted to the creation of self-contained nature reserves, successional events may well hasten the extinction of valuable species. In the evolution of ecosystems (ecological succession), an increase in biomass takes place, largely due to the growth of a number of individuals from a few dominant species. The closure and increased density of the canopy most often lead to the disappearance of species that require more sunlight and a measure of disturbance or grazing for their survival. For many species, grazing may directly affect individual plants and yet be beneficial overall; such is the case with many Gramineae and other tufted or tufflike plants where grazing causes a rejuvenation and stimulates new growth.

The preservation of climax ecosystems may thus be desirable for a number of ethical, scientific and educational reasons, but if we are to conserve endemic Mediterranean species we need to keep somewhat immature systems in approximately their present condition. These two objectives could be combined, as the semi-natural belts surrounding climatic reserves could be advantageously used to conserve a number of endemics.

Ecological stability usually increases with maturity of the system, and only attains a maximum in the final stage or climax. For the non-climactic stages to which we have referred to this chapter, stability is used in the sense of the continued existance of conditions necessary for the maintenance of that particular stage, so that this necessarily involves a certain degree of disturbance and exploitation. In a previous paper I made suggestions concerning a natural Mediterranean silviculture (Ruiz de la Torre, 1977) from which the following summary of the components of stability is taken.

- The autochthonous character of the components; the expansion of invasive species of an aggresive nature should be avoided.
- Good balance between production and explotation. This can be achieved through adequate corrective measures, subjected to periodic revision as necessary.
- Maintainence of the natural exploitation level.
- Decomposition and recycling of surplus debris.
- Maintainence of the biomass accumulation rates and the rates of activities at the different trophic levels.
- Prevention of dispersal of toxic contaminants that exceed harmful levels. The levels at which pollutants cause perturbation should be determined if contamination is presumed.
- Maintainence of divergence limits in general (see Section 6).
- Conservation of the genetic diversity.
- Conservation of the 'controller' organisms.
- Control of diseases, epidemics and parasites when strictly necessary through an integrated biologically-oriented programme.
- Prevention of year-to-year fluctuations in the water-table levels.
- Prevention of irreversible reductions in the summer reserves of water accessible to the ecosystem.
- Conservation of man's disturbance and levels of pressure. If thought necessary or advisable, their substitution by other equivalent but more natural forms of exploitation can be studied and applied.
- Attenuation of fires, or controlled burning where fires form part of the stability or rejuvenation conditions of the system with its endemics.
- Prevention of climatic or micro-climatic disturbances, such as thermal discharges, etc.
- Control of soil erosion and the stabilization of topsoil and/or substrate.

Among many other possible factors, some mentioned above, we should emphasize also the following destabilizing factors:

- Differences between production and exploitation levels.
- Arbitrary changes in the programme of management.
- Total or partial destruction or division of areas

of the ecosystem, possibly resulting in the isolation of parts, as with public works, industrial development, building estates, communication lines, fencing etc.
- Establishment of abrupt transition contacts amongst areas with very different degrees of artificialization.

Suitable treatments to stabilize instable conditions in immature ecosystems can be easily derived from the enumerations above.

For very mature stages the rule should be one of no treatment, always accompanied by the prevention of human influences. The same policy could be used with success in situations where the immaturity is a product of natural limiting factors as in alpine environments. But to a large extent our success in protecting a large number of endemics will be dependent upon our ability to maintain the present situations of inmature ecosystems (Figs. 2 and 3). Such a balance will be obtained by quantitatively and qualitatively maintaining the pressure of the exploitation that has led to the continued existence of the ecosystem.

In many instances, however, it may be positive to replace the present programme of management by other more 'natural' means, once specific preliminary studies and specific trials have been carried out to test the various possibilities. For example, sheep herds might be substituted by ibexes at the higher altitudes in the Sierra Nevada and other Mediterranean mountain ranges. Whenever possible, the replacement of artificial treatments by exploitation forms with an internal control, will lead to a more feasible and more economical conservation programme, and will give rise to an increase in the biological diversity and scientific value of the system. Only in situations leading to abandonment where irreversible losses of information may occur, will artificial means become necessary.

5. Demographic aspects

In the conservation of ecosystems, attention should be paid to population levels (Harper, 1977; Solbrig, 1980), especially those species which being essential, rare or endangered, are the primary objectives of our conservation efforts. Thus, importance should be attached to the mechanisms of pollination and to the means of establishment and dispersal of plant taxa. In cases where natural means and agents prove insufficient, artificial enlargement of populations should be resorted to.

The insects and animals responsible for bringing about pollination should be identified and their numbers protected by avoiding actions harmful to them. Thus, the use of insecticides or other toxins in the area, upstream or upwind, would be avoided. In the case of wind-pollinated taxa such as *Pinus sylvestris* var. *nevadensis* Christ., artificial pollination becomes necessary to extend existing populations, since the remnants of former stands have been surrounded by new plantations of exotic species.

The efficiency of dispersive vectors, depends on their presence at the right moment, and this may be ensured through proper monitoring and eventual control of the fauna involved. The conservation of the ecosystem infrastructure will assure the dispersion of the flora whether they be wind- water- or self-dispersed.

Among the germination and implantation agents we should note the roles played by the macrofauna (as for seeds that are scarified and distributed by passing through the digestive systems of herbivorous animals), the mycorrhizal fungi and the root-nodule bacteria. In cases where sharp reductions occur in taxa that depend on certain of these factors, it may be necessary to resort to special soil or seed treatment or innoculation. In any event, any type of treatment to the ecosystem should be compatible with the subsistence of such agents.

In the case of insufficient dispersal, seed viability, vegetative reproduction or seedling vigour, it may be necessary to resort to artificial renovation of populations by seed sowing or seedling transplantion. The use of exotic ecotypes, breeds or varieties is to be stronly avoided. Proper management of the ecosystem will in such cases make the necessary space available for these introductions and avoid allelophaties or other harmful phenomena.

The ecosystem may change, independently of the maintenance of the endemics, if essential non-endemic taxa are sharply reduced or disappear. To conserve endemics it is therefore important to know the causes for regression of other types of taxa. Some are due to large scale changes of climatic or even geological origin with or without the intervention of man. The whole stormy history of the Mediterranean region is a succession of such changes, which before the advent of human influence can be considered natural. The extent to which man-induced changes during historic times are 'natural' is a subject of controversy, at least from a quantitative point of view. To give a recent example, the advance of the desert along the northern edge of the Sahara combined with human pressure, is reducing the margin of life for many organisms.

Intrinsic causes of regression, as such loss of vegetative or reproductive vigour, have been mentioned above. Other causes are inherent in the ecosystem itself or are imported from neighbouring systems: diseases or plagues, inducers of allelophathies, new phytophagous species, excavators or enemies of collaborating organisms (as controls to pollinating insects). Most important nowadays are aggressive adventitious plant species that may alter the ecosystem and compete with rare species upon their arrival, by reducing the availability of water, nutrients, aerial or underground space, and perhaps casting shade or being a source of allelopathic substances.

6. General guidelines for a defence programme

Above we have briefly summarized the most important concepts to have in mind during the formulation of a defence programme. At this point we will proceed the enumerate the successive steps of that programme, including the necessary preparatory studies. It is evident that a plan of this kind should be duly co-ordinated with other conservation organizations (integral reserves, national parks, etc.) whose main concern is not necessarily the conservation of endemic species.

A. *Preparatory studies*

1. Preparation of a catalog of systems to be defended:
   a) Determination of the localities of endangered endemics (using the criteria explained in Chapter 11 of this book) not excluding those endemic taxa which are international and local at the same time, such as many Pyrenean endemics.
   b) Study of the plants' plan in ecosystems and classification of the ecosystems containing threatened endemics (see Dasmann, 1973a, b; Braun-Blanquet, 1965).
   c) Selection of ecosystems containing the highest concentration of endemism.
   d) Elaboration of the catalog of systems to be defended.
2. Study of auxiliary measures with regard either to the systems to be defended as a whole or to individual cases:
   a) Access and availability of the sites, including the legal background, such as the existence of legal defence statements, need for expropriation, present restrictions on the use of land, etc.
   b) Study of the alternatives to permanent installations (planned or existing) such as communication lines, polluting industries, quarries, etc.
3. For each site or sufficiently homogenecus group of sites, the actual study would consist of:
   a) Study of components. Identification of essential and special component taxa.
   b) Study of life conditions such as soil and climate, and including exploitation, especially by man, with its recent dynamics and trends.
   c) Determination of the degree of maturity, stability, and balance between production and exploitation.
   d) Study of the response to treatments directed to the maintenance of current conditions or to their readjustment.
   e) Study and test of possible 'more natural' alternatives
   f) Demographic studies on essential and

special taxa. Causes of regression and possible modes of renovation.

B. *The defence programme itself*

1. Preliminary measures to be taken:
   a) Official declaration of the need for protection of the systems catalogued (including cases of ecosystems to be reconstructed from lost ones).
   b) Appointment of supervisors to oversee the management of the protected sites, to prevent alterations and to ensure the study of possible alternatives to necessary and/or inevitable installations or development.
   c) Organization of administrative and scientific units in charge of the defence programme.
   d) Proposal of alternatives for permanent installations or developments that would affect the stability or subsistence of any of the catalogued systems.
   e) Establishment of funding for the areas to be defended by management.
2. The programme itself includes activities that tend to protect the ecosystems through one or a combination of strategies. Below, we list these activities under headings indicating the possible strategies:
   a) The defence of ecosystems through conservation of the groups of components.
      i) Lost or disappeared ecosystems: It will first be necessary to reconstruct the system by re-assembling the essential components in plots selected from a former, known area or from a potential area. Much care must be taken in the selection of the genetic material for the missing taxa, either from the wild or from cultivated collections, since they should correspond to the right ecotypes.
      ii) Deteriorated systems: When their composition has been affected and essential and/or endemic taxa are missing, the maximum possible diversity should be re-established, as far as this is compatible with the level of maturity of the ecosystem. Seeds may be either directly sown or plants raised from them in nurseries and transplanted.
      iii) Systems showing a sharp reduction in the population of endemic and/or essential components: An extension and strengthening of the present populations is advisable and, if necessary, the establishment of additional populations. As above, seeds and propagules need to be collected in advance. If material from seed banks or botanic gardens is used, the existence of the correct genetic characters is to be verified.
      iv) Well-conserved systems: Though a minimum of intervention is recommended, it may be necessary to stimulate certain interesting taxa.
      v) From an economical point of view it may be of interest to complete the representation of the existing endemics within the set of catalogued ecosystems. Restoration of the maximum possible concentration of endemics will in general be an acceptable goal. However, many limitations should be self-imposed to avoid major disturbances of present biogeographic patterns, to avoid unnecessary introductions into unsuitable ecological niches, to avoid unnecessary artificialization of the floristic composition and so on.
   b) Defence of ecosystems through the conservation of the prevailing life conditions:
      (In each case, the defence should imply the maintenance or reconstruction of the overall optimum life conditions of the ecosystem and the taxa, by providing the kind of treatment that leads to an increase in the strength and vitality, and assures the normal propagation of the essential and endemic plants. Fundamentally, the production/exploitation balance should be kept or restored.)

i) Exploitation overpressure: When this occurs, the intensities of the treatment need to be reduced. The populations of essential and special taxa will have been affected at least partially, so that we will be in one of the situations included under the heading B.2.a. Therefore, suitable action should be accordingly taken to restore the distribution of components to their previous importance and vigour.

ii) Balance between production and exploitation: The maintenance of the traditional actions should follow, with periodic revisions to adjust their intensities if necessary. Diverse stimuli to strengthen the numbers of the taxa to be defended may also be needed.

iii) Exploitation deficit: It will be necessary to ad controlled forms of exploitation, including one or more of the following:

- Extraction of biomass. The possible extension of the present range of land-uses should be considered.
- Extraction of surplus waste materials of high persistence or resistence to decomposition.
- Controlled burning.
- Management of the macrofauna, with emphasis on the wild species whenever possible, with proper adjustment of the different kinds, i.e. herbivorous, lignivorous, carnivorous, omnivorous.
- Adjustment and stimulation of controller organisms such as parasites, symbionts, commensals, saprophytes, etc.

b) We can be faced by situations where the traditionally used methods to conserve the structure of the ecosystem have long since disappeared. Thus, it will be necessary to adopt combined substitutive treatments, the efficiency of which should have been previously tested.

c) Defence of ecosystems through the regulation of the interspace dynamics.
Two aspects need to be considered:

i) In every case, proper care should be taken with regard to the nature of the boundaries of each ecosystem with the neighbouring ones, trying to achieve a predomination of 'divergent limits' (in the sense of Westhoff, 1971), that is, gradual transitions in the maturity of the ecosystems' boundary areas, avoiding sharp contrasts (Figs. 4 and 5).

ii) When the basic treatments for the subsistence of the ecosystem are interannual and/or itinerant with cyclic shifts, a mosaic distribution of periodically renovated plots will be adopted with maximum contact between pieces with vegetation of different ages. This arrangement should be sought when the aim is to conserve juvenile systems that contain pioneer or second-colonization endemics, such as the many short-lived, herbaceous, heliophilous and xerophilous taxa which are abundant among the endemics of the Mediterranean region. Various spatial models could be presented. With the irregularities and modifications that will often be made compulsory by the relief, the distribution of such sub-units should be brought as close as possible to equilateral triangles, squares or regular hexagons, in that order for increasing renovation periods. For juvenile systems, renovation will ordinarily mean tilling of the land, either alone or combined with the extraction of biomass or controlled burning.

d) Apart from the predominant strategy which is used, some general measures would be:

i) Regulation or delimitation of transit

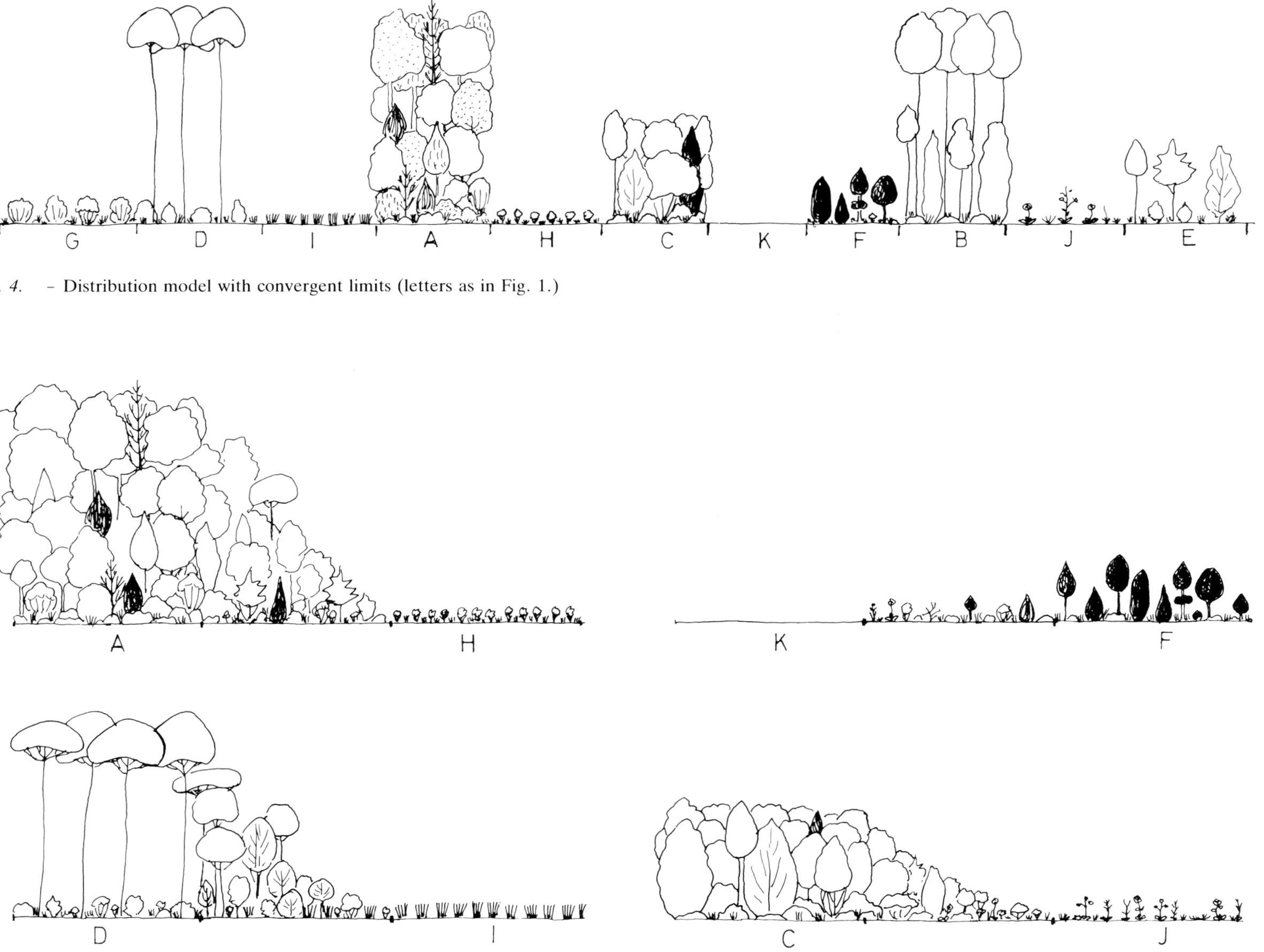

Fig. 4. – Distribution model with convergent limits (letters as in Fig. 1.)

Fig. 5. Distribution models with divergent limits (letters as in Fig. 1).

through and use of the plots to be defended.

ii) Limitation or suppression of the collection of components of the biocoenosis, substrate or soil.

iii) Prevention of catastrophic accidents such as fires, cattle invasions, mechanical destruction or damage by the access of toxic substances.

iv) Elimination of aggressive adventitious species.

v) Protection of pollinating and disseminating animals and insects.

e) Finally, proper monitoring and periodical revisions will be made to control the efficiency of the defence programme and to introduce the necessary modifications. The frequency of these assessments will decrease as the maturity of the ecosystem increases, but they should never be less frequent than once every ten years. For young systems with itinerant treatments, the results can be followed from one year to the next.

3. A complementary measure of the utmost importance will be to disseminate information concerning the defence programme. This information will consist of a clear explanation of motives and individual descriptions of the systems to be protected. It would aid in improving the knowledge, interest and support of the general population in regard to conservation problems. Also, it would draw possibly helpful criticism and aid from other scientists and naturalists.

7. Some examples

Below we present a group of type cases including those which have the greatest incidence and pertinence in the Mediterranean region with regard to the conservation of endemic species within their natural habitats. For each, the general guidelines that may be appropiate are briefly indicated.

7.1 Rocky environments

In these are found a large proportion of the Mediterranean endemics, within a diversity of niches which is larger than it appears (Heywood, 1953). As the prevailing form of exploitation is natural (limitation of space and available soil), they can usually be conserved with a minimum of human intervention. The site should be declared a reserve and control of possible collecting should follow this, even if these are carried out with a scientific objective. Re-introductions may be necessary in certain cases to complete the ideal grouping of species.

The establishment of quarrying activity will always be a potential problem and suitable alternatives should then be put into practice. In the case of rocks in a humid environment, any disturbance that could affect the water supply would also need to be foreseen and contingency plans made. In the case of easily accessible rocks, where plant populations are partially vulnerable to the action of grazing species (mainly goats), control of animal populations will be necessary to obtain the correct balance.

Summit rocks are particularly rich in endemic plant species (La Sagra, Yebal Qraa, Azrou Akchar, etc; see Chapter 6 for Greece). They are subjected to strong winds and general conditions of drought and nutrient stress. To conserve special taxa it may be necessary to exert a certain control on the macrofauna, such as ibex and goat, and to treat such places as integral reserves. In situations where the herbivorous species have been removed or have dissapeared, extractions of biomass, clearings, and even the occasional use of controlled burning may be advisable to avoid unwanted evolution of the vegetation.

7.2 Gypsaceous and saline areas

These types of habitats are abundant in the Mediterranean Region, where they contain important levels of endemic plants. The conservation of worth-while systems will in general involve the control of pasturage and livestock (Fig. 6) with eventual tillage of earth and controlled burning in localized areas. Tilling stimulates the renewal of many low woody shrub and herbaceous annual plant populations, whereas grazing and fire favour rebudding and rejuvenation of stocks of many vigorous herbaceous plants.

7.3 Steppes

The most common type of steppe in the Mediterranean area are dominated by small, often dwarf, woody plants together with perennial and some annual herbs. The traditional treatment is usually the pasturing of animals, extraction of the larger woody plants and occasional firing, with various combinations of pressures and cycles. Here, as in other cases, pasturing should be done with the same species and breeds as was previously the custom. Any change, even a desirable substitution of wild fauna for domestic stock, should be tested first (Harper, 1971). Wood extraction could be eliminated if lignivorous animals such as deer and goat are selectively encouraged.

7.4 Nitrophilous or ruderal pastures

Particularly in the semi-arid and arid parts of the Mediterranean region, this type of system contains abundant endemics. The expansion of urban areas, the itinerant night-stabling of cattle and intensive agriculture often endanger these systems. Conservation usually involves the control of domestic livestock and their solid wastes, and eventual cultivation of special areas.

Dry pastures in many North African countries are very often overgrazed (El-Hammoumi, pers. comm.) to the clear detriment of both productivity and conservation. It remains to educate the shepherds in the adoption of the proper loads of animals.

7.5 Permanently cultivated lands

The existence of threatened endemic weeds poses a difficult problem. Their survival is ensured by the maintenance of stocks of seeds in tilled land (soil 'seed banks') (Cook, 1980), so that for an effective protection it would be necessary to maintain a limit to the intensity of tilling and to avoid the use of herbicides and other selective chemicals. This is obviously at odds with modern agriculture, but the preservation of such obsolete methods could well be done in 'agriculture reserves' purposedly established to preserve endangered weeds.

Nonetheless, many weeds are well preserved as they grow in a variety of ruderal habitats. Many others survive in hedges, amongst low vegetation, or wherever the effects of herbicides are not too strong.

7.6 Humid zones

The humid zones of the Mediterranean region are of great importance with respect to their faunas, but contain relatively few endemic plants. Conservation efforts would mainly be directed towards the maintenance of the flow and water level, as well as the quality and trophic level of the water, the prevention of desiccation, toxic pollution and unfavourable eutrophication. The introduction of exotic aquatic animals, particularly grazing ones, is to be avoided.

7.7 Spiny cushion-plant formations (Xeroacanthion)

This brushwood partly corresponds to levels above the timber-line, where they may represent the climax vegetation, and partly the cleared or degraded facies of the higher levels of forest. It is characteristic of Mediterranean mountains at a certain altitude (approx. 2,000 m in Sierra Nevada or the Great Atlas) and usually contains a high level of endemic plants. Their spiny cushioned habit is the result of a long selection for protection against herbivores (in the Canary Islands where large herbivores were absent, they are substituted for by thornless woody brooms).

Suitable treatments (Fig. 7) should include the presence of lignivorous species such as ibex, goat and deer at controlled levels. In their absence, extraction of biomass and eventual tilling may be of help. The intensities and periodicities of these actions shall be determined through preliminary tests, whose results will be subjected to later cor-

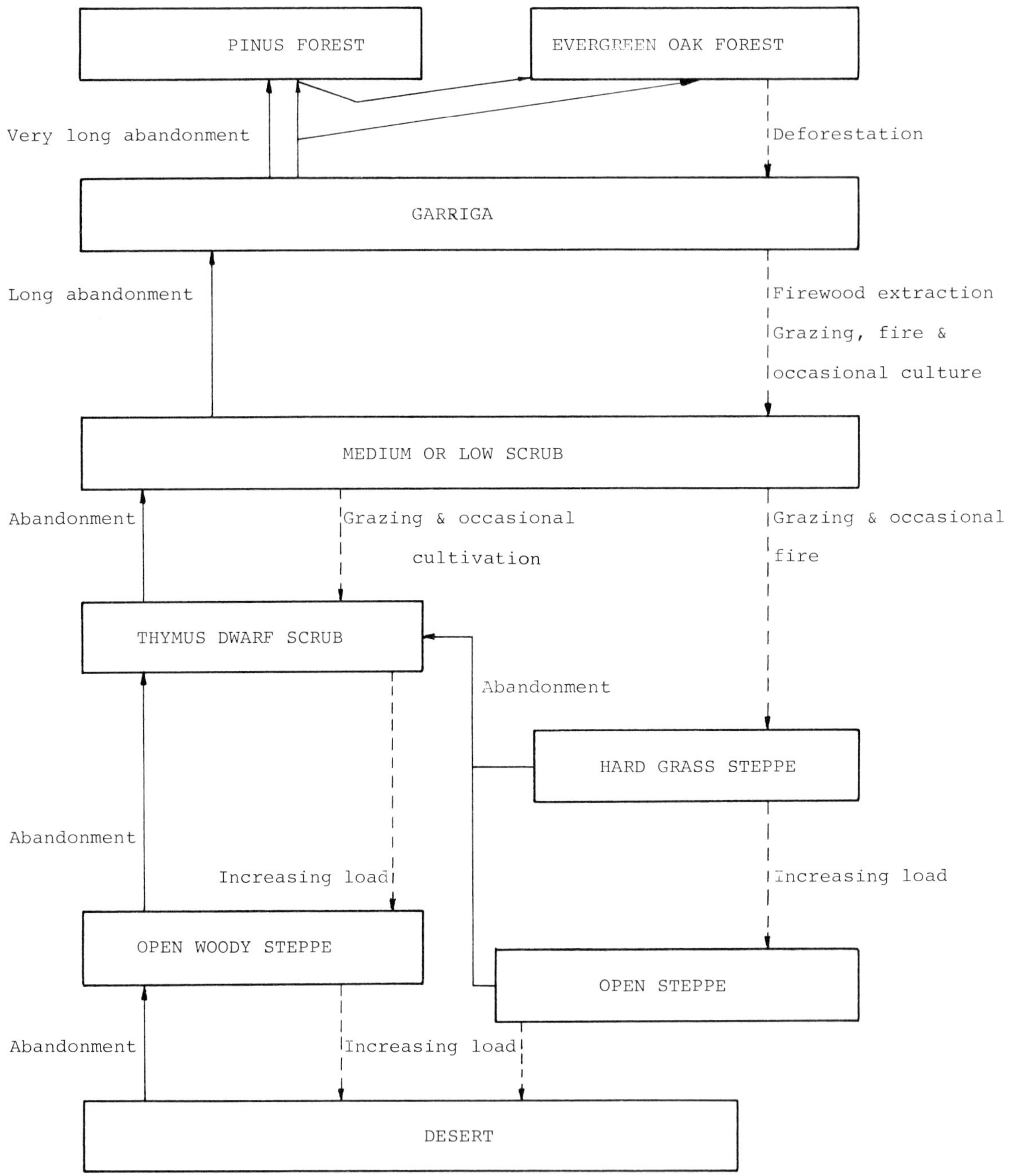

Fig. 6. A. A simplified dynamic model of succession on gypsaceous sustratum under accentuated Mediterranean conditions.
B. As Figure 6A, but under attenuated Mediterranean conditions. →

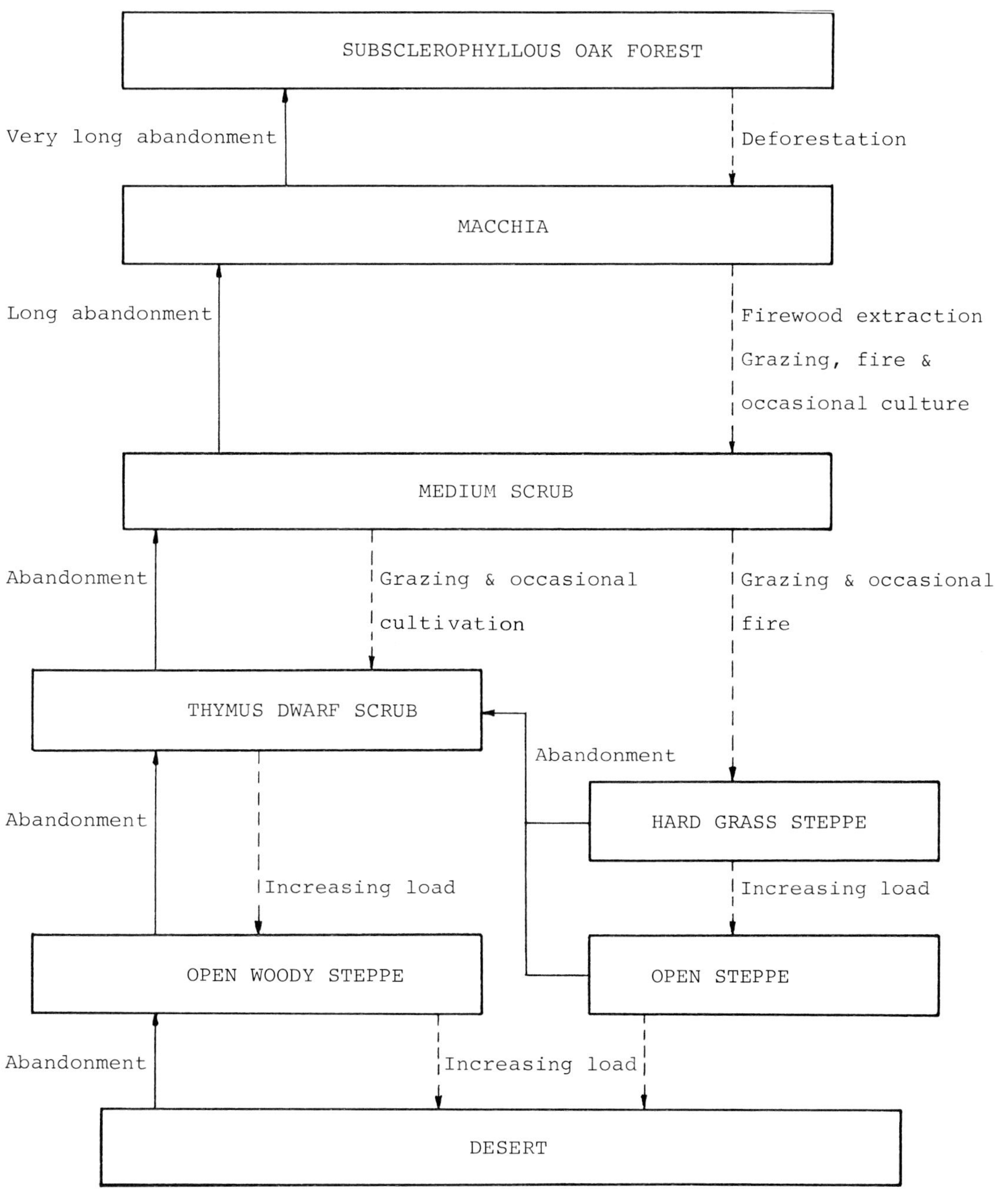
SUBSCLEROPHYLLOUS OAK FOREST
Very long abandonment
Deforestation
MACCHIA
Long abandonment
Firewood extraction
Grazing, fire &
occasional culture
MEDIUM SCRUB
Abandonment
Grazing & occasional
cultivation
Grazing & occasional
fire
THYMUS DWARF SCRUB
Abandonment
HARD GRASS STEPPE
Abandonment
Increasing load
Increasing load
OPEN WOODY STEPPE
OPEN STEPPE
Abandonment
Increasing load
DESERT

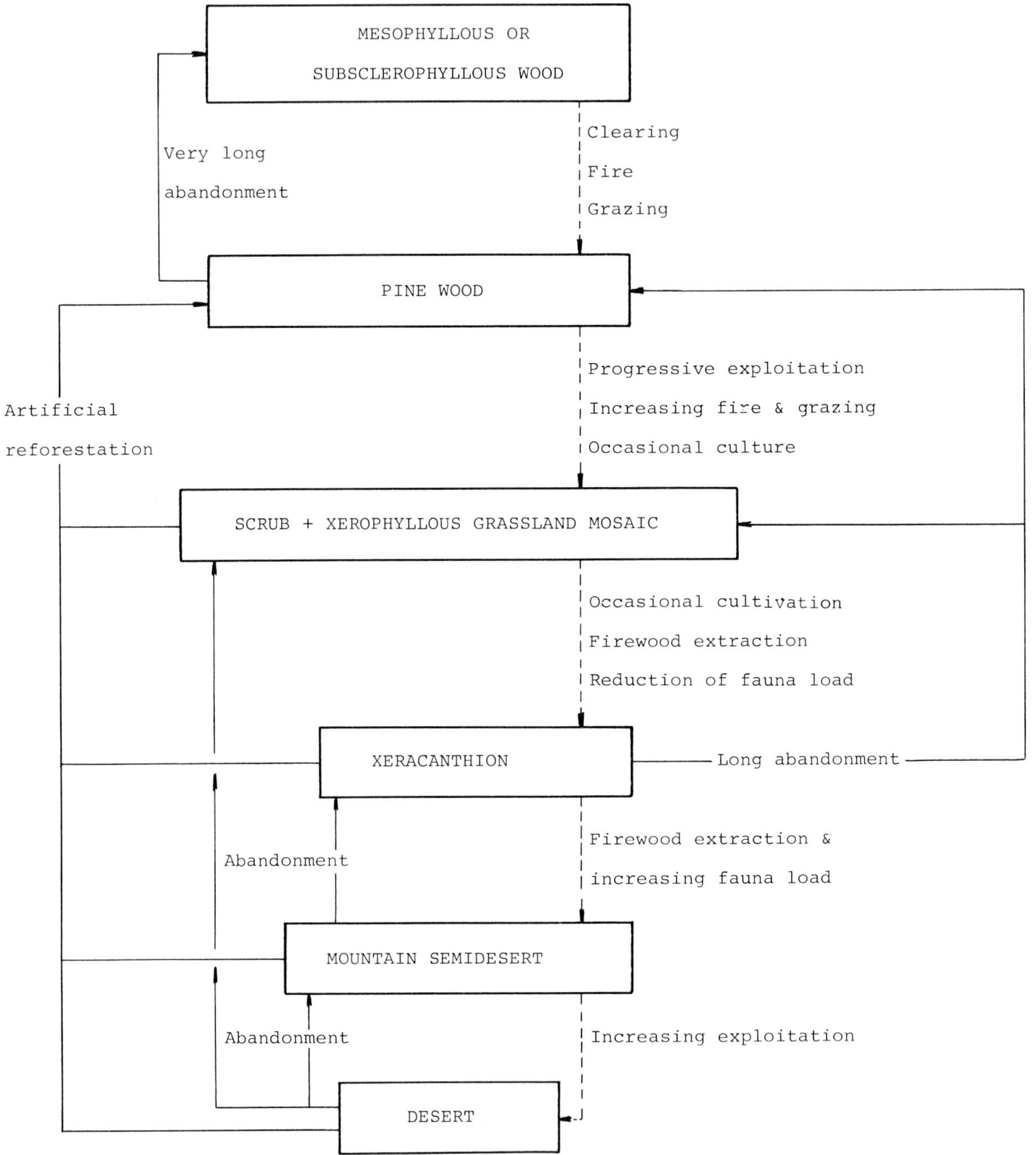

Fig. 7. A. A simplified dynamic model of succession in Mediterranean spiny cushioned montane vegetation (Xeracanthion) below timberline.
B. As Figure 7A, but above timberline. ⟶

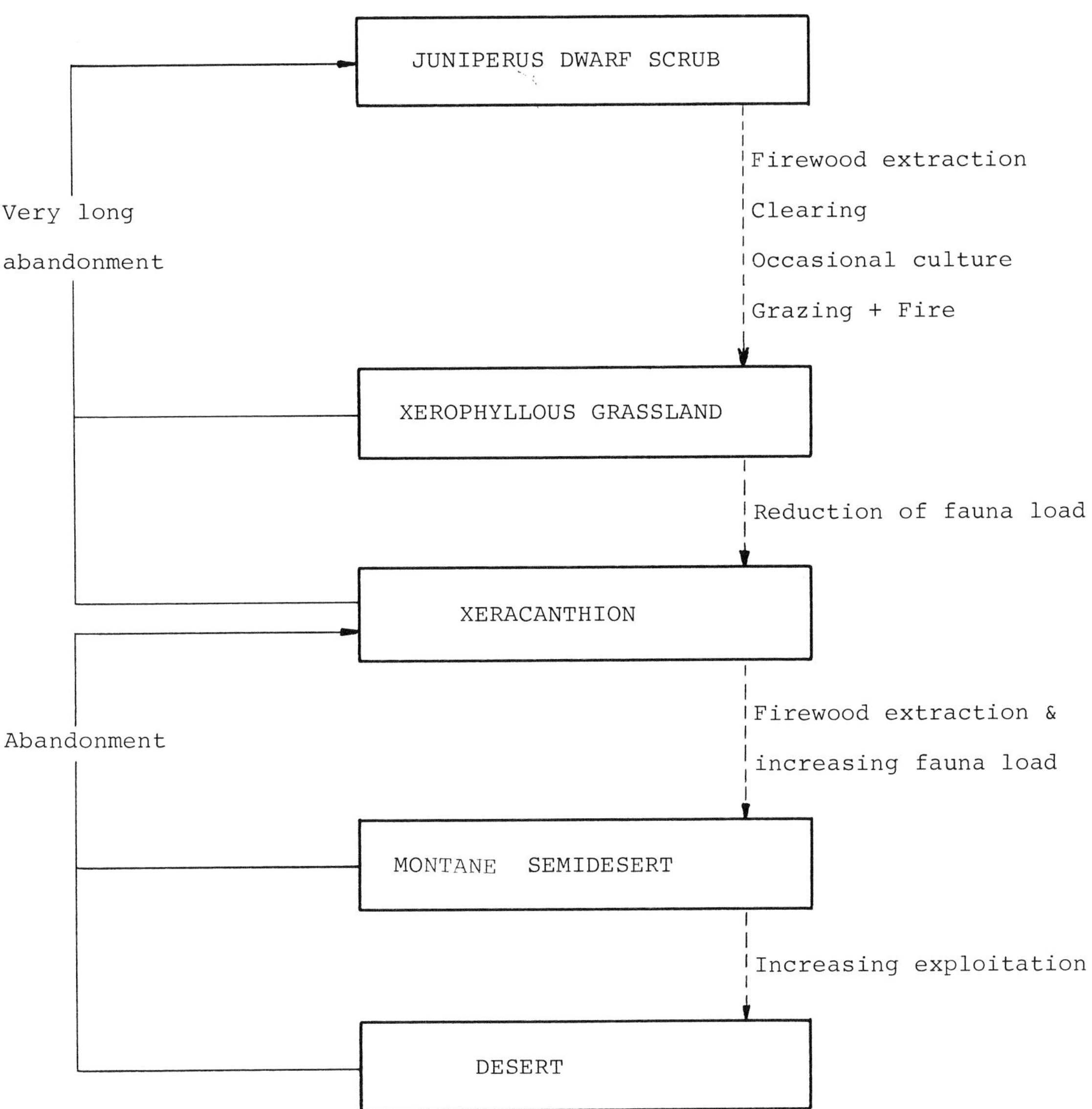
JUNIPERUS DWARF SCRUB
Firewood extraction
Clearing
Occasional culture
Grazing + Fire
Very long
abandonment
XEROPHYLLOUS GRASSLAND
Reduction of fauna load
XERACANTHION
Firewood extraction &
increasing fauna load
Abandonment
MONTANE SEMIDESERT
Increasing exploitation
DESERT

rections. Control plots should be held in reserve, to allow comparison of the evolution of their endemic populations.

7.8 Shrublands in the forest levels

Their conservation will involve the preservation of traditional methods or the establishment of new programmes of equal effect. What the latter are, need to be determined with proper tests and trials.

In the Mediterranean region, several types of shrublands can be recognized, but are usually grouped into four major divisions. In short two of them correspond to average maturity levels where pasture is combined with wood extraction and, very often, with bee-keeping; the other two correspond to low maturity levels and are the result of long rotation cycles in dryland agriculture with or without pasturing during the fallow periods. In every case the area to conserve will be broken up into a mosaic of pieces where the rejuvenation treatments (extraction of surplus wood, fire or cultivation) will be adapted according to a careful, gradual plan.

7.9 Arboreal species

Though arboreal endemic species are rather scarce in the Mediterranean area, we will briefly comment on this case also.

When the species is to be found in a stale mature systems, conservation may merely consist of the establishment of a reserve, and proper monitoring of the main species as well as of animal populations and wood surplus. At the opposite end of the spectrum, an heliophilous arboreal species may need to be introduced into an immature system. A stimulating treatment might then include the extraction of brushwood, soil tilling, etc.

If other races have been introduced which are able to cross with the endemic under consideration, the gradual elemination of the exotic ones should follow. The establishment of 'pure' populations of the endemic will be achieved from seeds obtained by articial pollination. Some breeds and special forms of great rarity and/or beauty can be propagated through controlled pollination, such as the silver cedars of Bel-lezna, Algeria, or local silver pinsapo firs in the mountains of Ronda, Spain.

In the case where there has been a sharp decrease of number in the area, it will be convenient to study and to recover some lost areas of the original distribution. A programme of this kind has been already undertaken with *Abies pinsapo* Boiss. an Ibero-Mauritanian species whose European area is now restricted to some locations in southern Spain.

References

Anonymous (1973). Conservation of natural areas and of the genetic material they contain.M.A.B. Report Series. UNESCO, Paris 12: 1–64.

Anonymous (1983) List of rare threatened and endemic plants in Europe. Kew: TPU, IUCN/CMC and Strasbourg: Council of Europe.

Bazzaz, F.A. (1979). The physiological ecology of plant succession. Ann. Rev. Ecol. Syst. 10: 351–371.

Braun-Blanquet, J. (1965). Plant Sociology: The Study of Plant Communities. London: Hafner.

Cain, S.A. (1950). Life forms and phyto-climate. Bot. Rev. 16: 1–32.

Clapham, A.R. et al. (1980). The I.B.P. Survey of Conservation sites: An Experimental Study Cambridge: Cambridge University Press.

Cook, R. (1980). The biology of seeds in the soil. In: Solbrig, O.T. (Ed.), Demography and Evolution in Plant Populations, pp. 107–129. Oxford: Blackwel.

Dansereau, P. (1958). A universal system for recording vegetation. Contrib. Inst. Bot. Univ. Montreal 72: 1–58.

Dansereau, P. (1961). Essai de représentation cartographique des éléments structuraux de la végétation. Methods de la cartographie de la végétation. Paris: H. Gaussen.

Dasmann, R.F. (1972). Towards a system for classifying natural regions of the world and their representation by natural parks and reserves. Biol. Conserv. 4: 247–255.

Dasmann, R.F. (1973a). A system for defining and classifying natural regions for purposes of conservation. I.U.C.N. Occas. Paper No. 7, Morges (Switzerland).

Dasmann, R.F. (1973b). Classification and use of protected natural and cultural areas. I.U.C.N. Occas. Paper No. 4, 30 pp, Morges, Switzerland.

Downes, R.G. (1969). Conservation in relation to the land and its use. In: Webb, L.J., Whitelock, D. and Brereton, J.L.G. (Eds.), The Last of Lands, pp. 11–17. Brisbane: Jacaranda.

Elton, C.A. (1958). The Ecology of Invasions by Plants and Animals. London: Methuen.

Emberger et al. (1968). Relevé methodique de la vegetation et du milieu. Paris: Editions du C.N.R.S.

Forsberg, F.R. (1967). A classification of végetation for general purposes. In: Peterken, (Ed.), Guide to the Check Sheet for I.B.P. Areas, I.B.P. Handbook No. 4. Oxford: Blackwell.
French, H.H. (1965). The Goat in the Mediterranean Climate Zone. Roma: F.A.O.
Greuter, W. (1979). Mediterranean conservation as viewed by a plant taxonomist. Webbia 34: 87-99.
Harper, J.L. (1971). Grazing, fertilizers and pesticides in the management of grasslands. Ilth. Symposium British Ecol. Soc. Oxford: Blackwell.
Heywood, V.H. (1953). El concepto de asociación en las comunidades rupicolas. Anal. Inst. Bot. Cavanilles 11: 463-481.
Komarek, E.V. (1983). Fire as an anthropogenic factor in vegetation ecology. In: Holzner, W., Werger. M.J.A. & Ikusima, I. (Eds.), Man's Impact on Vegetation; pp. 77-82. The Hague: Junk.
Melville, R. (1970). Plant Conservation and the Red Book. Biol. Conserv. 2: 185-1-188.
Muller, C.H. (1966). The role of chemical inhibition (allelopathy) in vegetational composition. Bull. Torrey Bot. Club. 93: 332-351.
Naveh, Z. (1971). The conservation of ecological diversity of Mediterranean ecosystems through ecological management-Ilth. Symposium British Ecol. Soc. Oxford: Blackwell.
Rabinowitz, D. (1981). Seven forms of rarity. In Synge H. (Ed.), The Biological Aspects of Rare Plant Conservation, pp. 205-218. New York: Wiley.
Odum, E.P. (1969). The stretegy of ecosystem development. Science 164: 262-270.
Preston, F.W. (1962). The canonical distribution of commonness and rarity. Ecology 43: 185-215 and 410-432.
Raunkiaer, C. (1934). The Life Forms of Plants and Statistical Plant Geography. Oxford: Clarendon Press.
Rice, E.L. (1974). Allelopathy. New York: Academic Press.
Rietz, G.E. du (1931). Life forms of terrestrial flowering plants. Acta Phytogeogr. Suec. 3: 1-95.
Ruiz de la Torre, J. (1977). La selvicoltura naturalistica nell'ambito dell'assestamento ecologico della regione mediterranea. Gollana Verde 4: 77-122.
Ruiz de la Torre, J. & Ruiz del Castillo, J. (1977). Metodología y codificación para el análisis de la vegetación espaņola. Madrid: Escuela T.S. de Ing. de Montes.
Schmithusen, J. (1968). Allgemeine Vegetationsgeopgrahie. Berlin: Walter de wgruyter.
Solbrig, O.T. (edit.) (1980). Demography and evolution in plant populations. Bot. Monogr. 15. Oxford: Blackwell.
Trabaud, L. (1981). Man and fire: impacts on Mediterranean vegetation. In: Di Castri, F., Goodall, D.W. & Specht, R.L. (Eds.) Mediterranean-type Shrublands, pp. 523-535. Amsterdam: Elsevier.
Westhoff, V. (1970). New criteria for nature reserves. New Scient. 46: 108-113.
Westhoff, V. (1971). The dynamic structure of plant communities in relation to the objectives of conservation. Ilth. Symposium British Ecol. Soc. Oxford: Blackwell.
Witkamp, M. (1971). Soils as components of ecosystems. Ann. Rev. Ecol. Syst. 1: 85-105.

Departamento de Botánica
Escuela T.S. Ing. de Montes
Universidad Politécnica
E-28040 Madrid, Spain

CHAPTER 13

The role of Mediterranean botanic gardens in the maintenance of living conservation-oriented collections

M. AVISHAI

1. Conservation-oriented collections in the Mediterranean region: Introductory remarks

One of the cradles of modern civilization, the Mediterranean region, is well known as such, but the price the Mediterranean environment had to pay for this development is less well known. The advent of agriculture, industrialization, urbanization, mass tourism with their side effects and more than three millenia of warfare have completely changed the Mediterranean landscape. Whole plant communities have been wiped out, many unique habitats have been destroyed and plant life of vast areas is now irreversibly changed. While rightly proud of their heritage and contribution to contemporary civilization, the Mediterranean people have only recently awakened to the magnitude of the environmental destruction and its impact on their life. Unfortunately, many of these people face other serious problems and conservation and rehabilitation of the environment is seldom a high priority.

In this state of affairs a heavy burden of responsibility rests on those institutions and individuals which may be at least partially able to change the direction of events. Botanical gardens and their staffs thus have a major responsibility in this respect and a crucial function uniquely suited to their special abilities. While national governments and regional or international organizations share this responsibility in many other wider environmental aspects, in matters related to plants and plant communities, botanists and their gardens are in a unique position to help; they face a tremendous challenge.

Although the Mediterranean region contains, in Italy and France, some of the oldest existing botanical gardens (see Table 1), in the rest of the region there is a grave shortage of this kind of educational or research institutions (see Table 1 and Fig. 1). Many countries have no botanical gardens at all, or have gardens that must survive under conditions which very seriously limit their ability to fulfill their role. Thus, the social and economic difficulties of the involved countries are reflected in the state of their gardens, and it is very difficult to see how the few existing gardens can play any role in conservation, unless a great change in their status occurs. Surely, the number of botanical gardens must increase and their location in relation to the threatened plant materials' (TPM) native habitats improved. This is necessary because successful conservation is only possible under similar ecological conditions, and the highly diversified climatic, edaphic, and topographic conditions of the Mediterranean region. The richness of the flora and the high rate of endemism make this a task of international urgency (Raven, 1976). To assume that the few existing gardens can solve the problem is, therefore, unrealistic. However, there are some good examples of what can be achieved (Bramwell, 1979) if local resources and support are properly mobilized.

One very serious shortcoming of the existing gardens in the region is their layout, their basic

Gómez-Campo, C. (ed.), Plant conservation in the Mediterranean area.
© 1985, Dr W. Junk Publishers, Dordrecht. *ISBN 90 6193 523 7.*

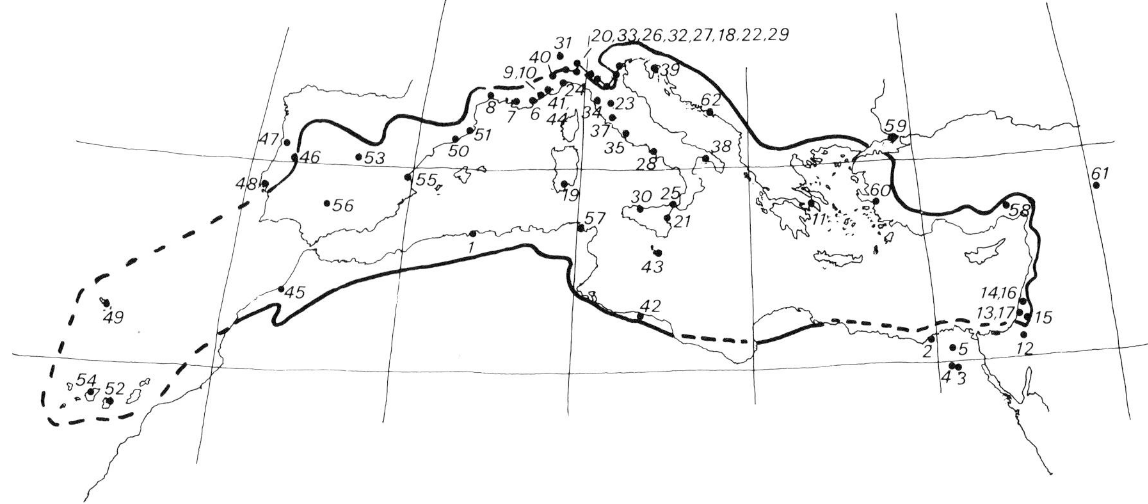

Fig. 1. Distribution of botanical gardens in the Mediterranean region (see Table 1 for numerical references).

design concept, and their public image. Many are located in densely built areas with very limited space available, and a serious danger of pollution. They are frequently considered to be a public show place, a nice luxury or ornament, a tool in environmental education, and not a sizable scientific institution. Their status as living museums of exotic plants conflicts with their true and potential functions, and local policy-making groups are simply unaware of the problem or unable to change things. Sometimes botanical gardens are just an exotic tourist attraction and tolerated as such by the supporting local governments or academic institutions.

Thus, to play any role in conservation-oriented work, existing botanical gardens must change, new ones must be created, and proper resources and support for this kind of work must be provided. Finally, conservation must become one of their primary scientific functions; it must be included in their master plan and layout, and it must become part of a policy in which both long-term and short-term goals need to be clearly defined.

The small number of botanical gardens able to contribute to the maintenance of threatened plant materials is especially important when the Mediterranean environment and climate are considered. The flora of many Mediterranean countries is rich in species for which this region represents the southern or northern distributional limits. Such populations show often extreme, adapted characters. One frequently finds populations of species which survive only as relicts or small 'islands' in an area not suitable for their survival. The special 'island' character deserves to be represented in collections as it often contains unique ecologic, cyto-genetic, or morphological traits. Conservation of these traits at a reasonable cost is only feasible under Mediterranean conditions, and it will be most effective within botanical gardens of the region (see Agreed Resolutions of the 1975 Conservation Conference, Anon., 1976). Snogerup (1979) describes efforts to maintain Aegean endemics in an artificial environment away from the region where they were collected. He concludes that only seed banks are a suitable solution.

This region has a highly seasonal climate with short transition periods. The summers are long and hot and solar irradiation is one of the highest per unit area on earth. Winters are cool and rainy, but they are interrupted by periods of warm sunny weather during which growth and development of plants is highly accelerated. It is highly improbable that these conditions could be recreated on a small scale anywhere else. Moreover, the climatic conditions are superimposed on one of the most diversified and fragmented geologic and edaphologic areas on earth. Sand, limestone, granites, or hills and cliffs replace one another over very short dis-

Table 1. Mediterranean botanical gardens. The date of foundation is given for most cases (compiled from Henderson & Prentice, 1977 and updated).

1. Algeria
   (1) Hamma: Jardin d'essais
2. *Egypt*
   (2) Alexandria
   (3) Cairo
   (4) El Saff, 1902
   (5) Qubba
3. *France*
   (6) Antibes, Villa Thuret, 1860
   (7) Marseille, 1880
   (8) Montpellier, 1593
   (9) Saint Jean Cap Ferrat, 1926
   (10) Nice
4. *Greece*
   (11) Athens
5. *Israel*
   (12) Beer Sheva, 1952
   (13) Holon, 1930
   (14) Ilanot, 1950
   (15) Jerusalem, 1931
   (16) Ruppin, 1949
   (17) Tel Aviv, 1964
6. *Italy*
   (18) Bologna, 1568
   (19) Cagliari
   (20) Camerino, 1828
   (21) Catania
   (22) Ferrara
   (23) Firenze
   (24) Genova, 1803
   (25) Messina, 1638
   (26) Milano, 1600
   (27) Modena, 1758
   (28) Naples
   (29) Padua, 1545
   (30) Palermo
   (31) Pallanza, 1930
   (32) Parma
   (33) Pavia, 1558
   (34) Pisa, 1543-44
   (35) Roma, 1883
   (36) San Bernardino di Trana
   (37) Siena, 1856
   (38) Tarento, 1938
   (39) Trieste, 1828
   (40) Turin, 1728
   (41) Ventimiglia, 1860
7. *Libya*
   (42) Tripoli
8. *Malta*
   (43) Floriana
9. *Monaco*
   (44) Monte Carlo, 1933
10. *Morocco*
   (45) Rabat, 1952
11. *Portugal*
   (46) Coimbra, 1772
   (47) Lisbon, 1874
   (48) Porto
   (49) Funchal
12. *Spain*
   (50) Barcelona
   (51) Blanes, 1951
   (52) Tafira Alta, 1952
   (53) Madrid, 1781
   (54) La Orotava, 1788
   (55) Valencia, 1802
   (56) Córdoba, 1983
13. *Tunisia*
   (57) Ariana
14. *Turkey*
   (58) Adana, 1977
   (59) Istanbul, 1935
   (60) Izmir-Bornova, 1969
   (61) Diyarbakir, 1978
15. *Yugoslavia*
   (62) Dubrovnik, 1959

tances. The Mediterranean flora is tightly bound to this mosaic of conditions and is perfectly adapted to the fluctuating character of the seasons. Plants revive from summer dormancy in a very short time and only gardens in the Mediterranean region can provide the conditions for the rapid and profund transitions that are characteristic for much of the area's plant life. This is the case with many bulbous species and especially the leafless flowering autumn plants such as *Biarum davisii* Turril, *Pancratium parviflorum* Decne or *Sternbergia clusiana* Ker-Gawl ex Schult. Even slight changes in growth conditions will cause a very rapid decrease in the viability and flowering pattern of these plants - all of which are on the threatened plant materials lists. Conservation-oriented collections of such plants must, therefore, be the sole responsibility of botanical gardens in the Mediterranean since no other institution can fulfill this role better.

The three basic actions needed when plants are to be conserved in a botanical garden (Brenan, 1979) are: First to obtain the plants; second, to establish and to maintain them; and third, to distribute propagating material as widely as possible. Simple as this statement is, each of its three mentioned stages has many aspects and the aim of this paper is to discuss these.

1.1 Sources of plant material

Three main sources can contribute to the existence of such collections in botanical gardens:

1.1.1 Surviving or surplus materials from completed or discontinued studies in the academic establisment

These constitute a major part of collections and must be considered of primary importance in view of the vast amount of research that is usually undertaken and published. Unfortunately the habitat information for the grower is too often incomplete. Data about temperature tolerance, humidity, soil type, exposure, special habitat conditions, etc. are frequently missing, and much is left to the intuition of the grower at the garden. The danger of the introduction of artificial selective pressures is thus very real and the threat of genetic erosion (see Esser, 1976) a constant factor.

The continued maintenance of such materials is one of the most common problems faced by botanical gardens and one of the most important aspects in their work, since very often new studies or continuations and elaboration of previous studies must rely on the verification of earlier results. Proper identification and correlation of the original samples with the corresponding results is crucial. This in turn requires a change in the routine work of supervision and research student. First of all the conservation status of any plant species of the garden should be assessed. All perennial plants included in one of the four I.U.C.N. categories should be incorporated into the threatened plant material (TPM) conservation-oriented collection with the active participation of the research student, and all pertinent data should be recorded and each sample given the proper accession number. Whenever other than qualified horticulturalists are involved, the collection of information important for cultivation should be stressed. Data about light conditions, soil drainage, water requirements, air movement etc. of collection sites should be recorded. Enforcement of this principle must be part of the responsibility of the supervisor and included within the approved curriculum as well as in the research budget for the study.

Remnants of discontinued studies constitute another very important source, as very often sizable collections of valuable material are assembled prior to the establishment of a project only to be rejected as unsuitable for various reasons.

1.1.2 Rescue collections by various official agencies or volunteers

This kind of TPM is becoming especially important with time as a result of urban development, road construction, afforestation, or other rapid changes in the environment. Though in many extra-Mediterranean countries such materials are considered as evidence for the failure of proper conservation measures, the present reality is that many sites are endangered by this kind of 'modern progress' and that botanical gardens are the only practical site for the deposit of such rescued, valuable TPM. The rapid pace of destruction in many regions makes cooperation with volunteers a crucial element in conservation-oriented work of botanical gardens. For example, when recently a track was widened into a road in the Judean Desert during the spring season, volunteers saved numerous samples of an *Iris atrofusca* Baker population by transplanting them to the Jerusalem Botanical Gardens. Transplanting to other locations would have caused heavy losses since no rainfall exists after winter in the Judean desert, and several circumstances prevented artificial irrigation of the plants in any possible nearby site.

1.1.3 Active conservation-oriented field collections

This kind of work can have different goals, but it is always characterized by the fact that collections are initiated by the botanical gardens and conducted under their sole and complete control. Very often legal permits are required for this work from governments authorities who are in charge of the enforcement of nature conservation laws and in such cases, often active assistance is offered. The very nature of such planned and conscientious efforts makes them one of the most efficient tools in the conservation and reconstruction of threatened plant populations, if they are part of an ongoing effort fully coordinated with the functions of the botanical garden, with the proper resources allocated, and if they are critically revised from time to time. Sporadic, uncoordinated and improperly funded efforts can be just as disastrous as no efforts at all.

1.2 The representive value of TPM samples

In conservation work every plant may be important. But it is obvious that the real scientific importance of this fact is often limited. Often genetic diversity is manifested not only in physical features of the plant, but also in its physiological requirements. Such distinct characteristics are sometimes recognized and known as special ecotypes, but most often they remain simply undiscovered; nevertheless, they exist and must be considered when the representative value of a certain TPM sample is assessed. Thus, in addition to the minimum sample-size dictated by the breeding system of each species and its variations, additional samples are usually required to present the ecological diversity of each species. This is not always the case as many TPM have a very narrow distribution pattern. But it must be borne in mind that these patterns of diversity may be secondary, i.e., the result of habitat destruction; all around the Mediterranean region this will often be the case. However, this region is characterized by great ecological diversity and this should be given due consideration.

1.3 Sample size

The sample size of any accession for explicit conservation purposes needs, therefore, careful consideration. In emergency situations when the rescue of the plants from their habitat is necessary, and in studies of inbred morphological divergence, the sample size must be large. For plants in the top two threat categories, collections can be compatible with the need for a lack of significant damage to natural populations and the specific life strategies of the species (see Bradshaw & Doody, 1978). If morphological divergence is the object of study and materials can be sent to the botanical garden in a relatively good state, seeds, cuttings or other propagation material should be preferred to any other sampling method as they, obviously, decrease the damage to the population. Propagation equipment maintained in the botanical garden and the proficiency of its staff make this approach possible and recommendable.

According to one important view (Frankel, 1970) preservation is seen as the persistence of genes in time. This raises the question of the genetic level with which conservation in the botanical garden should be concerned (e.g., species, subspecies, variety forms, hybrids or unique genotypes). Often the loss of unique gene pools or the reversal of direction of genetic drift or shift due to human or other environmental stresses is of prime concern to those involved in conservation work. This aspect, as well as the question of quantity, are examples of situations which require the explicit definition of the aims of conservation. In the Mediterranean region the danger of extinction of taxonomic species is probably the most urgent threat and aspects related to the genetic structure of populations can probably be seen as secondary, though not less important. Hawkes (1976) discusses the sampling of gene pools and the danger of genetic erosion as related to threatened plants.

The size of a collection has a tendency to increase. Soon a serious policy problem develops; how long should garden-grown stocks be maintained? What is their optimal size? The best solution to this problem is probably an aggressive program of reintroduction. The merit of this policy is clear, and whenever successful reintroductions of species or taxa into their original or secondary habitats are accomplished, the size of TPM conservation-oriented collections can be adjusted accordingly. The follow-up and continuous monitoring of the reconstructed populations are a necessary sequel that can be either directly implemented through the botanical garden or through another public or government agency. Often volunteer conservationists can help. In any case, the coordination of such steps with conservation authorities is crucial.

1.4 Living, conservation-oriented collections: their character and requirements

Conservation-oriented collections of plants that include threatened plant material (in short, TPM) can be of help in conservation only if certain conditions are met.
These are:

a. All the materials included are properly numbered, mapped, labelled, checked, and their names verified by a competent person, and retraceable to an original wild population, existing or extinct.
b. The collection represents a more or less significant sample of the morphological, physiological, and genetic diversity of the population it is said to represent.
c. Seeds of such TPM result from direct original collections in nature or fully controlled production in the garden or laboratory, conducted under the supervision of competent and properly trained staff, from well-grown disease-free stocks.
d. Data on all aspects of collection, storage, or

cultivation are properly recorded and readily accessible.

To fulfill these conditions, conservation-oriented collections of TPM in botanical gardens require the strict observance of some basic guidelines by all persons responsible for their creation, development, and maintenance. The establishment of conservation-oriented collections is a costly business, to be undertaken only if the righ conditions for complete success can be ensured. Allocation of the necessary funds is, therefore, required to implement the establishment of any program of this kind. However, as limitations are present everywhere, conservation-oriented collections should have a limited scope to start with, and be gradually expanded as more experience and resources become available.

In order to comply with the first condition concerning threatened plant material (TPM), specimens should be recorded with a permanent label attached to the plant or plants included in each sample. This requires the establishment of an accession system if it does not already exist in the garden, similar to those that are used in Herbaria (see also Hawkes, 1976). A single series of numbers should be used by each botanical garden with each accession recorded in a central file. As is usual for herbaria, each collector should use one series of continuing numbers during his entire career, fully corresponding to those used in his herbarium collections. The cost of labels and their lettering needs special attention in view of the extensive amount of work invested at this stage, but the special, often crucial, importance of threatened material in conservation-oriented collections should not be disregarded. The mapping of such collections is highly desirable as it decreases the chances of inadvertent, yet frequent, transfers of labels. Mapping of TPM collections also greatly facilitates the continuous checking required to ensure survival of the samples both in plots and in storage, while dormant or actively growing. To make sure that the sample is of significant size to represent the diversity of the sampled population, a competent person familiar with the group can be consulted. However, this is not practical in many instances, and, alternatively, sample size can be graded according to the seven categories designated by the Survival Service Commission of I.U.C.N. or to special conditions that are outlined elsewhere in this paper.

2. Management of living conservation-oriented collections

The best basis for an efficient and successful establishment and maintenance of conservation-oriented collections of TPM in botanical gardens is one combining close approximation of ecological (climatic and edaphic) conditions with an as complete as possible understanding of the breeding system of the particular taxa under natural conditions. The approximation of natural ecological conditions is the most important basis for successful cultivation of plants and the cost of the necessary compensating steps fully determines the cost of long-term maintenance of TPM in botanical gardens. Thus, the best place for the cultivation of Mediterranean plants is within this geographical area. The cost for creating, say, in Central Europe aestival dormancy would be prohibitive as one cannot reproduce either the temperature or other factors at a reasonable cost in Central Europe. Thus, an effort to maintain conservation-oriented collections of TPM from the sect *Oncocylus* of the genus *Iris* in Germany would be doomed to failure. The reason is the highly specialized character of these rhizomatous *Iris* species: their continued survival absolutely depends on a period of summer dormancy under conditions of high temperature, very low (less than 30%) relative humidity and alkaline soils. Also, since this section of *Iris* includes species of obligate outcrossers, their cultivation under one roof without any consideration given to their breeding system would result in a great mass of hybrid seed and genetically different progenies. Therefore, for conservation purposes, these plants require controlled conditions of flowering (such as paper-bagged flowers), aritifical or hand pollination, and bagging of the pollinated flowers.

The close approximation of ecological conditions may be very important but there are many other aspects. Among these the most important are the following:

2.1 The establishment of TPM conservation-oriented collections as a gradual process

The conservation of TPM is an urgent problem, but in view of the great difficulties involved, the establishment of collections must be planned as a gradual process, a process in the course of which trial and error play their fair share. Thus, every decision concerning the establishment and maintenance of TPM conservation-oriented collections should be based on strict priorities, and the efforts coordinated in accordance with the threat categories proposed by I.U.C.N. However, ornamental, economic, or scientific importance should also be given due consideration when the priorities are decided. As pointed out by Frankel (1976) botanical gardens must be highly selective in the choice of material to be preserved. The best solution would be a gradual process of building up a collection, not along the lines of general and species but for endangered taxa first of all, then for those defined as vulnerable, then for those defined as rare, and finally, those considered indeterminate. These four stages in the build-up should and could overlap, but they should be centered primarily on the natural habitat, and only secondarily should TPM materials be transferred to artificial growth conditions in a garden. This requires the involvement of botanical institutions also in the conservation process on a national or regional level as occurs, for example, in the Canary Islands (Bramwell, 1979).

The role of botanical gardens in the maintenance of TPM or other conservation-oriented collections should be an active one and not just one of providing services. Incoming materials should be handled according to priorities and incorporated into the existing collections, but stress should be directed towards the development of a gradually implemented program based on the recommendations of conservation organizations or well-informed scientific panels (see Fig. 2). Whenever such policy making sources are missing, guidelines must be initiated by the scientific and administrative management of the garden, according to the conditions in each case. In this respect, cooperation with other botanical gardens can be of great importance, and the Botanic Gardens Conservation Coordinating Body (Anon., 1979) set up in 1979, can play an important role in providing guidance, information, and other types of help.

2.2 Incorporation of TPM conservation-oriented collections into existing and new botanic gardens

In addition to the problem of accessibility of TPM in the garden and their protection from theft and damage, the incorporation of any special collections into a botanical garden is not always easy and often requires careful consideration. In many cases, mostly in existing and well-established gardens, the incorporation may be inconvenient and require profound changes in the design and concept of the garden. It is difficult to see how a garden designed in formal symmetric style, such as Versailles, can play any role in conservation. In new botanical gardens, or those under construction. In new botanical gardens, or those under construction, its role in conservation and the maintenance of conservation-oriented collections should become an integral part of the master plan and the scientific concept. Due consideration should be given to all the aspects and problems of the maintenance of such collections and the technical solutions for these aspects should be included in the planning process and construction of the garden.

The importance of the role of new botanical gardens in conservation will greatly increase in the future as new gardens are created all around the lesser developed countries. Conservation should be one of their first tasks, as is stated in one of the main resolutions of the last XII International Botanical Congress (Anon. 1975). In the design of these gardens more and more consideration is being given to the creation of small-scale samples of plant communities and biotopes, and it is through this approach that the role of botanical gardens in conservation can be enhanced. The alternative to this approach, often found in established gardens, is one that turns the botanical garden into a botanical research institute with a garden. This often leads to a situation where one is faced with row upon row of potted or container-grown plants –

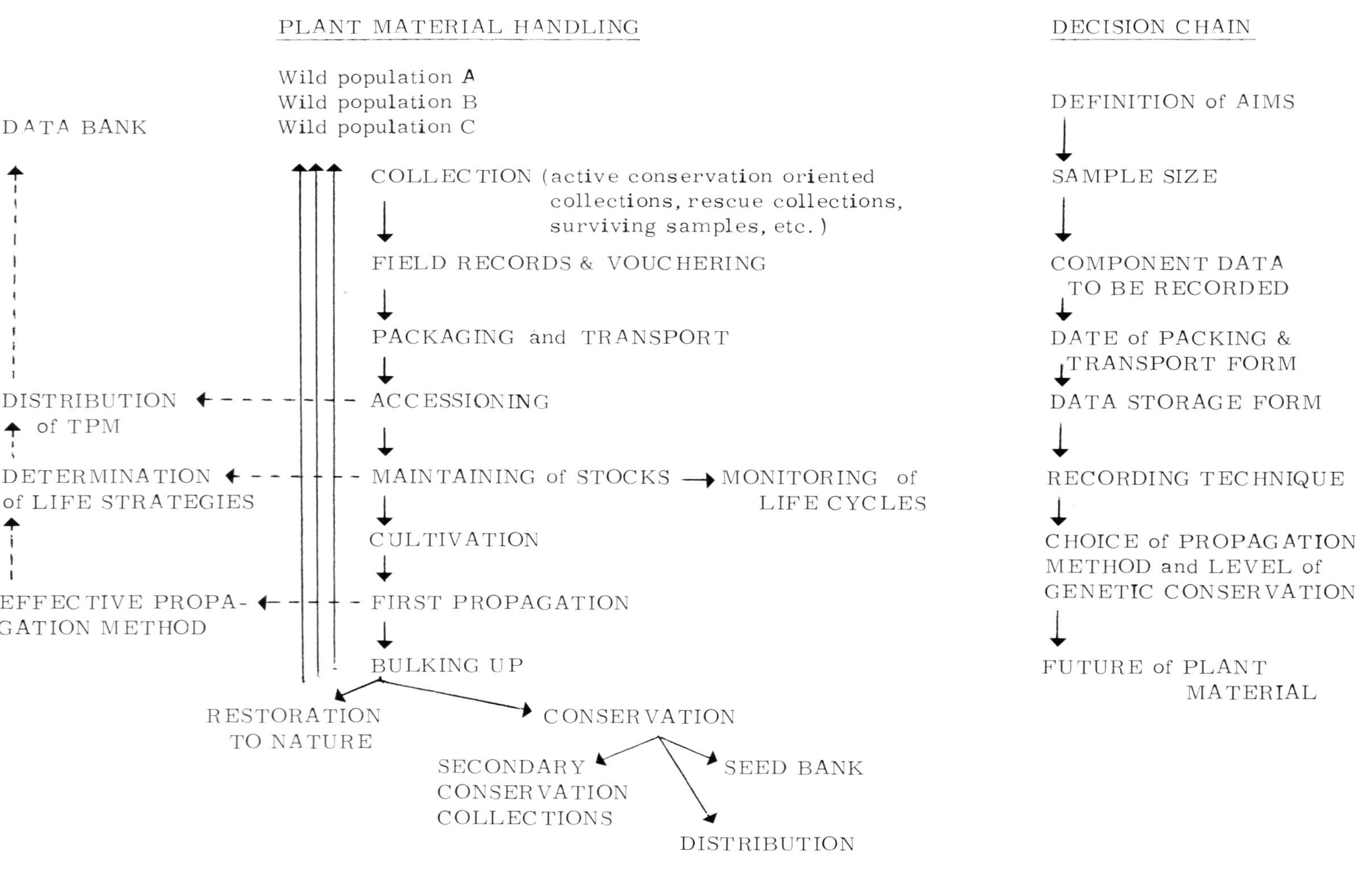

Fig. 2. Flow chart for management of living conservation - oriented collections. (–) TPM and decision flow; (---) data flow.

something that can hardly be called a garden.

In gardens designed with conservation as one of their important functions, the movement of visitors should be limited, and some 'off limits' areas should be separated from the open grounds. This does not impair the recreational and educational function of the garden, indeed the opposite is true. Rockeries, bog gardens, fern dells, sand dunes, or alpine gardens, while introducing valuable diversity to the botanic garden and creating centers of interest, should create the facilities for an unimpaired cultivation of TPM and yet keep that material at safe distance from the public. In this way, the apparent conflict between public display and maintenance of TPM can be resolved. By proper design and landscaping, what might have been considered a disadvantage can be turned into a most pleasing esthetic feature of prime importance in the teaching of conservation. Smith (1979) describes some of the problems encountered.

The creation of micro-biotopes in the garden will facilitate maintenance as different plants will be kept under similar conditions. This will decrease the amount of specialized attention required by each species and introduce a safety factor in their cultivation that does not exist in pot-grown collections.

In many cases the public display of the plants cultivated in the garden is an integral part of the public service of the garden. With TPM, this is practical for only a few carefully chosen examples with clear-cut educational purposes. The rest of TPM in conservation-oriented collections must be fully protected from the general public or from unqualified personnel, so that loss due to theft or damage to plants will be reduced.

Thus, although implementation of conservational measures in botanical gardens should be given publicity and stressed whenever possible, the actual sites should be as well concealed as possible and protected from intruders or hobbyists who may be attracted by the special aura of threatened plants.

2.3 Species accessibility in the botanical garden

There is one aspect of conservation work where the botanical garden plays a unique role and where no other institution could compete. This is the proximity at which plant material is held. This makes the conducting of many scientific studies easier and greatly facilitates any other work with TPM. Of course this in turn depends on the ready accessibility of the plants in the garden. This problem increases as the scope of research work is extended and with the increase of collections in relation to the size of the garden. Anyone familiar with botanical gardens knows from personal experience the problem of locating material: 'we do have that plant... somewhere!'. One obvious solution is the concentration of all TPM in one area of the garden, but this creates problems of maintenance of closely grouped plants with different requirements and problems of uncontrolled cross-pollination. It also removes all work with TPM from the visitors' eyes into a secluded corner. The opposite solution consists of the dispersal of TPM in their respective ecological communities throughout the garden, with special distinctive labelling and careful mapping and recording. Cultivation in this way, along with grouping plants of the same ecological requirements together, makes maintenance of such collections more economical and the additional special care required negligible. The ultimate solution probably depends on local conditions and resources, but this is a problem which now demands the attention of any garden management much more than ever before.

2.4 Conservation in the botanical garden

As it is known by anyone concerned with collections in botanical gardens, the fact that a plant is cultivated in the garden does not ensure its survival. Loss of plants is a common, even if lamentable, aspect of the work. Theft, damage, pests, or disease can be the most obvious reasons, but often cryptic reasons exist whose effect on plants can be disastrous. With TPM conservation-oriented collections, therefore, strict observance of the essentials of cultivation and management is necessary. Protective and preventive precautions must be implemented and all the details of the demands of the plants must be studied and recorded. Very often this will not be possible because of insufficient knowledge of the requirements of many wild plants. In order to alleviate this, conservation efforts should be coordinated with specialists in the ecology and biology of TPM and the advice and research by other botanical gardens or research institutions incorporated. Since many plants are susceptible to the effects of pollution and often succumb to them in a short time, the establishment of such collections should be made in areas where this kind of threat is minimal, and if this is impossible, equipment for environmental surveillance must be acquired or arrangements made for its availability from other agencies or institutions. Prominent, distinguishing labels are very useful and can aid work. Frequent checks or regular tests can be of great importance. Quick recognition of any kind of damage, such as pests or diseases, is necessary for the prevention of many disasters. Routine checking of root systems or other concealed areas of the plant body should form an integral part of the work of those responsible for the maintenance of TPM conservation-oriented collections.

In many commercial horticultural establishments, the use of growth-controlling chemicals such as germination inhibitors, selective or general-purpose herbicides is a normal part of the cultivation of crops. This practice, which is of great importance in the reduction of labor costs and in increasing crop productivity, can and should be applied in botanical gardens whenever possible. However, it cannot be used in the maintenance of TPM unless small-scale tests have been conducted previously and it is proved beyond any doubt that short- or long-term deleterious effects will not result from the use of such materials. It may be of interest here to give an example of damage caused in Jerusalem in this way. It is a general practice to use germination retardants like 'Ron Star' (2 tertio butyl-4 (2,4 dichloro 5 isopropyl oxyphenyl)) to improve the performance of many bulbous crops, and so this chemical in granulated form was used at the Jerusalem Botanical Garden's nurseries to combat weeds in a collection of TPM cultivated in pans. At first no negative signs appeared. But later in the season, as more of the chemical was leached into the root system from the upper crust of the soil, foliage yellowing appeared, growth was retarded, and finally, quite a few valuable plants succumbed to the effect of this germination retardant. Obviously, this can be true also for sprays that can drift in the air, especially when germination retardants are used on paths in nursery grounds, etc.

2.5 Conservation and administrative or professional staffs

In the day-to-day work, awareness of the importance of conservation is essential for all levels of staff. The problem is serious and deserves attention. Beyer (1979) describes the solution adopted at Kew. Every employee should understand the importance of the work being done by him and by others, and should be continuously aware of his contribution to the successful implementation of management policies and procedures. If this attitude is achieved, feedback is created which enables management to benefit from the full support and initiative of the staff. This is especially important in the maintenance of conservation-oriented collections of TPM as the very routine nature of such work can create conditions of absentmindedness, careless work, or even drudgery. Thus, to sustain interest, keen observation of procedures and all that concerns plants, is of great practical importance. Consistantly good performance should be encouraged through acknowledgement and reward. In line with what is common practice in scientific work, due acknowledgement in publications or other means of publicity should be given to the work of technical staff and even, in some cases, to the work of unskilled laborers. Enthusiasm and devotion which are crucial in the practical care of plants must and can be maintained. Staff and information exchange programs and up-to-date contacts can improve the awareness of technical personnel to the successes of other botanical gardens, and improve their motivation. The establishment of some sort of material reward or even prizes to be awarded for outstanding work or achievement are also highly commendable. Deep intellectual and emotional involvement in the work can thus be created and the goodwill of employees ensured. When every employee is aware of the challenge, he will be able to use to the best effect the resources put at his disposal.

The processes of administration with threatened plant material require constant review and reassessment of objectives. Implementation checks, follow-up at all levels of policy enactment, and long-range commitment of labor and funds, are all essential. By their very nature these tend to encourage centralized organization. But centralized organization limits initiative and often hampers independence, and thus can have a negative effect on the day-to-day functioning of the system. Therefore, follow-ups and reassessments should become part of a decentralized organization, one which increases flexibility and promotes initiative and the integration of volunteer help in the work. Though this requires some paperwork it gives conservation the essential continuity.

Successful management and maintenance of conservation-oriented live collections of threatened plant material depends on the technical know-how and dedication of staff. Thus, attraction and retention of the right kind of people is of great importance. Though dedication, enthusiasm, and goodwill are all very important, a proper staff policy is the basis for success in the attainment of objectives. The encouragement of initiative, continuous training, and improvement of the technical proficiency of the staff at all levels is of paramount importance. A garden can establish, maintain, and develop TPM conservation programs only if all levels of staff identify with the policy of the management and cooperate with it.

2.6 The role of botanical gardens in maintenance of TPM and data banking

Full scientific reliability of TPM requires complete information about each sample, data about provenance, collector, form, and quantity collected, the ontogenetic follow-up, etc. Storage of these data, their quick and exact retrieval upon demand, are essential, and their compilation is a costly and integral part of the work of botanical gardens. Few botanical gardens in the more developed industrial countries have the facilities required for this function, and none in the Mediterranean region has facilities with automated data bank and retrieval systems. There are two major approaches to the solution of this difficulty. The most common and easy solution is the use of punched cards as a simple mechanical recording system. A different solution is the establishment of an international 'regional plant records' data center. This is currently implemented in North America through the A.A.B.G.A. (American Association of Botanical Gardens and Arboretums), and one should also be developed for the Mediterranean region by one of the regional organizations such as O.P.T.I.M.A. (Organization for the Phyto-Taxonomic Investigation of the Mediterranean Area), as a joint project for botanical gardens of this region. This center would, within a short period of time, give a great impetus to conservation efforts and their coordination in the region, and provide great support to the work of botanical gardens in the region. The advent of a new generation of computers, the introduction of computers accessible over telephone lines, etc., make this a tangible aim. The methods described in Brenan et al. (1975), Perring (1976) or Hofman (1976) can be adapted to local conditions.

3. The ontogenetic follow-up, obscure factors in TPM maintenance and genetic contamination hazards

The maintenance of TPM is an inherent and natural part of the work and role of botanical gardens, and it is this very fact that makes the collection of data on ontogenetic development possible. For many plants, their origin or habitat character (such as cliff endemics) precludes this kind of study *in situ*, and the assemblage of phenological data and cultivation under garden conditions is the only way to gain this information. Often obscure or cryptic stages in development or in the pollination biology can be elucidated only under the conditions of prolonged observation in the garden (see Cook, 1976). This is not a perfect solution as the very removal of the plant from its ecological niche may exclude vital elements such as pollinating or dispersing agents. Whenever possible, therefore, the work in the garden requires complementary observation in the native habitat.

In addition, some genetic factors have to be considered. Esser (1976) gives a lucid discussion of these. He argues that conservation is impossible and only preservation can be achieved. This demands the attention of those in charge of collections. So do the problems of genetic contamination.

Genetic contamination is defined as the uncontrolled transfer of inherited characters among originally isolated populations. It is a byproduct of man's interference with natural ecosystems and, through introgressive hybridization, it may cause a deep transformation in the involved organisms. Since botanical gardens constitute an artificial habitat, genetic contamination poses a serious threat to conservation work. It is not only present at the level of individual plants, but can also change characters at the population level and create a false picture of the population parameters (Snogerup, 1979). The transfer of characters occurs both within the limits of the garden and out of them, between plants concentrated in the garden and those living in the surrounding areas. This was the main reason to discard conventional seed banks and reassess the seed lists of botanical gardens. The net effect depends on the existence and nature of pollen-transferring agents and the existing natural or artificial isolating barriers. Thus, for example, it will be extremely difficult to maintain 'true to type' collections of anemophilous species of grasses or woody species such as oaks unless intrinsic sterility barriers support these efforts. The cultivation, in close proximity of different genera, species, subspecies, etc. which is characteristic of botanical garden conditions, often requires very careful consideration and assessment of actual conditions at each particular site. A perfect understanding of the breeding system and the ecology of recombination systems (Grant, 1975) should, therefore, be a precondition to any conservation-oriented maintenance policy of threatened material in botanical gardens. Naturally, the information is often not available, and the staff has here a special role to play, which is beyond the passive maintenance of collections.

The threat of genetic contamination should not be overlooked, but under garden conditions, technical solutions (such as insect-proof structures) can be found to decrease this threat to successful conservation. Whenever possible, seed banks maintained in close conjunction with living plant collections will serve as important complementary tools.

4. The study of rareness

The key to a successful conservation effort in the botanical garden is a comprehensive understanding of the reasons that make certain plants in a floristically rich region like the Mediterranean rare and, therefore, liable to become the first victims of extinction. Why is a rare, endemic plant in its native home like *Pinus radiata* D. Don, the Monterey Pine, one of the best commercial timbers of the southern hemisphere? What are the specific reasons for rareness that are directly related to the plant and what is the limiting influence of the wild indigenous habitat? Obviously these are still some of the most perplexing questions, requiring a wide scope of proficiencies, and it is very likely that the answers will require cooperation with many people and organizations. However, the competence of technical and scientific staffs should be a good reason to locate this kind of scientific endeavor in the botanical garden. The following is a list of themes directly related to the maintenance of conservation-oriented collections in the botanical garden:

a. The assessment of the impact of environmental factors in the wild habitat like pests and diseases, as well as a comparison of the susceptibility of the plants to various threats from phytopathogenic of toxic agents under cultivation.
b. Life strategies, the rate of attainment of maturity, mortality and their relevance to conservation (see Bradshaw and Doody, 1978).
c. The study of mycorrhizal relationships between threatened plants in the natural, wild habitat and their impact on the cultivation of TPM.
d. The biology of flowering, dispersal and propagation in the wild and cultivated parts of the

gene-pool of TPM, fertility barriers, etc. (see Nettancourt, 1977).

e. The elucidation of breeding patterns and systems, and their cytogenetic basis; the natural isolating factors of the threatened species such as sterility caused by genomic mutations (aneuploidy or polyploidy, for example) or chromosomal mutations such as deletions, inversions etc. (see Esser, 1976).
f. Requirements for normal seed germination, seed viability and storage, and seedling establishment of threatened plants.
g. Allelopathic relationships between plants and the mechanism of mutual exclusion as related to rare threatened plants.
h. Limiting factors for the distribution of plants, in relation to the soil.

The collection of these data about the TPM maintained or to be established in the garden will make the scientific contribution of the botanical gardens much more meaningful. These data should be the basis for an efficient management policy of the conservation-oriented collections in the botanical gardens of the Mediterranean region. The information collected should made available to other gardens through regional publications such as the OPTIMA Newsletter.

These data are also important in contributing to scientific knowledge, in view of the fact that most of our information concerning the demography of plants is limited to annuals, weeds or plants of economic importance (see Harper and White, 1974).

5. The public role of botanical gardens in conservation

The role of botanical gardens in the maintenance of threatened plant material does create some secondary functions of importance. The first is the establishment of the garden as an information center for all kinds of technical methods and applications for government and volunteer organizations concerned with conservation. The highly technical character of many aspects of TPM maintenance requires technical and horticultural proficiency that is beyond the range of many agencies. The experience gained in the botanical garden and the sophisticated horticultural techniques developed, can help in the solution of many of the problems faced by these agencies.

The maintained collections turn the garden into a distribution center, a primary and often the only source of TPM, both as seed and living plants. This enables the botanical garden to initiate on its own, or to support initiatives for, the regeneration or restoration in nature of depleted populations of TPM from collections existing in the gardens. Under Mediterranean conditions, as in other less developed countries, conservation is often seen by the local population in rural areas as an interference in their conventional mode of life, a threat to their interests. This can have an adverse effect. I remember with sorrow a wonderful population of *Sternbergia clusiana* Ker-Gawl ex Schult, discovered recently in the Judean Mountains. This population was under threat of complete destruction as a result of the interest created and the large numbers of strangers that started to visit the site. Unless such situations are handled with utmost care and consideration for local traditions and fears, or unless local pride is 'cultivated', grave dangers for plant populations may be created. The garden technical staff, often themselves of rural origin, can help to overcome such fears. On the contrary, the mobilization of local aid can have far-reaching positive repercussions.

Beyond any doubt the public attraction exerted by botanical gardens helps in the creation of an awareness to the problems of threatened species. This interest can be cultivated and developed into a readiness to share in the efforts of conservation. Through the organization and instruction of volunteer efforts in conservation at all levels of the educational system (especially at school), the cause of conservation is furthered. Often it is possible to integrate projects such as development of special cultivation methods into highschool syllabi, and thus not only help conservation but add a new dimension to education. The cultivation of *Dracaena draco* L. seedlings by school children in Viera y Clavijo garden in Gran Canaria is a good example.

Of course, the character of botanical gardens as publicly supported institutions also creates a public commitment to conservation, and the funding through government or volunteer organizations can be a tool to deepen that commitment in the maintenance and enlargement of conservation work. Projects on a regional basis, such as the Cambridge University Botanical Garden's 'Eastern England Rare Plant Project' (Walters, 1979), is a move in the right direction. Where nature conservation agencies do not exist, botanical gardens should strive to establish small habitat reserves, or try to become involved in their maintenance as scientific consultants. Once again the Viera y Clavijo Garden in Gran Canaria (Bramwell, 1979) can be used as an example. The establishment of networks of small habitat reserves can be a practical and positive initial step in conservation. Whenever government or public agencies with conservation functions exist, close cooperation with these should be actively pursued.

In most Mediterranean countries effective conservation of large numbers of threatened species from different areas in their habitat is unfortunately a long-term goal, and botanical gardens with their facilities (seed banks, etc.) and know-how, can provide an excellent short-term back-up insurance.

The strengthening and deepening of contacts among botanical gardens concerned with maintenance of TPM, play a vital role. The rate of progress of conservation efforts in less developed countries can be greatly accelerated by the support of gardens from the more developed countries. Preferably this should be between gardens of similar ecological conditions, but this is not always necessary. Often the very nature of support from abroad increases the 'respectability' of conservation efforts with the political and social establishments of the supported countries, and creates valuable goodwill. In efforts of this type, international agencies can be mobilized, and new funds made available. An additional effect is the commitment of local government, which is very important especially for the less developed countries.

In conclusion, the public function of botanical gardens in the maintenance of conservation-oriented collections is a new, vital dimension in their work. A dimension whose basic and applied promise for society in the future will increase. This will no doubt enhance the value of botanical gardens for the community as a whole.

6. The distribution of TPM and conservation through cultivation

The distribution of propagation material as widely as possible is one of the most important aspects of the contribution of botanical gardens to conservation. Because we deal with rare plants (at least initially), a list of priorities for the distribution of such materials must be established. What material do we have for distribution, and where and to whom should the material go and why - are questions to be considered and answered.

The first destination for bulked-up material is the original wild habitat, or if this has been destroyed or otherwise been made unsuitable, neighboring reserves or other conservation-oriented areas. The next in line should be other interested institutions like botanical gardens in close proximity. The next step should be a sharing program between centers of research and environmentally suitable sites, to ensure the continued survival of such plants. This way strong mutually beneficial relationships could be built up between research centers with original Mediterranean plants located in the temperate regions of central and northern Europe and environmentally suitable gardens in the Mediterranean region. Problems like those listed by Snogerup (1979) concerning the cultivation of Aegean endemics, or by Poppendieck (1976) concerning South African succulents could effectively be resolved by such a sharing program.

The extensive, often lavishly landscaped campuses of the new universities currently under development in the Mediterranean region should be another recipient of such plants. On such sites quite a few of the suitable rare and threatened plants of the region could be introduced for cultivation. This scheme, proposed also by Raven (1976), would serve an educational function and ensure survival without continuing cost to the botanical garden.

The cultivation of rare and threatened plants and their introduction to commerce is of course not easy and simple, and requires the consideration of various legal and administrative aspects and is justified *only* after the successful restoration of wild populations. But it is (to quote Raven, 1976) certainly 'better to have a few genetically depleted individuals of a given species remaining in cultivation than none at all.'

The sale of rare or threatened plants through volunteers or school children could, if successful, generate public support for conservation and the gardens themselves.

Successfully maintained collections of threatened plants in the botanical garden produce seed. This should be collected and conserved. But even better is seed of threatened plant species that is collected and distributed through the Index Seminum channels to interested scientists and institutions. Though it is not easy, such valuable seed should be traced and a follow-up program maintained to determine the eventual fate of the seed and the plants raised from it. In the list of seeds offered for exchange TPM should be marked, and probably the best method would be an index system that marks the species according to the threatened category to which each of the respective species belongs.

In conservation, the maintenance of representative samples of natural diversity is extremely important. But the capacity to bulk-up species is not less important. Botanical gardens have this capacity. Often some threatened plants are ornamentally promising or proven species, or plants of economic importance as, for instance, many of the highly aromatic herbs or spices that are common around the Mediterranean. Mass production and sale through volunteer organizations, or other distribution channels, can provide an efficient source of funds for conservation work. Cultivation of such plants in private homes or gardens can have a positive impact on relations with people, and greatly reduce the pressure on many threatened plants. *Origanum dayii* Post and other species are well-known examples. Though the genus includes a few originally very narrow endemics, quite a few of them, until recently, were common plants of the Eastern Mediterranean, but populations of these valuable spices were so severely decimated by collecting for consumption that they now must be included in the TPM categories.

Cultivation of threatened material could also change the unfair balance in the distribution of profits between exporting and importing countries in the trade of endangered material (see also TPC-Newsletter No. 4, July 1979).

The distribution of plants from existing collections of botanical gardens could aid conservation by decreasing the size of collections of easy-to-maintain, highly reproductive species, freeing resources and space for more demanding or threatened species - see also Shaw (1976).

A basic prescription for maintenance of conservation-oriented collections in botanical gardens:

1. State aims of conservation explicitly, inform all staff and coordinate with related organizations and individuals.
2. Identify and list stocks to be conserved in order of priority.
3. Restrict numbers so as to be as compatible with space and resources, with emphasis on ecologically suitable taxa.
4. Maintain in conservation-oriented collections and propagate only threatened stocks which are (a) taxonomically authentic, (b) with accurate provenance information, (c) of adequately representative sample size, and (d) well grown and disease-free.
5. Set up decision-making system and consultant list to assist the resolution of conflicts.
6. Set up monitoring program and review periodically results against aims.
7. Identify and place problems for research in order of priority.
8. Carry out maintenance-oriented research.
9. Identify and locate support potential and mobilize for conservation and restoration.
10. Establish policy on display and begin distribution and sales of threatened plants.

References

Anonymous (1975). Resolutions of the XII International Botanical Congress, Leningrad. Taxon 24: 701-704.

Anonymous (1976). Agreed resolutions of the 1975 Conservation Conference. In: Simmons, J.B. et al. (Eds.), 'Conservation of Threatened Plants', pp. 311-312. New York - London: Plenum Press.

Anonymous (1979). Agreed conclusions of the 1978 Conference about the practical role of botanic gardens in conservation. In: Synge, H. & Townsend, H. (Eds.), 'Survival or Extinction', pp. 6-7. Kew: Royal Botanic Gardens.

Beyer, R.I. (1979). The appreciation of the conservation role by staff of botanic gardens. In: Synge, H. & Townsend, H. (Eds.), Survival or Extinction, pp. 171-174. Kew: Royal Botanic Gardens.

Bradshaw, M.E. & Doody, J.P. (1978). Plant population studies and their relevance to nature conservation. Biol. Conserv. 14: 223-242.

Bramwell, D. (1979). A local botanic garden: Its role in plant conservation. In: Synge, H. & Townsend, H. (Eds.), Survival or Extinction, pp. 47-52. Kew: Royal Botanic Gardens.

Brenan, J.P.M. (1979). Introduction. In: Synge, H. & Townsend, H. (Eds.), Survival or Extinction, pp. 1-5. Kew: Royal Botanic Gardens.

Brenan, J.P.M., Ross, R. & Williams, J.T. (Eds.) (1975). Computers in Botanical Collections. New York - London: Plenum Press.

Cook, C.D.K. (1976). Autecology. In: Simmons, J.B. et al. (Eds.), Conservation of Threatened Plants, pp. 207-210. New York - London: Plenum Press.

Esser, K. (1976). Genetic factors to be considered in maintaining living plant collections. In: Simmons, J.B. et al. (Eds.), Conservations of Threatened Plants, pp. 185-198. New York - London: Plenum Press.

Frankel, O. (1970). The time scale of concern. In: Simmons, J.B. et al. (Eds.), Conservation of Threatened Plants, pp. 245-248. New York - London: Plenum Press.

Grant, V. (1975). Genetics of Flowering Plants. New York -London: Columbia University Press.

Harper, J.L. & White, J. (1974). The demography of plants. Ann. Rev. Ecol. and Syst. 5: 419-464.

Hawkes, J.G. (1976). Sampling gene pools. In: Simmons, J.B. et al. (Eds.), Conservation of Threatened Plants, pp. 145-154. New York - London: Plenum Press.

Henderson, D.M. & Prentice, H.T. (1977). International Directory of Botanical Gardens II, Regnum Vegetabile 95.

Hofman, M. (1976). Index systems for special collections as illustrated by the Orchidaceae. In: Simmons, J.B. et al. (Eds.), Conservation of Threatened Plants, pp. 119-122. New York - London: Plenum Press.

Nettancourt, D. de (1977). Incompatibility in Angiosperms. Berlin - Heidelberg - New York: Springer Verlag.

Perring, F.H. (1976). The future and electronic data processing. In: Simmons, J.B. et al. (Eds.), Conservation of Threatened Plants, pp. 113-117. New York - London: Plenum Press.

Poppendieck, H.H. (1976). Mesembryanthemums and their cultivation. In: Simmons, J.B. et al. (Eds.), Conservation of Threatened Plants, pp. 55-60. New York - London: Plenum Press.

Raven, P.H. (1976). Ethics and attitudes. In: Simmons, J.B. et al. (Eds.), Conservation of Threatened Plants, pp. 155-179. New York - London: Plenum Press.

Shaw, R.L. (1976). Future: Integrated international policies. In: Simmons, J.B. et al. (Eds.), Conservation of Threatened Plants, pp. 39-47. New York - London: Plenum Press.

Smith, M. (1979). Creating specialized habitats in a garden. In: Synge, H. & Townsend, H. (Eds.), Survival or Extinction, pp. 219-221. Kew: Royal Botanic Gardens.

Snogerup, S. (1979). Cultivation and continued holding of Aegean endemics in an artificial environment. In: Synge, H. & Townsend, H. (Eds.), Survival or Extinction, pp. 85-91. Kew: Royal Botanic Gardens.

Walters, S.M. (1979). The eastern England rare plant project in the University Botanic Garden. In: Synge, H. & Townsend, H. (Eds.), Survival or Extinction, pp. 37-46. Kew: Royal Botanic Gardens.

Jerusalem Botanical Gardens
Jerusalem, Israel

CHAPTER 14

Seed banks as an emergency conservation strategy

C. GÓMEZ-CAMPO

1. Introduction

Simple techniques are now available to conserve the viability of seeds over long periods of time, possibly hundreds or even thousands of years. These techniques have been used in the past few years to conserve vanishing cultivars of crop species, and more recently, they have started to be used for wild species as well. Such 'seed banks' are primarily to avoid possible extinctions, but can also serve as a source of valuable genetic material for research or plant-breeding purposes. The use of long-term conservation techniques make the need for re-collection or multiplication less important and makes seed collection an essentially cummulative task. The time and effort which is saved can then be devoted to many other aspects of the same activity.

It is important to state that storing seeds of wild plants in a seed bank should never be a substitute – but rather a complement – to the conservation of the species in nature within their natural habitats. But the steps to establish protected natural areas are very often too slow in the Mediterranean region. In part, this is because the very diversity and richness of the flora contributes to make more difficult the application of floristic criteria to the selection of suitable areas. In part it is also because public awareness to this problem is still poorly developed. Conservation of wild species in their Mediterranean habitats will often need some management, and we need time to acquire the necessary know-how before it can be properly done. If we add financial and legal difficulties, we must conclude that a satisfactory situation in this respect may be still many years ahead. In the meanwhile, a number of species may become extinct. Seed banks provide an emergency method to avoid such events and the necessary source of genetic material to restore lost taxa or populations.

2. Long-term preservation of seeds

Aging of seeds is generally thought to be a consequence of cummulative poisoning by certain products of their retarded metabolism. Basically, seeds are adapted to a long life-span – to facilitate their dispersal in space and time – and this is achieved by slowing down the metabolic activity through desiccation. But desiccation is necessarily accompanied by unusual reactions and the accumulation of unusual metabolic products that are often toxic or mutagenic and constitute the ultimate cause of seed death. The rate of seed metabolism is very dependent upon environmental conditions, and so is seed longevity. The lower the metabolic rate of the seed, the slower the accumulation of undesirable products and the longer the life-span. By using the proper environmental factors (fundamentally, low temperature and low humidity) we can take the metabolic rate to near zero and thus obtain considerably extended longevities.

Some early estimations of seed longevity have been summarized by Barton (1961). Two principal

Gómez-Campo, C. (ed.), Plant conservation in the Mediterranean area.
© 1985, Dr W. Junk Publishers, Dordrecht. *ISBN 90 6193 523 7.*

sources for old seeds of known age were used: herbaria and garden collections. Though 'normal' longevities were found to be of 5–25 years, it was also noticed that the seeds of some species (mostly hard-coated Leguminosae) could live 200–300 years. Also some instances with very short life-spans, such as a few weeks in some *Salix* and *Populus* species, were detected. Obviously there is an important genetic component, but the decisive factor is the environmental conditions under which the seeds are stored. Early measurements of seed longevity largely neglected the storage conditions, and, therefore, their value is very relative.

Serious long-term experiments devised to determine the influence of storage conditions began early in this century. Consequently, their conclusions are at least valid for a period of 70–80 years. Though different species may exhibit somewhat different responses, some general rules that are followed by most can be given. The relative roles of temperature and humidity are summarized by Harrington (1972) as follows:

a) For each 5°C of decrease in the temperature at which the seeds are stored, their life-span is approximately doubled.
b) Independently, for each 1% of decrease in the seed water-content, the life-span is also approximately doubled.

The effect of reducing the oxygen in the atmosphere of conserved seeds has been also investigated, but here the results are more erratic. It is felt, however, that the use of anaerobic conditions may be beneficial.

From Harrington's figures, it becomes apparent that the most extended longevities can be obtained if low temperatures, low humidities or a combination of both are used. As an example let us take a seed sample which under normal room conditions (20°C and a relative humidity equilibrated to approximately 10% of seed water-content) can conserve its germination ability for ten years. If it is stored at 0°C with its water content being maintained, this would result in extending its life by a factor of 2^4, i.e. to 160 years. If instead, we desiccate it down to 5% of water content, its life would be extended by a factor of 2^5, i.e. to 320 years. If we use both methods simultaneously, the life-span of that seed would become $10 \times 2^4 \times 2^5 = 5{,}000$ years approximately. Through similar computations we could conclude that storage at -20°C with 3% of water would result in the conservation of viability for more than 300,000 years!

As experiments on actual seed longevity have been conducted for only 70–80 years, the reliability of these optimistic calculations depends upon how the experimental conclusions can be extrapolated to periods of hundreds or thousands of years. Can they? There is enough evidence that, at least to a certain extent, the answer in yes. Several instances of naturally preserved seeds in properly dated strata are now available and they give figures that reach 10,000 years in the case of some *Lupinus* seeds from Greenland (Porsild, Harington & Mulligan, 1967). Even if extrapolations are not quite exact, at least we know the procedure to extend the life-span of the seeds over long periods of time. If a given seed sample is expected to live 5,000 years and only lives 500, it is obvious that the main proposed objectives would be satisfactorily accomplished anyway. A practical upper limit for the life-span of preserved seeds is posed by the action of cosmic rays.

There are some limitations to the use of low temperatures and low humidities (Roberts, 1975). For instance desiccation of seeds below 2% of water content is technically difficult and also biologically dangerous, since irreversible damage can be produced. Practical minimum values could be 3% for oleiferous seeds and 4–5% for other seed types. On the other hand, minimal temperatures to be attained should always be in relation to the water content of the seed in order to avoid freezing damage. Moistened seeds, for instance, would rapidly die by freezing if they are stored at only a couple of degrees below zero. In general, the drier a seed is, the lower the temperature at which it can be conserved (Fig. 1). Only very rapid freezing, as by immersion in liquid nitrogen, can be done at any water content, the success also being a function of the use of cryo-protective substances. However, very low temperatures introduce a forbidding economic factor when long periods of time are involved.

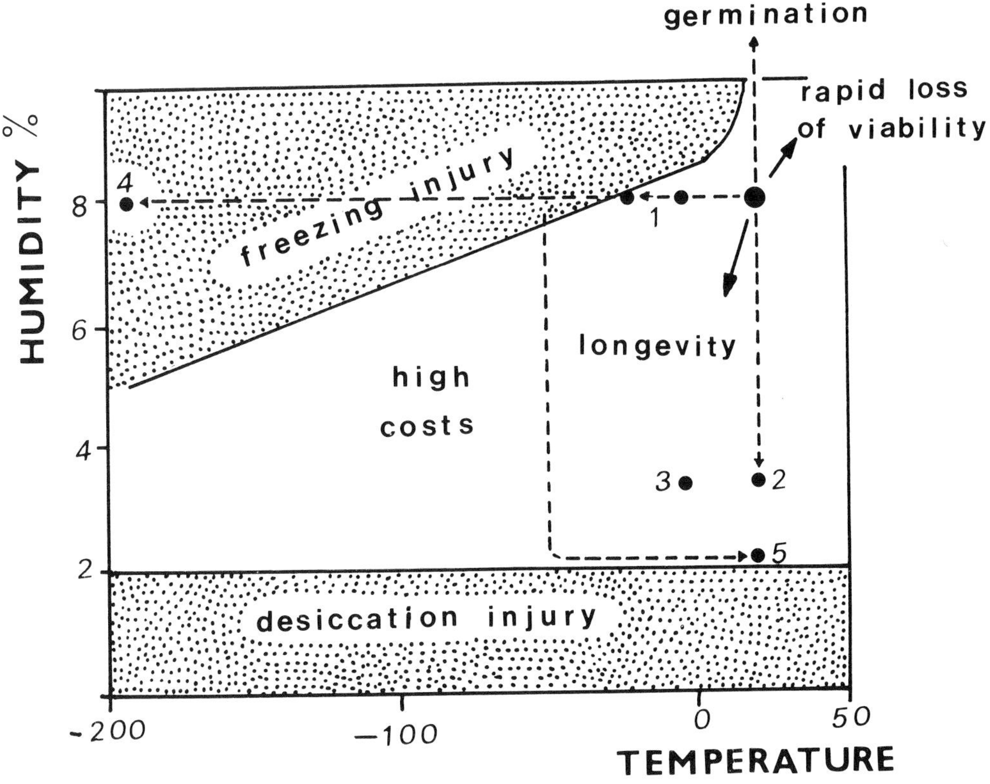

Fig. 1. Temperature-humidity diagram where several conservation situations are expressed: 1. Storage at cold temperature. 2. Desiccation (with silica-gel) only. – 3. Low temperature and desiccation combined. – 4. Storage in liquid nitrogen. – 5. Liofilisation (at present only an unuseful possibility). (from Gómez-Campo, 1976).

Some large-sized farinaceous seeds, such as those of chestnuts and walnuts, do not allow desiccation without permanent damage. For these 'recalcitrant' seeds the above conservation methodology is limited. Low temperatures certainly extend their life-span, but no extreme longevities can be artificially obtained. However, it should be noted that among Mediterranean endemics, the proportion of 'recalcitrant' species is minimal.

In practice, the use of a simple domestic refrigerator permits us to extend the life of any seed sample by at least a factor of ten. Similar results can be obtained if the seeds are dried and stored at room temperature in a hermetic container. The combination of both methods is highly recommended because, as they act independently, the life-span becomes enormously expanded. Seed drying can be easily done by solar or artificial heat or by the use of chemical desiccants such as $CaCl_2$ or silica gel. Heat must be used properly because too high a temperature might damage the seed. Chemical methods have many advantages in this respect. The desiccants absorb humidity from the atmosphere, and the seeds reduce their own water content to 3–5% by becoming equilibrated with the dry atmosphere after only a few days.

It is desirable to allow a period of several weeks or months to elapse between collection and permanent storage, because this helps to avoid possible interference by post-harvest dormancy. The same lapse can be simultaneously used to obtain a gradual drying. A good practice consists of temporarily storing the collected seeds with silica gel while awaiting transference into permanent containers. The desiccant can also be included within the permanent container, so that its action is ensured for a very long period of time. Silica gel is sold with a cobalt indicator showing a blue colour. When it absorbs humidity it becomes pink and thus helps detect any crevice, hole or other air leak.

It is unnecessary to emphasize the importance of hermetic sealing in the containers to be used for permanent conservation. Sealed glass capsules are by far the safest system, but good screw-cap bottles can also be used. Sealed tin cans have also been used, but it is now believed that seeds can be damaged by metal vapours in the long run. Plastic containers should have properly tested impermeability to water and oxygen.

The choice of procedure depends on the value of the material preserved and also of the objectives that are to be pursued. For valuable endemic material to be conserved over long periods, sealed glass capsules or ampules with some silica gel inside are recommended; some permeable cotton or a piece of paper should be placed to avoid its mixing with the seeds themselves. The low temperatures to be used are only limited by the available equipment and the cost of its maintenance. For shorter conservation periods, a number of simplifications are possible. In fact, when storing endemic material, a dual objective often exists: (a) long-term preservation to avoid possible extinction, and (b) conservation over shorter periods of material to be distributed. In practice, two different procedures might be used in parallel.

3. Seed collection

As a botanical activity, seed collection for conservation in a seed bank shows a number of differences from the usual practice of collecting specimens for a herbarium. On the one hand, seed collection is to be undertaken comparatively late in the growth season. Some of the earliest taxa can certainly be collected in April or May, just when the collection of herbarium specimens is at its peak in the Mediterranean region. However, the best months for seed collection might be June–July in the plains and August to October or even later in the mountains. On the other hand, to recognize rare species when their fruits are at the ripe, shattering stage may be difficult for collectors who are used to seeing the species in its flowering or early fruiting stages. Therefore it is comparatively frequent that unidentified samples are collected on a probability basis and given a number for *a posteriori* checking of their botanic identity. A voucher specimen of the plant itself should always be taken, but some necessary characters of the flower or leaf may not be present when the plant is collected. A small part of the seed sample can be sowed and identity determined on cultivated material the next season.

Collecting expeditions should be carefully planned (Bennet 1970) by determining in advance which areas have a maximum concentration of the species of interest, and when is the proper time for most of them to show the right maturity. When the programme becomes more advanced, trips need to be planned to visit not one but several, often distant, localities where many species are known to be mature at that time. Economically, this may be less productive in relation to the costs, mostly when rare, difficult-to-find species are involved. In such a situation it is suggested to combine seed collection with other botanic activities as the collection of ecological or demographic information (see Chapter 11) on rare species. The collaboration of local botanists is another important factor in increasing efficiency and cutting down the costs of seed collection. Success in mobilizing local contacts may be of great importance to the whole programme.

Within each population, seeds should be taken from as many different individuals as possible in order to sample a maximum of genetic diversity. Second or third collections from other geographically distant populations provide a good method to improve that variability, but such collections should be kept separately in the seed bank. Plant populations from which the seeds are collected should be free of pest and diseases and at the right stage of ripeness.

A small or medium-sized plastic pan in the hand plus a pack of paper envelopes in the pocket may be the most useful tools for the seed collector. The pan is not only very practical for the collection itself but it also makes a preliminary cleaning very easy. Paper or cloth envelopes are much preferable for transport to plastic ones, because moulds could develop in the latter when the seeds are not properly dry. A plastic bag on the belt may serve to carry plant specimens. Most species keep their turgidity for a few hours within a plastic bag. A glove may be used when dealing with thistles and other spiny plants. Other equipment such as screens can remain at the car or campsite for subsequent use.

It is very important to take proper care that seed collection itself does not threaten the survival of the taxon involved. This is unlike to occur, but it could happen in certain extreme cases. Partial collection in several succesive seasons and/or multiplication in the garden may be what is called for in this type of situation. In very extreme situations (some cases exist where a single surviving individual has been reported), vegetative multiplication, tissue culture or even anther culture might be tried.

The sampling techniques described in the literature for crop species are largely unapplicable to the collection of rare wild species. The size of the sample, for instance, cannot be fixed in advance because it is subjected to a number of circumstances. Some endemics can be easily and abundantly collected in five minutes at the roadside, while others need several hours of search on difficult, nearly inaccessible, rocks. One should just aim to do the best possible without inflicting significant damage to the population.

As mentioned above, it is convenient to store the seeds temporarily in a cool dry place, if possible a hermetic box or chamber with some silica gel or Cl_2Ca in it. This has the twin effect of allowing time to overcome possible postharvest dormancies, and of obtaining the necessary desiccation prior to permanent storage by the mild method that is provided by chemical desiccants. The preparation of the material for permanent storage thus becomes a task to be carried out in late autumn or in winter, without any interference with summer field-work. However, aged seeds, i.e. more than one year old, should be rejected as material for long-term conservation.

4. Germination tests

There are at least two reasons why it is convenient to conduct germination tests on collected threatened plant material before it is placed into permanent storage. Firstly, tests supply reference data that serve as control, to evaluate the conservation method in the future. But, most importantly, the tests tell us whether or not the material being stored has any dormancy problems.

Dormant seeds are perfectly viable, but they are affected by some limiting factor that prevents germination even when they are placed in the right temperature-humidity-oxygen conditions. Such a limiting factor has to be removed if we wish to have those seeds germinate. If dormant seeds are stored and sub-samples are sent away in the future, the recipient will be unable to produce plants and the result would be exactly the same as if we had sent unviable seeds. Our effort, time and research money would have been wasted.

Fortunately, only about 15% of the Mediterranean endemics exhibit serious dormancy problems (Ayerbe & Ceresuela, 1981). The most frequent instances refer either to hard-coated seeds of Leguminosae or to mountain plants which are naturally subjected to a cold season and need some chilling treatment. In the first case, mechanical scarification or chemical treatment with sulphuric acid may be enough to facilitate germination. In the second case, 2–4 weeks storage at 1–3°C is usually sufficient to obtain satisfactory effects. According to Thompson (1971), the seeds of Mediterranean plants tend to germinate at relatively low temperatures, an adaptation to prevent premature germination under the high soil temperatures of summer, when some moisture derived from occasional storms is intermittently present. Natural times for germination are autumn and spring.

There is some evidence that dormancy can be abruptly interrupted under the conditions of long-term preservation. The author of this article tried to germinate conserved seeds of *Caulanthus hallii* (a Californian annual) without success for three seasons, but obtained almost 100% germination in the fourth year. Whether this type of behaviour is common or exceptional is not known.

When a dormant sample has been detected, the next step would consist of checking the type of dormancy it exhibits. Figure 2 shows a simple standard test proposed by Thompson & Brown (1972) to obtain this kind of information. The number of seeds recommended there for each test may need to be reduced when dealing with rare species of which we have been able to collect only small amounts. Even if the statistical significance is negatively affected, our reluctance to consume too large a proportion of material in the germination tests is understandable.

Once the method of removing dormancy from a seed is known, the material can be treated in the proper way before it is placed into permanent storage. An alternative method would be to store the seeds in their dormant condition and give the proper recommendations to prospective users, so that they can employ the treatment themselves before sowing the material.

Species which remain ungerminated after passing the whole set of Thompson's assays are either unviable or have some obscure dormancy mechanism that would constitute an excellent subject for laboratory study. If seeds have been collected from mature and healthy wild populations they should be viable in a majority of cases. Unviable seeds can sometimes be detected with X-rays or through the use of tests proposed for this purpose, such as that based on tetrazoilum salts.

In fact many dormant seeds will readily germinate if they are treated with a sufficient dose (1,000–2,000 p.p.m.) of exogenous gibberelic acid; this seems to be a good substitute for a wide range of environmental factors and an excellent neutralizer for the action of possible inhibitory substances. According to the experience in our laboratory, the doses recommended by Thompson are apparently rather low and lead to an underestimate of the practical value of gibberelic acid in breaking dormancy (Durán & Retamal, 1983).

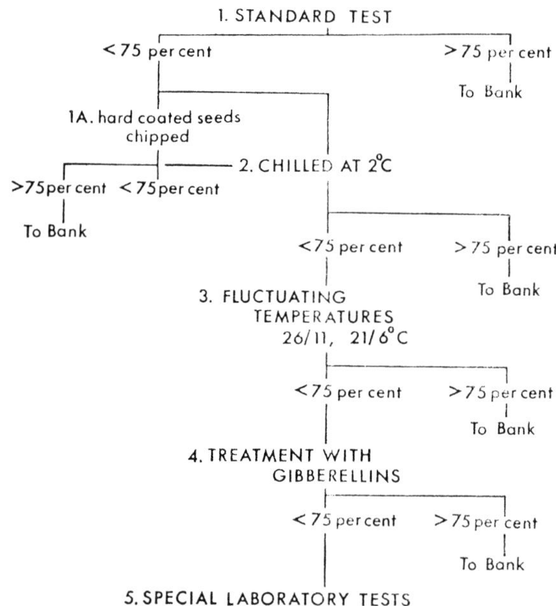

Fig. 2. Series of tests proposed by Thompson and Brown (1972) to screen seed before storage in the bank.

5. Setting up a seed bank of endemics

Whether a given species is threatened or not is often very difficult to determine. Thus, a seed bank restricted to threatened plant material would meet continuous problems in the selection of the taxa to be stored. To establish a better defined objective becomes necessary, and it is believed that the collection of endemics provides a satisfactory upper limit. However, this might be only one part of a project or programme with much wider scope, as for instance a seed collection by the staff of a botanic garden. Widely distributed species are seldomly threatened, and the problem may consist of detecting the few cases which are.

The proportion of threatened taxa in the so-called narrow endemics or geographically restricted plants, becomes high or very high. The term 'endemic' needs to be referred to a defined geographic territory, and it is practical to use it with reference to single countries since protective legislation, capacity for action, and so on is actually based on these political units. Nonetheless there is a basic fallacy in this approach because political boundaries do not always correspond to biogeographical limits; this should be properly understood and compensated for. To give an example, many narrow endemics grow on mountains that form the border between two countries and they might be incautiously left out of the programme. Such is the case of many Pyrenean endemics between Spain and France. Plants with very narrow ecological requirements or living in very fragile ecosystems should also be included, even if they are present in two or three countries. Since seed banks often need to be started before the full necessary information can be assembled, wide flexible criteria are better suited to the situation than narrow strict ones.

As the establishment of crop seed-banks (see Hondelman, 1976) was started in the decade of the 60s, often with liberal funding, this created the idea that all seed banks are necessarily very costly installations, requiring advanced technology, large investment and costly equipment. This has proved to be misleading and probably has retarded the initiation of programmes to preserve the endemic flora in a number of countries. Actually, wild endemic species have much smaller seeds, so that the samples to conserve are also smaller. The total number of samples would also be much lower in general, because they are basically referred to species or sub-species; for crop species, as many varieties, populations and genotypes as possible are usually welcomed so that the number of samples soon climbs to many thousands. To store seed samples of endemic plants requires much less space and much cheaper installations (Gómez-Campo, 1979). Actually, seeds of the whole endemic flora of a country with, say, two hundred endemics, could well be accommodated within a standard domestic refrigerator.

The development of a seed bank of endemics could be promoted in at least three possible alternative ways: (1) As an initiative of a botanist or a group of botanists working closely with problems of plant conservation, (2) as a part of large-scale programmes developed by existing crop seed-banks, or (3) as an evolution of present seed collections in botanic gardens. In my opinion, the most logical procedure is the last. But botanic gardens are so often affected by budgetary shortages – especially in the Mediterranean region – that we feel obliged to suggest other possible alternatives.

Seed banks of crop plants are partially interested in wild species because these can be a source of useful genes transferrable to cultivated plants. But their interest is commonly restricted to taxa that are genetically closely related to well-known crops. This area of interest may be now widened, after the advent of modern techniques of protoplast fusion.

But the objectives of conserving vanishing cultivars and conserving vanishing wild species are still too far apart from a practical point of view. The first is mostly a task for geneticists and agronomists, the second for botanists. Nonetheless, if the collaboration of some botanists is secured, the protection of wild species could be incorporated as an appendix to the programmes being carried out in crop seed-banks. Only a small space would be occupied in the cold rooms. The conservation of endemics would additionally benefit from the better economic situation that usually accompanies the seed-banks of major crop species.

Seed catalogues from botanic gardens have long been an important source of living plant material for basic or applied botanic research. By 1968 there were at least 150 of these catalogues published in Europe alone. In the past few years this number has decreased sligthly due to the economic crisis, but is still very high. In general, there is a sharp contrast between the cost of labour involved in making such large collections of seeds, and the real value of this material to the institutions or persons who receive and use it. This is the reason why the compilation of such seed exchange lists has long been criticized (Heywood, 1964, 1976; Howard et al., 1964; Böcher & Hjerling, 1964). The evolution of botanic garden seed collections towards the style provided by seed banks is suggested by Thompson (1970, 1974) as a means to improve their efficiency and quality.

A quantitative analysis of the weaker aspects in the performance of botanic gardens in this respect (Gómez-Campo, 1969) was based on a comparison of the contents of crucifer seed samples from 111 European catalogues with the listing of the existing crucifer flora according to *Flora Europaea* (Tutin et al. 1964).

On the one hand, errors in the identification of samples were as high as 30%, in a total of 200 samples received and checked. Many of the misnamed samples corresponded to weedy species thus suggesting the probable mechanism which lead to such a situation: weeds had grown in the place of ungerminated samples – either because these were unviable or dormant – and were collected together with the original label without a proper check of its identity. The same errors are reproduced by seed exchange afterwards, and they are often found in several catalogues.

On the other hand, only 45% of the existing European crucifer flora are accessible by this procedure. The remaining 55% of species were not listed in any of the catalogues. By contrast, some species such as *Arabis alpina, Matthiola incana* or *Isatis tinctoria* were included in almost every catalogue – the so-called 'Botanic Garden flora'. For rare local endemics, the performance was much poorer. Only 15% of the 'endemic to a single country' category was included in some of the 111 catalogues. The remaining 85% was completely inaccessible by this means. The availability of species which are of interest from the point of view of conservation was extremely low. As the seeds become unviable after a few years, the real value of these samples for conservation is restricted to their use to establish living collections in the garden.

Obtaining seeds from cultivated garden material is linked to serious inconveniences, apart from being a continuous source of possible misidentification. The genetic diversity is artificially reduced by genetic drift or by unwanted selection, and cross-pollination with other nearby related species is favoured. While living plant collections can be easily obtained from seed collections or seed banks, the reverse action – taking seeds from living cultivated collections – is not recommendable. As an example, let us think of the genes responsible for the dormancy mechanisms in wild populations. Diversity with respect to these genes is maintained in the wild because dormant and non-dormant seeds all eventually germinate after a few years. By contrast, in every garden generation we always collect from germinated material and inadvertedly introduce a strong selective pressure, not only against such genes but also against others that might be linked to them. When not from garden origin, the seeds offered in the catalogues of botanic gardens have been obtained through exchange with other similar institutions. Probably for economic reasons, collection in the wild seems to be rather limited. This is in turn reflected in a general failure to give proper data on the geographical or ecological origin of the seeds.

In spite of all this criticism, botanic gardens are best suited to deal with seeds and can make enormous improvements in a short time through the incorporation of modern seed preservation techniques. A slow but clear evolution in that direction has already taken place in the past decade. The Royal Botanic Gardens, Kew took an initiative by establishing a seed bank and physiology unit at Wakehurst Place (Thompson & Brown, 1972) and now we have noticed at least half a dozen gardens which are involved in seed banking activities. The Wakehurst staff have developed several useful concepts about seed banking (Thompson, 1971, 1975, 1976) and have done a good deal of work on seed germination and dormancy in wild species.

The adoption of long-term seed preservation techniques directly results in the disappearance of all the above-mentioned problems. Even if expensive, expeditions or trips to collect seeds in the wild are much more rewarding if each sample can be expected to live several hundred years. Thus every effort brings a straightforwards enrichment in the quantity and quality of the collection. Periodic multiplications to maintain the material alive becomes completely unnecessary, and all the time which is saved can be used instead to take care of identification, to search for rare and valuable plants and so on.

A seed collection of crucifers established in the laboratory of the author may be an example of how these activities are easy and simple enough to be undertaken by a reduced group of persons (Gómez-Campo, 1972). The first mimeographed seed list with preserved material of natural origin was published in 1967 and it was probably a pioneer in the application of seed banking techniques to wild species. The method of conservation used is summarized in Figures 3 and 4. The most recent printed catalogue (Gómez-Campo, 1978) contained 500 entries of crucifers including many rare and threatened taxa. The next issue will expectedly contain near 600 entries plus an appendix with a number of seed samples of n = 9 Mediterra-

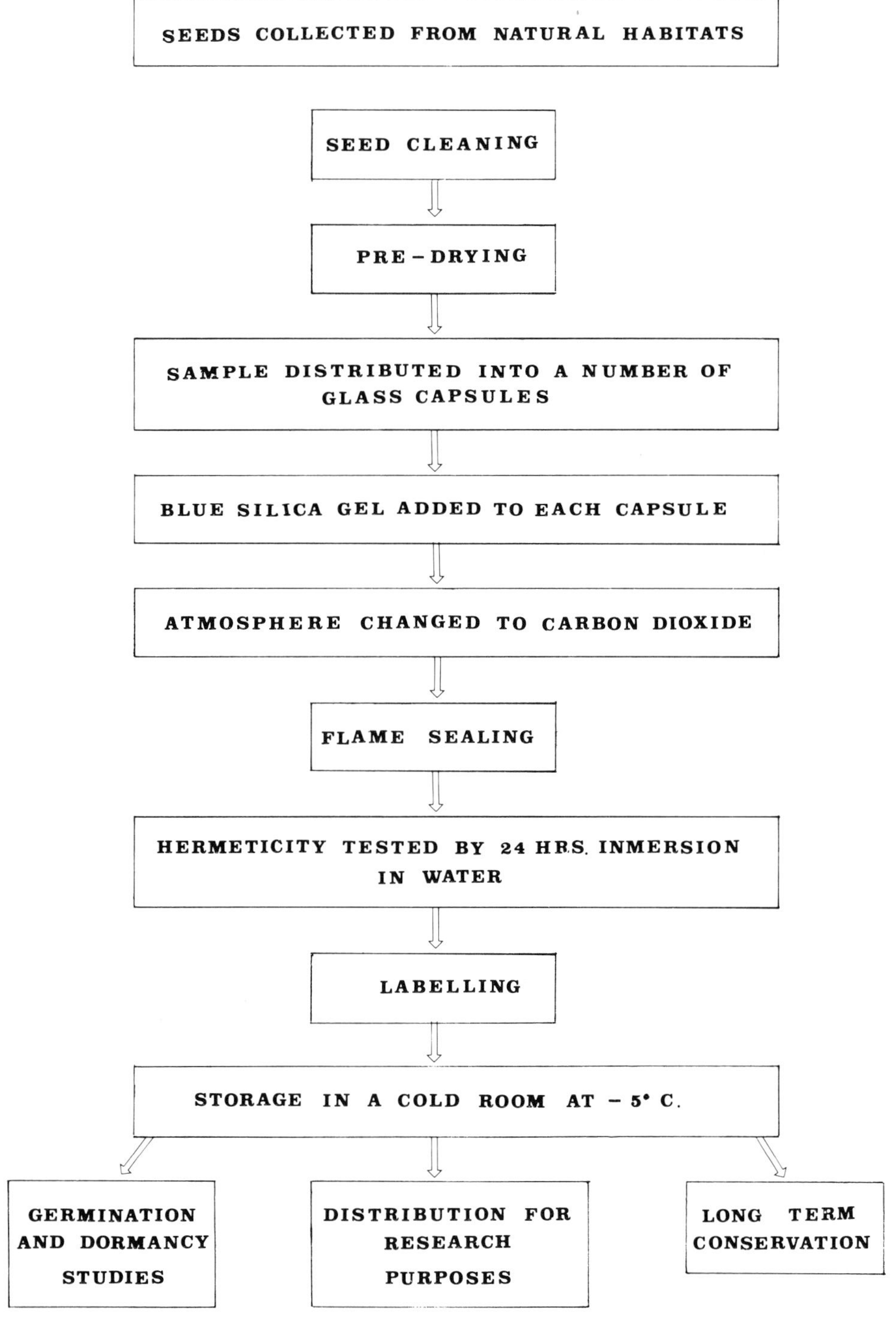

Fig. 3. A scheme of operations in a seed bank of wild taxa, as exemplified by current programmes in the author's laboratory.

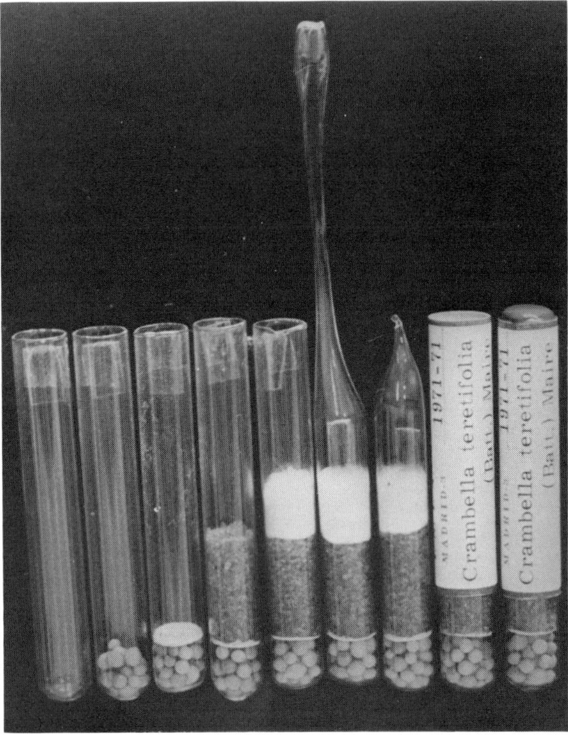

Fig. 4. Detail on the elaboration of encapsulated seed samples in a seed bank of crucifers.

nean *Brassica* species, subspecies and populations, collected under the auspices of IBPGR-FAO (Int. Board of Plant Genetic Resources) (Gustafsson, Gómez-Campo & Zamanis, 1983). Specialization in a particular taxonomic group has a positive influence on the correct identification of the samples, but collecting expeditions may become too expensive unless they are combined to achieve other additional objectives.

A second programme aimed at Iberian and Macaronesian endemics was started in 1973. Seed collection is done either through expeditions from Madrid or through the collaboration of a number of local botanists. By January 1983, the bank contained more than 1,000 samples (Fig. 5) corresponding to about 900 different taxa. Only about 2,5 square metres of shelving in the cold room were occupied. From 1981 onwards, the project has been extended to cover the whole Mediterranean area under the auspices of OPTIMA (Organization for the Phyto-Taxonomic Investigation of the

Fig. 5. Iberian and Macaronesian endemics stored in a cold room at -5°C.

Mediterranean Area) and with the expected collaboration of botanists from every Mediterranean country. 'Ad hoc' expeditions are also being organized. A duplicate of each collected sample will be eventually returned to the country of origin. This strategy has been planned to stimulate the establishment of a net of local seed banks around the Mediterranean region.

It is important to note that possible duplication of work arising from the overlapping of local, national or international programmes is not only admissible but desirable in this type of activity. Keeping duplicates of seed samples of the same taxon in different banks, in fact constitutes the safest way to minimize losses of valuable material.

References

Ayerbe, L. & Ceresuela, J.L. (1982). Germinación de especies endémicas. Anales Inst. Nac. Invest. Agrarias, Madrid 6: 2-41.

Barton, L.V. (1961). Seed Preservation and Longevity. London: Leonard Hilld and New York: Interscience.

Bennet, E. (1970). Tactics of plant exploration. In: Frankel, O.H. & Bennet, E., Genetic Resources in Plants: Their Exploration and Conservation, pp. 157-180. Oxford-Edinburg: Blackwell.

Böcher, T.W. & Hjerling, J.P. (1964). Utilisation of seeds from botanical gardens in biosystematic studies. Taxon 13: 95-98.

Durán, J.M. & Retamal, N. (1983). Efecto del ácido giberélico en la germinación de semillas de mostaza silvestre (*Sinapis arvensis* L.). Anales I.N.I.A. 24: 11-54.

Gómez-Campo, C. (1969). The availability of crucifer seeds from European botanic gardens. Plant Introduction Newsletter (FAO) 22: 25-32.

Gómez-Campo, C. (1972). Preservation of West Mediterranean members of the cruciferous tribe *Brassiceae*. Biol. Conserv. 4: 355-360.

Gómez-Campo, C. (1976). Conservation techniques of crucifer seed banks. Arabidopsis Inf. Service 13: 18-21.

Gómez-Campo, C. (1978). A germ plasm collection of crucifers. Catálogos I.N.I.A., Madrid 8: 1-40.

Gómez-Campo, C. (1979). The role of seed banks in the conservation of Mediterranean flora. Webbia 34: 101-107.

Gustafsson, M., Gómez-Campo, C. & Zamanis, A. (1983). *Brassica cretica* Lam. germplasm collection in Greece. EUCARPIA Cruciferae Newsletter 8: 2-4.

Harrington, J.F. (1972). Seed storage and longevity. In: Kozlowski, T.T. (Ed.), Seed Biology, Vol. 3, pp. 145-245. New York - London: Academic Press.

Heywood, V.H. (1964). Some aspects of seed lists and taxonomy. Taxon 13: 94-95.

Heywood, V.H. (1976). The role of seed lists in botanic gardens today. In: Simmons, J.B. et al. (Ed.), Conservation of Threatened Plants, pp. 225-234. New York: Plenum Press.

Hondelmann, W. (1976). Seed banks. In: Simmons, J.B. et al. (Ed.), Conservation of Threatened Plants, pp. 213-224. New York: Plenum Press.

Howard, R.A. et al. (1964). Comments on seed lists. Taxon 13: 90-94.

Porsild, A.E., Harington, C.R. & Mulligan, G.A. (1967). *Lupinus arcticus* Wats grown from seeds of Pleistocene age. Science 158: 113-114.

Roberts, E.H. (1975). Problems of long-term storage of seed and pollen for genetic resources conservation. In: Frankel, O.H. & Hawkes, J.G. (Eds.), Crop Genetic Resources for Today and Tomorrow, pp. 269-298. Cambridge: Cambridge University Press.

Thompson, P.A. (1970). Seed banks as a means of improving the quality of seed lists. Taxon 19: 59-62.

Thompson, P.A. (1971). Research into seed dormancy and germination. Proc. of the Int. Plant. Propagators Soc. Ann Meeting, pp. 211-228.

Thompson, P.A. (1974). The use of seed banks for conservation of populations of species and ecotypes. Biol. Conserv. 6: 15-19.

Thompson, P.A. (1975). The collection, maintenance and environmental importance of the genetic resources of wild plants. Environmental Conserv. 2: 223-228.

Thompson, P.A. (1976). Factors involved in the selection of plant resources for conservation as seed in gene bank. Biol. Conserv. 10: 159-167.

Thompson, P.A. & Brown, G.E. (1972). The seed unit at the Royal Botanic Gardens, Kew. Kew Bulletin 26: 445-456.

Tutin, T.G., Heywood, V.H., Burges, N.A., Valentine, D.H., Walters, S.M. & Webb, D.A. (1964). Flora Europaea, Vol. 1. Cambridge: Cambridge University Press.

Departamento de Biologia
Escuela T.S. Ing. Agrónomos
Universidad Politécnica
E-28040 Madrid, Spain

Part IV

Indexes

Subject index

Systematic index*

* These taxa being the object of a 'case history' or which are otherwise treated more extensively in the text, are written in italics.